Expedition agroparks

TransForum

Wageningen UR
For quality of life

Expedition agroparks

Research by design into sustainable development and agriculture in the network society

Peter J.A.M. Smeets

Wageningen Academic
Publishers

ISBN: 978-90-8686-163-7
e-ISBN: 978-90-8686-719-6
DOI: 10.3921/978-90-8686-719-6

First published, 2011

Wageningen Academic Publishers
P.O. Box 220
6700 AE Wageningen
The Netherlands
www.WageningenAcademic.com
copyright@WageningenAcademic.com

Table of contents

Preface 9

Acknowledgements 11

Summary 17
1. Context, aim and method of this book 17
2. The network society 18
3. Delta metropolises 20
4. Agriculture in the network society 21
5. Theoretical starting points in research by design 21
6. Research by design applied to agroparks 23
7. Discussion 33

1. Context, aim and method 37
1.1 Scientific objective 41
1.2 Method 42
1.3 Reading guide 44

2. The network society 47
2.1 The emergence of the network society 47
2.2 Time and space in the network society 51
2.3 The environmental movement' as a synthesis of spaces of flows and spaces of places 56
2.4 Spatial development policy 58

3. Delta metropolises 65
3.1 The historical development of cities: centralisation and dispersal 65
3.2 Polycentric mega city regions 71
3.3 Spatial planning of metropolises in the network society 78
3.4 Delta metropolises 84
3.5 The Northwest European delta metropolis 88

4. Agriculture in the network society 93
4.1 Mega-trends in the urbanised network society 93
4.2 Agriculture in the network society 105
4.3 Spatial planning of agriculture in the network society 111

5. Theoretical starting points in research by design 119
5.1 Research by design 119
5.2 Theoretical production ecology and the De Wit curve 121
5.3 Landscape ecological theory and the three dimensions of landscape 127
5.4 The design process during design and implementation 132

6. Research by design on agroparks 145
6.1 Definition of agroparks 145
6.2 Deltapark 151
6.3 Agrocentrum Westpoort 161
6.4. New Mixed Farm 177
6.5 WAZ-Holland Park 200
6.6 Biopark Terneuzen 210
6.7 Greenport Shanghai 224
6.8 IFFCO-Greenport Nellore 240
6.9 Iterative testing of the resulting hypotheses 256
6.10 Conclusions from research by design 261

7. Discussion 271
7.1 Resource use efficiency of metropolitan foodclusters 271
7.2 Landscape theory 277
7.3 Methodical elaboration 281
7.4 Greenport Holland 288
7.5 The knowledge infrastructure of Greenport Holland 293

References 301

About the author 313

Keyword index 315

Preface

This book is the result of several years of expedition into uncharted territory by the author Peter Smeets. His fascination for the Dutch agricultural landscapes led him to the conclusion that improving the efficiency of agriculture is the most effective way to safeguard the quality of such landscapes. The wasteful modes of production developed in the past 150 years have led to a serious decline in both the surface area and the quality of the high valued landscapes. Closing the loops within the agricultural production system and increasing their productivity is therefore the best remedy to arrest this decline.

Closing loops and increasing productivity is something that forms the very foundation of agriculture, viewed at systems level. Agriculture is nothing more than making productive use of the natural processes of photosynthesis and nutrient-cycling. So, restoring these cycles within the agricultural production system and increaing their productivity became the aim of the expedition.

By combining different modes of agricultural production, i.e. by coupling plants and animals in cycles that at the end of the day have few if any leftovers, waste will be a thing of the past, energy consumption will be minimised, and the productive use of the land maximised. This is the basic concept underpinning the development of agroparks: new combinations of agricultural production in a confined region that make it possible to close loops and optimize the efficient use of inputs. By developing agroparks in urbanized or metropolitan areas, other forms of more extensive or recreational land use and landscape conservation become possible.

Smeets carried out a substantial part of his expedition while working with the TransForum innovation organization. His ideas fitted in closely with the objectives TransForum was working on: to show that the more sustainable development of agriculture is possible and to illustrate how current and future knowledge can contribute to that development. The agroparks that Smeets describes in this study are all examples of what one might call 'sustainable intensification'. It is about intensification in terms of 'doing more with less', but it is also a step towards more sustainable development by eliminating wastage and reducing the negative impact of agriculture on its physical, social and natural environment.

For everybody involved in the challenging endeavour of improving agriculture and opening up a new future, I very much recommend reading this book. It contains inspiring examples of new and promising pathways for agriculture that will help shape the future, not only of agriculture but also of our landscapes and green space.

Dr. Henk C. van Latesteijn
General Manager, TransForum

Acknowledgements

The plan for writing this book is now more than 12 years old. When, in 1996, the opportunity arose to work extra hours to save up for a sabbatical leave, I seized it with the ambition of being able to write this publication after 7 years' saving, and building up the knowledge required for it until that time. The basis for the content of the book had already been laid in the years before when I worked at the former *Rijksplanologische Dienst* (Physical Planning Agency) on the Rural Areas and Europe project. Research by design, co-design, action research, these are all perfect words for the methodology used in this thesis, but also for the way in which we dealt at that time with spatial planning under the inspirational leadership of Peter Dauvellier and Hans Leeflang and in the project group with Frans Bethe, Emmy Bolsius, the late Jan Groen, Yvonne van Bentum, Lilian van den Aarsen and Marcel Wijermans.

The plan was there, but in the daily practice of integral management of Centrum Landschap, it could only lie dormant. However, as a manager in the years from 1996 to 2004 I did succeed in inspiring on an intrinsic level, and working in a culture in which entrepreneurial researchers took responsibility for the content they wanted to develop.

An important stimulus for me was working in the Think Tank on the Pig Sector with Ge Backus, Jan Blom, Johan van Bommel, Arjen Bonthuis, Herman Bosman, Theo Coppens, Eric Daandels, Henk de Lange, Bennar Dirven, the late Jaap Frouws, Anton Hilhorst, Chris Hoeven, Theo Holleman, Chris Kalden, Jan Melis, Huub Nooijen, Ad Romme, Arjan Schutte, Bennie Steentjes, Eric Thijssen, John van Paassen, Dick van Zaane, Kees Veerman, Hans Verhoeven, Rene Vermunt, Peter Vingerling, Theo Vogelzang, Chris Wijsman, Marcel Zandbelt and Wijno Zwanenburg. In my experience the Think Tank was the first time that science stepped out of the academic world into society, to really get a system innovation going together with entrepreneurs, community groups and governments. In the Think Tank I learned first and foremost that despite all the criticism, which certainly at that time was aimed at livestock farming in the Netherlands, this latter is at the same time and precisely because of this criticism, the best in the world.

With the budget that Wageningen UR made available from 2000 for strategic knowledge development, we set up the Regional Dialogue and proceeded with the research by design in a practice of transdisciplinarity. In the background, Dick van Zaane stimulated the innovation, which we wanted to be involved in, first in Zeeland and then in North Limburg. The list of active participants in the report on the Regional Dialogue North Limburg fills two pages but Jan Ammerlaan, Frans Bethe, Jannemarie de Jonge, Klaas de Poel, Romé Fasol, Eric Frijters, Hans Hillebrandt, the late Jan Heurkens, Paul Kersten, Raymond Knops, Gé Peterink, Marcel Pleijte, Maarten Souer, Hans Sprangers, Paul Stelder, Jan van de Munnickhof, Madeleine van Mansfeld, Peter van Weel, Mark Verheyen, Marcel Wijermans and Annoesjka Wintjes still remain in my memory as inspirers. Together we stood in North Limburg on the site of what has since evolved into Greenport Venlo, a region which the whole world comes to see as an example of modern agrologistics and which will house the Floriade in 2012. When we began, the joint problem as perceived was a regional inferiority complex. Now we from our side are justifiably proud that we are still able to be part of the process.

In 2002 the *Innovatienetwerk Groene Ruimte en Agrocluster* (Innovation Network) arose from the former NRLO. The spiritual father of Innonet, A.P. Verkaik supported our work in North Limburg from the word go, and within Innovation Network Jan de Wilt was a first-rate advocate when it came to research by design on agroparks. He initiated the report 'Agroproduction parks: perspectives and dilemmas', within which we took on the task of designing Deltapark. Jan ensured that Innonet helped finance the work on Agrocentrum Westpoort and New Mixed Farm. By means of these projects the circle of researchers working on agroparks steadily grew: Jan Broeze, Arjen Simons and Marco van Steekelenburg leant their creativity and here too the collaboration was transdisciplinary. The dedication and willpower of Bram Breure in the Westpoort Agrocentrum project and the enthusiasm with which Simon Bijpost and Rene Overdevest carried out the cost-benefit analysis on the same project are unforgettable.

Inspired by the Regional Dialogue North Limburg, KnowHouse bv was set up in North Limburg. Rinus van de Waart and Trudy van Megen embraced the New Mixed Farm as a strategic spearpoint. Peter Christiaens, Huub Heijer, Martin Houben, Marcel Kuijpers, Huub Vousten and Gert-Jan Vullings helped out as entrepreneurs. For me, the tenacity of this couple in their long trek through the corridors of bureaucracy is more real proof of the power of the agropark concept. Frans Tielrooij, Chris Bartels, Arne Daalder, Evert Jacobsen, Leon Litjens and Joost Reus were invaluable coaches and inspirers for the strategy in the Steering Group.

The work in North Limburg showed those of us in Wageningen that the process management of research by design had to become more professional. The process management of transdisciplinary regional development initially became a craft competence, before we were able to start focusing on scientific research. Jannemarie de Jonge set the ball rolling by enrolling on a course at COCD. Madeleine van Mansfeld, Annoesjka Wintjes and I followed. It was the start of a collaboration with Helga Hohn, who continued to stimulate and coach us on how to inject creativity and multiple intelligences. The process work was given a boost by the work of Remco Kranendonk and Paul Kersten at Communities of Practice. With this they laid a scientific basis in the working environment for learning how to learn. The Agrologistics Platform has the honour of having invested in this, and thanks to the Agrologistics Community of Practice, the network of researchers, entrepreneurs and government personnel remained intact, precisely in those years when our projects were vulnerable. 'It can be done!' was the constant call of Frans Tielrooij that kept us going, and Lucy Wassink and Jochem Pleijzier defended this effort against all the scepticism of the bureaucrats in The Hague.

The crowning glory of this work on process management and for me a great help in the writing of this work, was Jannemarie de Jonge's thesis: Landscape Architecture between politics and science.

When I stopped working as an integral manager at the Landscape Centre of Alterra in 2004, I actually found time to work on my dissertation. With the creative efforts of Hein van Holsteijn we turned the sabbatical year into a project. Alterra therefore did not have to pay for this doctorate. Thank-you, Hein, for your patience: it took longer than budgeted for, external projects now get priority and by definition an expedition explores roads that later turn out to be dead ends.

The intrinsic work on agroparks was given a real boost by TransForum, to whom I have been seconded since 2004. Henk van Latesteijn, Jeroen Bordewijk, Johan Bouma, Evert Jacobsen, Sander Mager, Hans Mommaas, Lia Spaans, Jan Staman, Hans van Trijp and Tom Veldkamp: you helped me understand what sustainable development was all about. But even more importantly: together we were the third space *avant la lettre* that introduced the first agropark system innovation. As project director I was privileged to be able to work both at New Mixed Farm, Biopark Terneuzen and Greenport Shanghai. The project managers of these projects were the ones who actually saw the results of our contributions to sustainable development materialise as a consequence of their personal knowledge and experience: Maikki Huurdeman, Trudy van Megen, Rinus van de Waart, Madeleine van Mansfeld and Mark van Waes. Anne Charlotte Hoes and Barbara Regeer did the donkey work by monitoring the work processes in detail and deriving the learning experiences from them.

As a lone strategic scout at Alterra, Bert Harms prepared the way in the period between 2000 and 2004 for our projects in China and India. As account manager he brought in the WAZ-Holland Park project, and made the first contact with the Yes Bank.

In the autumn of 2004 we worked in Changzhou on our first overseas expedition. Within a week we had a concept, and in that week the skills and attitudes that we had built up along all the above-mentioned lines proved their worth. The people responsible for doing that were crucial: Jan Broeze, Chen Jianlin, Wibo de Graaff, Jan de Wilt, Pim Hamminga, Bert Harms, Jiang Jin Ming, Fransje Langers, Lu Yi, Bob Ke, Pieter Krant, Enrico Moens, Rik Olde Loohuis, Rinus van de Waart, Rene van Haeff, Madeleine Van Mansfeld, Marco Van Steekelenburg, Wang Qiang Sheng and Lucy Wassink. Parts of that design have since been implemented; whether the whole thing comes to fruition is too early to tell.

In 2006 Pim Hamminga put us into contact with Chonghua Zhang. For his part, Chonghua convinced the Shanghai Industrial Investment Corporation to design the agro-industrial zone to be developed by them on Chongming Island as an agropark, thereby laying the foundations for Greenport Shanghai. With support from TransForum we were able to get down to business, the first step being a delegation of entrepreneurs, managers, knowledge workers and innovation brokers, who went to China in October 2006 and reinvented the agropark. It is with great satisfaction but also with a certain wistfulness that I think back to the moment when we missed the fast ferry from Chongming Island to Shanghai and made a virtue out of necessity by performing the plenary presentations on the slow ferry boat. And a new concept unfolded from these presentations – not a supply but a demand-driven concept. In fewer than five days, Frans Balemans, Peter Christiaens, Ger Driessen, Martin Eurlings, Ge Lan, Lei Heldens, Martin Houben, Jan Janssen, Marcel Kuijpers, Frank Laarakker, Lu Hongmei, Henk Hoogervorst, Pierre Nijsen, Murk Peutz, Jan van Cruchten, Wim van de Belt, Wim van de Beucken, Kees van de Kroon, Rinus van de Waart, Jean van der Linden, Henk van Duijn, Henk van Latesteijn, Madeleine van Mansfeld, Trudy van Megen, Jérôme Verhagen, Jose Vogelezang and Chonghua Zhang had turned the agropark concept completely on its head. In March 2007 the accompanying Master Plan was created. The team of designers grew. Wijnand Bruinsma, Steef Buijs, Huub Heijer, Sander Mager and Leo Stumpel joined. On the Chinese side contributions were made by 40 Chinese experts headed by Ma Cheng Liang and Gao Gui Hua.

In parallel with and inspired by the results in China, the strategic collaboration between Wageningen-UR and Yes Bank in India emerged. Kees Slingerland and Rana Kapoor signed a detailed Memorandum of Understanding in March 2008 and shortly thereafter the late Chief Minister Rajasekhara Reddy gave proof of his personal involvement by laying the first stone of IFFCO-Greenport Nellore. The Conceptual Master Plan of this agropark has since been implemented down to the last detail. But it will never be finished, dynamic as it is, and Paul Bartels, Jan Broeze, Annelies Bruinsma, Steef Buijs, Peter Christiaens, Alwin Gerritsen, Else Giesen, Huub Heijer, Anton Hiemstra, Joke Hoogendoorn, Herco Jansen, Gopinath Koneti, Jack Kranenburg, Marcel Kuijpers, Chris Nab, Rik Olde Loohuis, Koen Roest, Janneke Roos Klijn Lankhorst, Edo Raus, Tons Schoonwater, Han Soethoudt, Pallavi Srivastava, Pierre Stals, Alex van Bakel, Michiel van Eupen, Frank van Kleef, Jetty van Lith, Shiva Vishnoi, Sunjay Vuppuluri and Jan Vorstermans continue to make their contributions.

The collaboration in the management team of the Indian expedition was very intensive and inspiring. Kalyan Chakravarthy, Raju Poosapati, Arjen Simons and Madeleine van Mansfeld, we keep on ploughing our furrow and our friendship keeps on growing.

ICT carries the network society and the technology that has made this thesis possible. The maintenance of hard- and software is never sufficiently recognised in the acknowledgements, but I have survived more than once the desperation of a crash and of a laptop which drank coffee, luckily without sugar. It is at those moments that ICT people demonstrate their real worth. Michiel Pieters, Jaap Spaan, Bertus van de Kraats, Jettie van Lith and Dick Verhagen: without your goodwill, this show would never have got on the road.

Thanks to Sandra McElroy and Katinka Horvath for their help with the English translation and publishing of this book.

Steef Buijs made a key contribution to this thesis by helping me to write a report on agriculture in metropolises that formed the basis of Chapter 3. Herman Agricola provided the basis for Figure 48. Henk van Latesteijn made invaluable contributions to the content and structure of this work.

The source of inspiration for the Rural Areas and Europe project, with which Expedition Agroparks began, was the 'Ground for Choices' report by the Scientific Council for Government Policy (WRR). No greater fortune has befallen me in my work since then, than that Rudy Rabbinge, main author of 'Ground for Choices', consented to be my promoter for a PhD-degree based on this book. A teacher in the true sense of the word, indefatigable and inspiring, challenging and sharp in discussion, extremely patient and, above all, ultimately transdisciplinary in the way in which he combines and holds up as an example science, politics, management and genuine involvement in the major problems facing the world.

Hanni Claassens, my love:

> '...If it be your will
> That a voice be true
> From this broken hill
> I will sing to you
> From this broken hill
> All your praises they shall ring
> If it be your will
> To let me sing...'

Stein, Lin and Janna:

> 'To propose sustainable development as intergenerational solidarity brings together healthy selfishness and systematic thinking in an evolutionary perspective'

Sustainable development is about your future. That is why this book was written. When I look at you, this future seems happy and full of promise.

Thank you, *mijn geliefden*, for making this possible.

Summary

1. Context, aim and method of this book

The world is undergoing a process of rapid urbanisation. Globalisation and the emergence of a worldwide network society are simultaneously a cause and consequence of this urbanisation process. It is to the cities that people turn in order to improve their economic lot, for improved schooling for their children and for better employment prospects. This generates an antithesis between spaces of flows, which jointly shape the worldwide network society, and spaces of place, which give each city its own local identity, history and uniqueness.

In few social activities does the tension between globalisation and local identity emerge quite so starkly as in modern agriculture. Agricultural production throughout the world is becoming an element in chains and networks involving the industrial supply of raw materials, primary production and industrial processing. At the same time agriculture is clearly evident at local level, not least in and around metropolises. Already now, the future for the most highly productive forms of agriculture (glass horticulture and intensive livestock farming) is being sought in much more far-reaching spatial concentration. In addition, the prospect of population growth and urbanisation has once again raised the question as to whether the current agricultural system is capable of continuing to feed the world population. This is not just a matter of the area under cultivation and productivity: precisely as a result of urbanisation, consumer demand is changing, there are fewer people producing agricultural products and the area of production is shrinking. At the same time the availability of critical growth factors such as water and plant nutritional substances is declining.

This publication centres around agroparks. An agropark is a spatial cluster of agrofunctions and the related economic activities. Agroparks bring together high-productivity vegetable-based and animal-based production and processing along industrial lines combined with the input of high levels of knowledge and technology. The cycles of water, minerals and gases are skilfully closed and the use of fossil energy is minimised, particularly by the processing of various flows of waste products and by-products. An agropark may therefore be seen as the application of industrial ecology in the agrosector. Agroparks are the outcome of a design process in which a new balance is sought between agriculture as it functions in global networks and the local environment of those same farms. It amounts to a system innovation, i.e. not just the innovation of agricultural production itself but also of other relationships among the stakeholders concerned. In this regard, the concept of sustainable development occupies centre-stage as a set of objectives that is simultaneously concerned with a reduction in environmental pollution, greater economic return and a better working and living environment for the people concerned.

The tension between the global network society and the local identity of rural areas and green areas outside cities provides the setting in which the design process I refer to here as 'Expedition Agroparks' takes place – at the interface between agricultural development and spatial planning. An expedition is a methodological quest, set up as a company, based around an ambition. Expedition Agroparks is the quest for sustainable development and the position of agriculture and food supply within that context, in metropolitan areas. The academic aim of this publication is to find answers to the questions whether agroparks contribute to sustainable development in metropolitan areas as well as how an agropark is developed and how its design should be arrived at.

The scientific aim of this project is to find answers to the questions whether agroparks contribute to sustainable development in metropolitan areas, how they are developed and how the design should be arrived at.

As a method, this publication combines an inductive with a deductive approach. Under the inductive approach I use working hypotheses, derived from various theories, as a guideline for testing the content and process of the designs. In the current projects the working hypotheses are used in order to intervene in the design and the process. In the deductive approach these working hypotheses are, in so far as they are confirmed by the examples, used to enrich these theories. This publication introduces this emerging theory into the scientific discourse, once again leading to the enrichment of the theoretical starting points and conclusions.

2. The network society

Agroparks fit into the context of the network and information society. It is the third development stage of humankind after the agricultural and industrial societies. Knowledge is one of the essential source of increases in productivity and power and the revolution in information technology (including gene technology) is the driver. A quarter to a third of people live in prosperity in this network society, in what is sometimes termed the 'Crystal Palace'. Under the current technical, energy-political and ecological conditions, the incorporation of all people within the Crystal Palace is impossible. The big question is whether this inequality between within and without is systematic. When it comes to agroproduction it is my premise that the productivity of agriculture can be increased to such an extent that good, sustainably produced food could be available for all people. Agroparks are put forward in this publication as a contribution to that goal.

In the network society, 'spaces of flows' form the physical organisation of social practices that divide up time and work via flows. They are without spatial contiguousness and are mutually interrelated in the network via flows of information, technology and organisational interaction, and work in timeless space. Spaces of flows are physically linked to spaces of places, but each space of place is at the same time a place where people live. 'Spaces of places' is not the antonym of the global space of flows. The global society is not just the sum total of the local spaces without boundaries. In both spaces fundamentally different experiences of time and distance arise. Processes such as reproduction, the raising and education of children and the handing down of culture are autonomous and remain the most closely related to the housing function of spaces of places.

In this contrast between the global space of flows and the local spaces of places we also find the philosophical core of the public debate about contemporary agriculture. Agroproduction has become globalised into an element of the space of flows. This process rests on an economic and ecological rationale, but for many people this is indigestible. Whereas as consumers they daily consume products from the agroproduction system, living beside a farm often entails inconvenience. In addition many citizens have difficulties with the industrial nature of modern agriculture since they have memories of how agriculture used to provide the backbone of village life in which many urban dwellers have all or some of their roots.

Communicative self-steering is the capacity of people to reflect on possibilities for holding their own in the network society. Only democratic systems have the ability to translate communicative self-steering to agencies with the capacity to regulate the space of flows. At the same time, however, this democratic system runs the risk of ossifying in an inhibiting context that is inherent in a modern metropolis and is the barrier that must be overcome in the implementation process and hence also in agropark design.

There is a different way of dealing with time in the current age from the concept of 'timeless time' that is dominant in the network society. The 'environmental movement' invokes glacial time, in which the long-term evolutionary relationship between culture as a product of human civilisation and nature is examined. Sustainable development as solidarity among various generations is the combination of healthy self-interest and systematic thinking in an evolutionary perspective. This characterisation of the environmental movement helps shape the design of modern agriculture. Such designs are driven in terms of the limiting conditions and new possibilities that arise in a world that is operating increasingly as a single global system. At the same time they have to be fully embedded in the local setting in which they function and must not make any concessions to precisely those characteristics to which local residents object.

The spatial development policy as developed by the Scientific Council for Government Policy provides a framework for dealing with the power to obstruct of self-aware citizens, whilst also setting aside a significant place for sustainable development.

3. Delta metropolises

Two dominant features of historical development of cities are centralisation and dispersal. These lead to urban hubs in agricultural areas with a tendency to extend their radius of action ever deeper into the countryside, and trading towns competing not for a grip over their immediate environment but for distant markets. Trading towns were therefore able to continue existing and developing in close proximity to one another. They depended critically for their existence on the transport infrastructure. By origin they were located on the sea or acted as meeting points between rivers and the ocean.

Since the industrial revolution this pattern has become less clear-cut with the advent of rail and road infrastructure. In addition the transport modalities have increased the scale on which cities operate. This has given rise to polycentric mega cities, i.e. metropolitan networks of larger and smaller towns, generally based around a big central city. All this makes the definition and description of metropolises in the network society particularly difficult: a metropolis is characterised by its place in the network and not by any clear-cut delimitation. An important frame of reference is provided by the advanced producer services supporting the characteristic service and knowledge enterprises in the network society: accountancy firms, banks and finance companies, the insurance industry, the legal profession, management consultants and the advertising world.

As a physical node in the network society, metropolises are places where managers and knowledge-workers live. At the same time metropolises are, viewed historically, generally places with stories and are therefore attractive to the tourist. This creates a huge demand for services, which is provided by cheap labour. But the same cities simultaneously act as a beacon of hope for all those people outside the 'Crystal Palace', who seek their fortune there in great numbers, legally or illegally. These are the urban nomads with nothing who populate the inner city areas and slums of the metropolises.

Since the Middle Ages the metropolises that have dominated the world economy have all been Delta metropolises: Venice, Genoa, Antwerp, Amsterdam, London and New York. Shanghai and Hong Kong are also port cities, located on the estuary of a river, functionally linked to a large hinterland. Delta metropolises are polycentric mega-city regions of a distinctive kind. Not only are they of historical interest but they also offer particularly favourable conditions for the development of highly productive agriculture. The Northwest European Delta metropolis comprises the area between Lille, Amsterdam and Cologne and is the cradle for the development of agroparks, with which this publication is concerned.

4. Agriculture in the network society

The development of agroparks is a phenomenon associated with the metropolises of the world. They centre around the contrast between the people inside and outside the 'Crystal Palace'. At the same time account must be taken of important shifts in power structures. Nation-states are declining in importance and the network society is made up of multilateral institutions, themselves in turn also networks. Individual people become disengaged and fall back on themselves and their primary networks. It is a context with a highly complicating effect on the system innovations at issue.

As against this there are also promising trends, which provide the basis for working on agroparks. There has been a continuous improvement in the productivity of both land and labour and, since the 1970s, in the input of fossil energy, combined with industrialisation and the far-reaching application of information technology. Agroproduction currently takes place in well-orchestrated chains and it is the clustering of these chains into networks that is at the core of the development of agroparks. The agrosector covers ever more products: apart from traditional food production there are also luxury products such as flowers, aromatic substances and flavourings and also industrial raw materials, energy and medicinal drugs.

Agriculture in the network society is a system of agroproduction with the ambition, through the new and intelligent connections inherent to the network society (between producers, sectors, raw materials, energy flows and waste flows, between stakeholders and between their value systems), of being able to satisfy the changing and competing demands of the urbanised population on a sustainable basis.

This agriculture is in the first place part of the space of flows and the Delta metropolis has traditionally offered advantages of location, which are both interrelated and mutually reinforcing. In terms of spaces of places, as a spatial challenge, the transformation to metropolitan foodclusters is more awkward since the dominant decision-making in the various chains and networks is located in the space of flows, over which regional and local governments have little if any control. Facilitation needs to take place in the spirit of spatial development policy, particularly by the provision of space for local and regional initiatives from the bottom-up. This publication discusses seven such initiatives.

5. Theoretical starting points in research by design

The scientific method used in this publication in order to analyse the various designs of agroparks is research by design or co-design. In the social sciences this method is closely related to Action Research. It is a form of engineering with regional designs as the end-products, where scientific research may take the form of feasibility and suitability studies, as well as process evaluations concerned with the generation of greater generic knowledge. The research is interdisciplinary; it covers both the natural and the social sciences, while also taking account of aesthetics, cultural history and communication.

The design produced in co-design is aimed at the generation of interventions and interventions, leading ultimately to the system innovations required for agriculture to link up with the new challenges of globalisation and the network society in the spatial planning of metropolitan areas – or, more specifically, to design and actually implement agroparks. Since this consistently involves practical spatial planning situations in which scientific knowledge is in an ongoing process of iteration with the practical know-how of the various participants in the concrete projects, it comes down to transdisciplinarity in practice.

Three theories are key for the research by design carried out on the basis of the various designs. The first of these is the resource use efficiency theory. An agropark is primarily concerned with production and the processing of vegetable- and animal-based products. The resource use efficiency theory holds that the efficiency of the agroproduction process in a chain increases the greater the yield per hectare. In this publication I investigate the hypothesis that this also applies in heightened form to agroparks.

The second theory concerns the three-dimensional landscape. This covers both physical aspects such as soil, water and vegetation (the matterscape) and the social sciences when it comes to the balance of power between people and groups in the landscape and the related economic aspects (the powerscape). Thirdly there are the subjective aspects such as aesthetics, history and communication forming part of the humanities (the mindscape). In terms of the theory of the three-dimensional landscape an agropark is regarded as a landscape in which matterscape, powerscape and mindscape each play an important role and must be specifically designed.

The third theory concerns the design process itself. What conditions must the design of a complex system innovation like an agropark satisfy for it to be a realistic prospect in present-day society? What are the steps from invention to implementation? Which parties need to be involved in the design and how do they cooperate?

A number of working hypotheses have been derived from these theories:

- An agropark realises lower costs, greater added value and lower environmental pollution per unit of output and space.
- An agropark can be arrived at only on the basis of an integral design of matterscape, powerscape and mindscape at both the global scale of Intelligent Agrologistics Networks and at the local scale of a landscape.
- An agropark is a knowledge-driven system innovation and makes a significant contribution to sustainable development

- The design and implementation of system innovations such as agroparks necessitates the participation of knowledge institutes, entreprises, NGOs and governmental organisations, (acronym KENGi). It is a transdisciplinary process in which the explicit knowledge of research institutes and the tacit knowledge of the other partners are developed in a process of continuous iteration. KENGi brokers act as the facilitators of this transdisciplinary process.
- In all the decision-making concerning the realisation of the integral agroparks design, involving matterscape, powerscape and mindscape aspects, arguments from the world of justice and trustfulness prevail over the arguments from the world of truth.

These working hypotheses are developed further on the basis of the various agropark designs and result in an answer to the question whether the development of agroparks does in fact make a contribution to sustainable development. Together they make contributions to the method used of research by design as a specific elaboration of the Landscape Dialogue or Co-Design method.

6. Research by design applied to agroparks

Deltapark

Deltapark is a theoretical design of an agropark in the port of Rotterdam. It provides for the regional clustering of production and processing of animal proteins, glass horticulture, waste separation and animal feed production, which are integrated with the chemical industry in the port. A feasibility study on the basis of the design worked out positively and pointed to substantial social benefits. The most important goal of the design was to place the public debate concerning agroparks on the agenda, but no communication plan was drawn up for this aspect. Deltapark certainly met this goal and initiated the public debate concerning system innovations in the future development of agriculture. The limited attention to communication strategy during the design of the plan elicited a sharp reaction on the part of various NGOs (focusing on the environment and animal welfare) following the presentation, as a result of which the plan became stigmatised (as a 'pig flat').

Agrocentrum Westpoort

The Amsterdam Port Authority, the Green Space and Agrocluster Innovation Network and Wageningen-UR worked between 2002 and 2006 on Agrocentrum Westpoort, a design for an agropark involving pig production, fish cultivation and vegetable-based production in modular, seven-storey buildings, combined with a slaughterhouse and co-fermentation plant. To begin with the design was intended for the Amsterdam docklands. The plan was for industrial-environmental relationships to be expanded with other companies in the docklands. A cost-benefit analysis conducted by Ballast

Nedam indicated the potential for major operational cost savings in comparison with conventional firms not working in a cluster. In addition it would be possible for a one-off saving to be made in the form of a discount on the procurement of the necessary pig-production rights. The environmental benefits were also convincingly demonstrated.

The design was the subject of intensive discussions with potential stakeholders. Ultimately none of these came forward as a pioneer. Subsequently the concept lost the support of the Port Authority.

The absence of primary producers and processors proved a particularly big handicap in the working process. Furthermore the participants in the design process had difficulty in gaining acceptance among their own grassroots supporters for the leap in innovation they were advocating. The design therefore had to explore the scope for compromises: there was no question of a free space.

Agrocentrum Westpoort was not implemented. It did however become the basis for an Agrologistics Community of Practice that produced new designs. The design assembled a great deal of basic knowledge, which was successfully used in later designs. The most important lesson, however, came from the design process itself. It became clear that active communication campaigns were needed in order to lift the public debate concerning intensive livestock farming out of its defensive mode and to do something about the social stigma that had become attached to intensive livestock farming. Since the design process itself consistently arose out of a broadly-based forum of KENGi parties, this stigma and the lack of effectively formulated counterarguments also seriously hindered the design process.

New Mixed Farm

Efforts to establish an Agropark had been made in North Limburg since 2001. This is now taking specific shape in the form of the New Mixed Farm, an initiative on the part of agricultural entrepreneurs in the Agricultural Development Area in the Municipality of Horst aan de Maas in North Limburg. In combination with a bioenergy power plant, New Mixed Farm is designed to provide sufficient space for 35,000 pigs and 1.2 million chickens. In 2003 the project was adopted by the Agrologistics Platform and by TransForum. This resulted in the setting up of a Steering Group to support the project group consisting of the entrepreneurs and the KnowHouse innovation broker. In 2004 Minister Veerman promised the project 'separate status'.

When the local planning procedures were instituted this gave rise to protests among local citizens, supported by a number of political parties and national action groups. Notwithstanding this, plans for the Witveldweg reconstruction area, including New Mixed Farm, were accepted by the Municipal Council on condition that the sustainability of the project could be demonstrated. To this end a supplementary scan was conducted, which reached a positive conclusion and was adopted by the Municipal Council. The most important contributions to sustainable development are a reduction in ammonia emissions, the internal energy-generation of the project, lower emissions of greenhouse gases and, at regional level, a significant reduction in stench nuisance. Veterinary risks are lowered by the reduction in transport, while New Mixed Farm makes a major contribution to the improvement of public health by minimising the use of antibiotics through targeted management. The project generates considerable employment and leads to an improvement in the working conditions in the participating enterprises.

The decision-making procedure in respect of New Mixed Farm has not gone smoothly. Despite the promised 'separate status' it was to have in the decision-making procedure, all sorts of additional tests and special rules were devised with which the project had to comply. The criteria laid down by the various parts of the government were divergent and sometimes downright contradictory.

The various KENGi partners provided the project group with intensive support and guidance in the form of the New Mixed Farm Steering Group. Heavy investments were made in external communication, in which sustainable development, innovation, the open structure and the socio-economic development of the area were consistently emphasised. Having made explicit their choice, the national and local governments have abided strictly by the formal requirements of the various procedures.

The entire working process in the research by design of regional dialogue and New Mixed Farm may be summarised as a transdisciplinary process, in which person-related knowledge concerning the participants from the KENGi parties is embedded by means of scientific tests in successive iterative rounds and internalised by all the parties, after which the process commences a fresh cycle.

As a system innovation, New Mixed Farm has not yet fully proved its right to exist. For this it still requires 'approval' in the various currencies of the KENGi parties concerned, which can be achieved only given by a realignment of their mutual relationships.

With regard to three of the four KENGi groups concerned, it is already clear at this stage that in terms of their own currency they regard the project as an innovation: apart from the entrepreneurs this certainly applies to the knowledge institutes. Similarly the support that the project has received – at least verbally – from politicians and civil servants may be interpreted as an attempt to allow this innovation to take root in terms of licensing policy and legal rules. The project is rejected only by a number of NGOs – although where it really counts, namely in the political deliberations of the municipal council, they are in the minority.

New Mixed Farm is the most advanced agropark in the Netherlands involving intensive livestock farming and in which all other sectors initially dropped out, partly on account of the image problems faced by intensive livestock farming in the Netherlands. The long development process has successively shown how hard it is for SMEs to invest long term in complex system innovations and how difficult it is for the government to facilitate such innovations in terms of legislation and regulations that are largely oriented instrumentally towards the regulation of the existing situation and which have been worked out in fine detail.

New Mixed Farm is first and foremost a regional design. It concerns a spatial concentration of and the development of synergy among existing farms in the North Limburg and East Brabant Peel region, with a dominant role being played by the Limburg provincial government, local municipal councillors and the KENGi broker KnowHouse, whose focus is primarily regional. New Mixed Farm is expected to be opened in 2012.

WAZ-Holland Park

Wujin Polder forms part of the hinterland of Changzhou in the Chinese province of Jiangsu and is being developed by the Wujin Agricultural Zone (WAZ) Authority. Alterra delivered a Master Plan for the polder in late 2004. The plan provided for a high-tech agricultural development zone designed to act as a regional exemplar in the form of an agropark involving both Chinese and foreign firms. The agropark would bring together animal production, vegetable-based production and mushroom production. Storage and co-fermentation of manure and biomass is to take place in a Central Processing Unit (CPU), where these will be converted into biogas for the generation of electricity, and CO_2 and heat for use in the park. The CPU also contains a composting facility. The Master Plan furthermore includes a recreational area with Chinese and Dutch agriculture as the theme; and a marketplace for the sale of agropark products; and a meeting place.

The WAZ-Holland Park Master Plan comprises a group of inventions. The two most important of these are the integration of agricultural production and processing in a CPU and the combination of highly productive agricultural production with recreational and educational facilities. The design has moreover been implemented as a complete landscape plan, including detailed water management and surface and underground infrastructure.

During the design of the process the KENGi network was not complete. No NGOs were involved and just one entreprise from the Netherlands. The design was arrived at through research by design, completed in China in the space of one week and later elaborated in the Netherlands. In this working process, transdisciplinary working took shape between Dutch designers and scientists, representatives of the Dutch government and an entreprise and Chinese government officials.

Taking the Master Plan as the point of departure, talks were held from May 2005 onwards concerning the implementation of the plan. Agriculture turned out to be unknown territory for the Chinese project developer. WAZ-A has proved to be less willing to undertake risky investments in this area than in industrial, non-residential and residential building, where it generally delivers turnkey projects.

For the development of agroparks the Chinese work on the basis of a comparable consortium structure to that used in industrial investments. They expect a foreign partner to take a significant share of the investment in China, thereby disregarding the fact that the potential investors in both China and the Netherlands are small and medium-size enterprises and that these have neither the presupposed commitment nor the capacity for such investments. Since the talks about joint implementation came to a halt, the WAZ-Authority has begun to develop elements of the plan for which Chinese investors could be found.

Biopark Terneuzen

An agropark is being built in the dock area of Terneuzen. Under the direction of De Bunt consultants and with the support of TransForum, the Province of Zeeland and the Municipality of Terneuzen, a coalition of knowledge institutes drew up a trend scenario, in which existing lines of development were projected forward, together with three agropark scenarios, in which activities would be clustered on an ever-growing scale. Between 2005 and 2007 a partnership scheme was set up among the existing firms in the dock area, and the development of a 240 ha glass horticulture area was set in motion. From 1 July 2007 this joint venture was formalised by all the stakeholders concerned under the name of Biopark Terneuzen. Under the plan an existing fertiliser producer will cooperate with glass horticulture firms in a new cluster covering over 200 ha. New industrial functionalities such as biomass processing, bio-ethanol production and the purification and production of various grades of water are being added to this. To begin with the project is concerned with linkages between the industrial firms already operating in the Sloe area, but the return becomes much greater with the addition of primary agricultural production. The biomass plant will draw 50% of its manure requirements from the intensive livestock farming in the province of Zeeland.

The exchange of waste products and by-products among these firms will result in lower costs, lower environmental emissions and a lower take-up of space. Biopark Terneuzen is expected to generate 2350 new jobs: 80% of these will be accounted for by new glass horticulture businesses. The extra added value on the basis of extra employment is estimated at € 42 m/year.

The Biopark Terneuzen project has created a new collaborative venture between the University of Ghent, which has been building up expertise in industrial ecology and process theory for a considerable time, and Dutch knowledge institutions concerned with agropark development.

Where it builds on the knowledge developed in the Agrocentrum Westpoort project and New Mixed Farm, the Biopark Terneuzen working process may be regarded as a form of co-design. The most important substantive learning experience from Biopark Terneuzen is undoubtedly that the non-inclusion of animal production has helped overcome obstacles to acceptance both at government level and among the industrial players. As a consequence Biopark Terneuzen did not become the focus of national environmental groups in the way that Agrocentrum Westpoort and New Mixed Farm did. The lack of intensive livestock farming has substantially reduced the implementation period for the project.

By way of extension of the original design, a new project, financed by the EU Region, the Flemish Region and the Netherlands, providing for the establishment of a pilot plant for bioenergy processing, was approved at the end of 2008.

Biopark Terneuzen is in brief a success story. The vigorous approach that has been adopted and the continuing increase in urgency of the energy issue also give the project an attractive forward-looking orientation, which can be worked on further. Furthermore the design in question enables the spatial concentration in Biopark Terneuzen of the widely distributed intensive livestock farming in Zeeland-Flanders, instead of further proliferation in rural areas. At least in the Netherlands, however, the selected development path of large-scale manure processing first and then only as a secondary element the addition of intensive livestock farming is much more skilful in strategic terms.

Greenport Shanghai

Chongming Dao is an island north of the city of Shanghai in the estuary of the Yangtze River. This is where Shanghai Dongtan Ecocity – intended as a model of sustainable urban development – is being developed. The project has been placed with the Shanghai Industrial Investment Company (SIIC). Dongtan Ecocity is divided into four zones, aimed at the development of a garden city, an area with offices and educational functions, a nature conservation zone with wetlands and a 27 km^2 area for modern agriculture. For the development of this agricultural area SIIC also sought collaboration with Tongji University and Wageningen-UR in a project being supported by TransForum: Greenport Shanghai.

In October 2006 TransForum and KnowHouse sent an initial mission to Shanghai. This included all the entrepreneurs participating in New Mixed Farm, plus colleagues involved in the North Limburg network of KnowHouse.

During the mission the group formulated a number of principles with which the design for Greenport Shanghai would need to comply. They proposed no longer working on the blueprint principle but first setting up a demonstration park in which the divergent aspects of industrial agriculture could be seen in practice. This demonstration park will enable research to be conducted into market and production conditions, while by adding a trade park to the demonstration park in a second stage, products will be able to reach that market quickly allowing the true market scale of those products to be established. Not until it is evident that there is a market on a sufficient scale for a particular product and that the product can also be optimally produced in Shanghai will the chain in question be set up in Greenport Shanghai on as integral a basis as possible.

This *Demo>Trade>Processing>Production* framework represented a radical transformation of the design principles from the previous supply-driven design to demand-led design. At the same time, however, the SIIC continued to seek a highly detailed Master Plan for Greenport Shanghai. That Master Plan accordingly lays down no-regrets specifications for the environment, water, zoning and the main infrastructure. In terms of this no-regret plan four scenarios for production and processing have been elaborated on the one side in order to clarify the internal requirements that this no-regret plan will need to comply with, and on the other side to set limits in later stages on the development of Greenport Shanghai in respect of the debate that will take place concerning trade, production and processing, In each scenario the various elements have been balanced in terms of their input to and output from, the CPU, the most important criteria at all times being a sufficiently large scale for economic activity and a minimal environmental burden.

The public debate concerning animal welfare in intensive livestock farming as conducted in Northwest Europe is virtually absent in China. In the design for Greenport Shanghai, the existing Dutch animal welfare standards have therefore been maintained and actively brought to the attention of the Chinese clients, in so far as a connection could be established between productivity and inconvenience. Otherwise the position has consistently been taken that the Chinese clients should specifically ask for the deployment of state-of-the-art technology from the Netherlands.

After the completion of the Master Plan, a brief audiovisual presentation with English and Chinese text was drawn up on the basis of a three-dimensional virtual simulation of the design. This laid down 'Greenport Shanghai' as the new brand name, whereby the generic trade name provides the means for fostering the concept of agroparks outside the Netherlands. The SIIC supports the idea of using Greenport not just as the designation for new agroparks but also as a brand name guaranteeing top quality.

Seen in that light Greenport Shanghai would need to obtain the services of a quality controller, for subsequent implementation as the Greenport network is extended.

The Master Plan Greenport Shanghai is an invention. During the Greenport Shanghai design process, the KENGi network developed a new planning methodology that was strongly inspired by the Dutch concept of spatial development policy as advanced by the Scientific Counsel for Government Policy (WRR). The Master Plan played its intended role in the formal Chinese planning system, even though the scenario method meant that it lacked the blueprint approach that would normally be expected of such a plan in China. The municipality and province of Shanghai accordingly sought and obtained experimental status for this departure from normal practice. In the meantime the methodology used has been drawn to the attention of the Chinese national Development and Planning Committee. In this sense there has been an innovation in the domain of the Chinese government.

On 1 July 2007 this phase was completed with a presentation of the Master Plan in Shanghai and the Master Plan was formally approved. Since then the implementation has continued along three lines. The SIIC has started to elaborate the no-regrets plan. TransForum has begun to set up an international knowledge network based on the collaboration between knowledge institutes and Greenports in the Netherlands. This Platform on Innovation of Metropolitan Agriculture was established in Beijing in October 2007. On behalf of the province of Limburg, KnowHouse began to organise and recruit entrepreneurs interested in investing in Greenport Shanghai. A relevant factor in this regard is that the target group of Dutch companies consisted primarily of small and medium-sized enterprises, most of which were too small to take the step of investing in China. A support network will need to be organised in the Netherlands in which government authorities, companies, financiers, knowledge institutes and other interested parties share the risks and potential benefits of the initial steps being taken at Greenport Shanghai.

The basis for cooperation is that the Dutch invest primarily in terms of know-how and time (and to a limited extent with money) and that the large-scale financial investments are made by the SIIC and via joint ventures with Chinese entrepreneurs.

The most important learning experience to emerge from the various evaluation and reflection meetings with participants is that to an even greater extent than in the Dutch situation, where KENGi networks seek to bring about an innovation leap, the creation of confidence is vital in an international, multicultural situation. This takes time. In the complexity of a multicultural environment, this means consistently and step-by-step enticing new participants by showing them just what has already been achieved.

IFFCO-Greenport Nellore

At the time of writing, the development of IFFCO-Greenport Nellore is in full swing. This example has been included on account of the powerscape development that led to the first agropark design in India. A critically important factor in this regard was a strategic partnership between Wageningen UR and the Indian Yes Bank from September 2005 onwards, which provides the basis for the various agropark projects in India. Partly on the basis of the experience gained with the WAZ-Holland Park project, the collaborating knowledge institutes of Wageningen-UR concluded that working in a multicultural setting would benefit in the long term from cooperation with local strategic partners, with which multiple projects could be tackled simultaneously.

The work in India also made us aware of the importance of the logistical infrastructure that has arisen in Northwest Europe over the course of several centuries and which forms an essential part of this system, for which reason the theme of agrologistics has been to the forefront of public attention in recent years. In India just 2% of all the food produced is processed. The rest is brought to market in fresh form by means of a totally underdeveloped infrastructure. The incorporation problems in the existing situation therefore call for much more attention than in the designs to date. The designs of agroparks in India accordingly start with Intelligent Agrologistics Networks (IAN), in which agroparks together with distribution and consolidation centres are embedded. An IAN ensures that large-scale and industrially produced products from the agrosector are brought to consumers throughout the world by means of the finely-meshed network of supermarkets and speciality stores in the metropolises. With a view to the producers wishing to bring their products to market via an IAN, the concept of Rural Transformation Centres (RTCs) was developed which, apart from the range of agricultural products, also cover activities concerned with rural development. RTCs have also been included in the IAN. RTCs provide the connection between the agropark and the local communities.

In India the agropark designs are in the first place aimed at the creation of processing capacity and provision of the associated storage facilities. Secondly they are directed towards the new market demand generated by the rapidly growing middle class and the out-of-home market in India. Here too the solutions consist primarily of new ways of processing existing products and of other logistics. This transition cannot succeed without the introduction of other forms of production, a large part of which can be added to the agroprocessing in agroparks. IFFCO-Greenport Nellore is therefore being set up as an element of an Intelligent Agrologistics Network that also includes other agroparks and in which consolidation centres provide the connecting link between the large-scale production in the agroparks and the demand for agroproducts in the metropolis.

Since November 2008, Yes Bank and Wageningen-UR have organised a number of business missions from the Netherlands and Israel, during which they have made a start on organising joint ventures with Indian firms that will be engaged in activities in the agropark, RTCs and/or the IAN. Notably lacking in this network are specialised KENGi brokers. This role has been taken over by the collaborating knowledge parties.

The working process in India began with the creation of a basis of support within the Federal government and among the various state governments. In a certain sense this may be compared with the role that the debate concerning Deltapark in the Netherlands played in the generation of public support for or alternatively opposition to agroparks. The design process of the concrete IFFCO-Greenport Nellore project was set up as a Landscape Dialogue, to which various scenarios were added. The scenarios were not however worked out in parallel but in an iterative process, in which various entrepreneurs were given the opportunity to inject their wishes.

The aforementioned working hypotheses have been developed further for each example in the research by design and where necessary reformulated or supplemented by other working hypotheses. Ultimately the research by design results in the following conclusions:

- An agropark as knowledge-driven system innovation realises lower costs, more added value and reduced environmental pollution per unit of output and surface area and in so doing makes a significant contribution to sustainable development.
- An agropark can only come into being on the basis of an integral design of matterscape, powerscape and mindscape at both the global scale of Intelligent Agrologistics Networks and the local scale of a landscape.
- The design and implementation of system innovations like agroparks necessitates the participation of knowledge institutions, enterprises, NGOs and governmental organisations and a positive outcome from an integral test on sustainable development. It is a transdisciplinary process in which the explicit knowledge of knowledge institutions and the tacit knowledge of the other partners are developed in a process of continuous iteration.
- The organised campaigns against agroparks as a system innovation by organisations such as Friends of the Earth Netherlands, the Socialist Party and the Animal Protection Foundation, all of which are concerned with environmental and animal protection, totally ignore the demonstrable improvements to the environment and animal welfare that these agroparks can bring both in the Netherlands and elsewhere in the world.
- Design of the agropark orgware with knowledge institutions, enterprises, governments and citizens from the local area, where the park is to be implemented, can only take place once sufficient trust has been built up between these parties on the basis of an appealing matterscape design, in which there are still many open options.

- Greenport seems to be an attractive international logo, under which a broadly formed network of knowledge institutions, governments and entrepreneurs, focusing on open innovation in the agrosector, can propagate system innovations and quality management in the global network. For this purpose Greenport must embrace the entire Dutch agrosector, and extensive collaboration and synergy between the existing and future Greenports in Dutch is essential.

7. Discussion

To conclude I would submit the conclusions from the research by design. First of all to the theory.

The first conclusion from the research by design is that the resource use efficiency theory will also apply in heightened form to the complete industrial ecological complex operating in agroparks. The resource use efficiency of the parks is greater the more complete the level of integration. This also applies to the agrosystems themselves, if they are integrated with one another, to the integration of elements in the chain, and to the combination of production, processing and trade. It even applies to the combination of integral chains in the associated logistics and to the integration of the supplying and processing industry and primary agricultural production. In doing so, the level of integration in the theory has been extended from crop plots to crop systems and land use and from there to complex industrial ecological systems or more accurately industrial ecological networks, part of which is spatially clustered. The list of resources expands correspondingly from physical inputs to factors such as logistics and market knowledge or in other words from matterscape to powerscape and mindscape.

In this way the theory generates a call for far-reaching clusters of industrial agriculture in agroparks and for the integral design of agroparks and the associated intelligent agrologistics network as a response to the worldwide movement towards a network society and urbanisation. In its classical formulation the resource use efficiency theory was a plea for multidisciplinary cooperation between the natural sciences, agricultural sciences and economics. In its application in the WRR report Ground for Choices it acted as an invitation for cooperation among politicians, policy-makers and scientists. On the basis of the examples described here, and based on the resource knowledge derived from the cooperation among the KENGi partners, it now invites transdisciplinary cooperation among scientists in the domain of explicit knowledge and entrepreneurs and other stakeholders in the domain of tacit knowledge. With the success of transdisciplinary cooperation among stakeholders, the resource use efficiency theory consequently becomes a plea for the participatory approach in powerscape.

In the Space of Place the design of the concrete agropark takes place in the three dimensions of the landscape: matterscape, powerscape and mindscape. In this 3D landscape the resource use efficiency theory fulfils its ever more far-reaching promises of higher productivity combined with the most efficient possible input of the factors of production. A learning system arises that is capable of adaptation and that is continually enlarging its responsiveness to its immediate environment. In the Space of Flows the agropark forms part of an intelligent agrologistics network comprising agroparks, supply areas with land-dependent agriculture and consolidation centres.

Together, these metropolitan networks form the global network society in which they each have their own unique position; on the one hand in relation to important centres of production of primary commodities, energy, animal feed, commodities from the biobased economy and waste products and by-products from agroproduction; and on the other hand in relation to other leading metropolises in which the consumers live, who in turn generate the demand for the products produced in the park.

When it comes to the landscape theory all the examples allow the generic conclusion to be drawn that in the absence of structural cooperation among, at the least, knowledge institutes, governments and industry, such projects do not get beyond a plan on paper. For the Dutch situation this can be taken further on the basis of the second and third conclusions in the light of spatial development policy. In the Netherlands there is no hierarchy between matterscape, powerscape and mindscape in the sense that truth, justice and trustfulness dominate at different points in the gestation process and can each be decisive at certain points.

The level of scale of the landscape of which an agropark forms part corresponds the most closely with the municipal level of government. It is at this level that the cooperation among knowledge institutes, government authorities, industry and citizens in the design process is required to take place. In terms of the operation of the resource use efficiency theory, an agropark reflects important generic principles of Dutch spatial planning:

- The concentration of urbanisation: agroparks redefine highly productive forms of agriculture as an urban activity (i.e. metropolitan foodclusters) and reorganise these spatially into the most compact area possible.
- Spatial coherence: agroparks localise the industrial agricultural activities in the urban area, close to multimodal logistical hubs, thereby combating congestion.
- Spatial differentiation: agroparks facilitate an exchange process whereby industrial agricultural activities with a small direct but large indirect claim on space disappear from rural areas, thus creating more space for other functions.

The methodological innovation emerges in particular in the different approach towards the working process developed in the various projects. System innovations are key, hence the demand for fundamentally different relationships among the KENGi players. The knowledge institutions make the switch from interdisciplinarity to transdisciplinarity. Government authorities do not just get around the table to hand out directional plans with the intention of testing the resultant designs later on but create an experimental space in which they themselves become co-designer. Businesses enter into joint ventures with knowledge institutions and also sell their knowledge. The KENGi partners must also concern themselves with the powerscape: what is the innovation ambition in the process; how are the experimental spaces organised; when do which parties have legal, political and/or publicity power and where are they prepared to forego executive power? The parties need, moreover, to be aware of their shared mindscape: how do they deal with the various kinds of knowledge and with emotion, aesthetics and education?

In addition, the resource use efficiency theory provides a number of points of reference during the process for the *ex ante* testing of, in particular, profit and planet aspects of sustainable development.

For the input of classical resources such as nitrogen, phosphate, biocides and fossil energy, it can simply be argued that the aim must be to minimise each party's consumption per unit product. The theory shows in particular this trade-off needs to be made on a coordinated basis.

Matters become more difficult with the powerscape and mindscape resources – such as space and knowledge. To some extent these can be expressed in money terms (land prices, the hourly rates of knowledge workers), but that value does not reflect everything. How should we deal with a stench circle when it comes to land prices? Even more difficult is the testing of aspects such as animal welfare or aesthetic aspects on the basis of objective norms. In a generic sense it may be concluded that the KENGi parties need to state their opinions in these areas, both during the design process itself and later in the form of a regular, recurrent discussion of these aspects.

Taking the projects as a whole, the methodological conclusion may be reached that an initiative undertaken from the bottom-up alone has less chance of success then if there are two orchestrated lines, one consisting of the realisation of top-down support and commitment which is then worked out and established on the basis of bottom-up designs of concrete projects.

Research by design therefore leads to insight concerning the wider applicability of the theories used. The applicability of resource use efficiency, previously applied to crop plots, crop systems and land use, was extended to industrial ecology but also helps shape the input of resources such as space and knowledge. With the latter this theory becomes a plea for participatory planning. The landscape theory gives expression to spatial development policy, particularly the way in which this should be applied at regional level. The Co-design theory is enriched with knowledge concerning the necessary partnership arrangements, sustainable development as a quality objective and transdisciplinary co-operation. The method of research by design itself may also be added as a combination of induction and deduction to the repertoire of Co-design as a design theory.

1. Context, aim and method

The world is undergoing a process of rapid urbanisation (Figure 1). Half of the world's population now live in cities and according to the United Nations this proportion is set to rise to 70% in 2050[1].

Globalisation and the emergence of a worldwide network society are simultaneously the cause and consequence of this urbanisation process[2]. Cities in this society assume many forms: in addition to the traditional central cities (Paris, Berlin, Moscow) there are extended mega cities (New York, Tokyo, Mumbai, Beijing) or yet bigger areas, sometimes deltas, which become completely urbanised, such as the Pearl River Delta in South China or the Southern Yangtze Delta between Shanghai and Nanjing. Our own environment is described by urban planners as a 'polycentric metropolis'; some people restrict this to the Randstad, others mean the Dutch city ring and still others

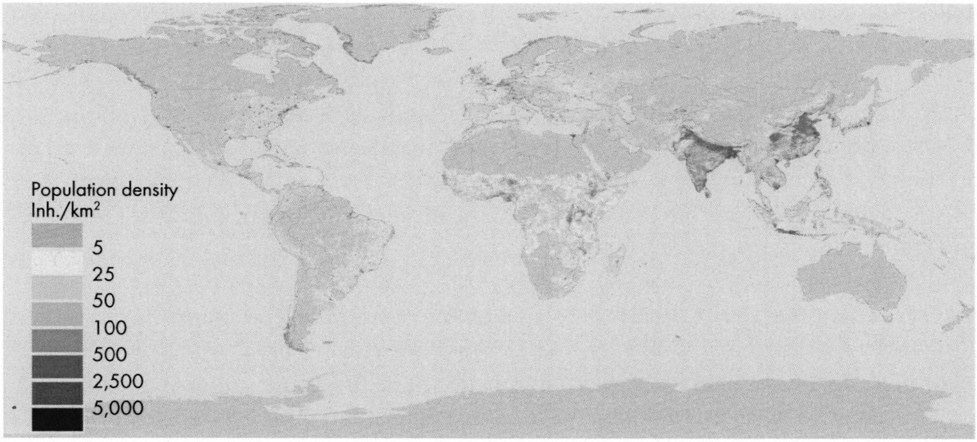

Figure 1. Global population density.
Stichting Onderzoek Wereldvoedselvoorziening van de Vrije Universiteit Amsterdam (2009).

[1] United Nations Department of Economic and Social Affairs/ Population Division (2008). *World urbanization prospects. The 2007 revision.* United Nations, New York, USA: 1: 'The (...) world population will reach a landmark in 2008: for the first time in history the urban population will equal the rural population of the world and, from then on, the world population will be urban in its majority. (...) Between 2007 and 2050, the world population is expected to increase by 2.5 billion, passing from 6.7 billion to 9.2 billion (...). At the same time, the population living in urban areas is projected to gain 3.1 billion, passing from 3.3 billion in 2007 to 6.4 billion 2050. Thus, the urban areas of the world are expected to absorb all the population growth expected over the next four decades'
[2] Castells M. (2000). *The information age: Economy, society and culture. Volume 1: the rise of the network society.* Blackwell, Oxford, UK.

have the Northwest European Delta in mind[3]. Urbanisation is highly evolved in Europe and North America. Figure 1 shows that the major population concentrations are in Asia and Africa, and it is precisely there that urbanisation is happening most rapidly. It is to cities that people look for financial security, a better education for their children and job prospects, and they migrate there in great numbers for these reasons. In cities an antithesis is emerging between spaces of flows, which jointly shape the worldwide network society, and spaces of places, which give each city its own identity, history and uniqueness. In terms of space many metropolises are a sequence of built-up places and open spaces, formerly known as countryside. Where there was, in the past, a sharp division between city and countryside, this distinction has disappeared in and around metropolises. In the polycentric metropolis of North West Europe it has certainly become increasingly difficult to identify places that are rural in the traditional sense of the word. The city is everywhere.

In few social activities does the tension between globalisation and local identity emerge quite so starkly as in modern agriculture. Agricultural production throughout the world is becoming an element in chains, involving the industrial supply of raw materials, primary production and industrial processing. These chains are all connected to each other in a global network. At the same time agriculture is explicitly present at the local level, not just outside cities but in and around metropolises and network cities. In an urbanised country like the Netherlands agriculture takes up about two thirds of the total space. But even this relatively extensive use of space, which urban dwellers are increasingly questioning, is under pressure. For decades agriculture has been squeezed of space that is used for urbanisation, road networks, nature reserves, and water management, etc.

Industrialisation of agriculture is barely taken into account in the spatial development of modern metropolises. From its own perspective agriculture is a process that still takes place in the rural area, is the main player there and wants more space in which to grow. But most other residents in the same place are inhabitants of the metropolis and while they live 'in the country' they live an urban life as regards work, health care facilities, culture and education. The demands made of agricultural production from the global network are leading to major tensions at the local level. Conversely, demands that neighbouring populations in a concrete location make of their farming neighbours, can often not be met because this weakens the farmers competitive position in the global network. Further development of agriculture, even if this takes

---

[3] Hall P. and K. Pain (2006). *The polycentric metropolis. Learning from mega-city regions in Europe.* Earthscan / James & James, London, UK, 256 pp.
Smeets P.J.A.M., W.B. Harms, M.J.M. Van Mansfeld, A.W.C. Van Susteren and M.G.N. Van Steekelenburg (2004). Metropolitan delta landscapes. In: Tress G., B. Tress, W.B. Harms, P.J.A.M. Smeets and A. Van der Valk (eds.) *Planning metropolitan landscapes. Concepts, demands, approaches, Delta series.* Wageningen University, Wageningen, the Netherlands, pp. 103-114.

place in the countryside as it used to, is becoming increasingly difficult to implement in most places in the metropolis. Already, the future for the most intensive forms of agriculture (glass horticulture, intensive livestock farming) is being sought in much more far-reaching spatial concentration, as was the case in the 1960s with the expanding industrial activities in cities.

In addition, the prospect of population growth and urbanisation has once again raised the question of whether the agricultural system is capable of continuing to feed the world's population. That is not just a matter of the area under cultivation and productivity. More people in the cities means more wealth and buying power, which includes more money to buy food. Precisely because of urbanisation, consumer demand is changing, there are fewer people producing these products and the productive area is shrinking. At the same time the availability of critical growth factors like water and plant nutrients is diminishing.

This publication centres around agroparks and the central question is whether they contribute to sustainable development in the metropolitan areas where worldwide urbanisation is expressing itself most explicitly.

An agropark is a cluster of agrofunctions and related economic activities on or around a location. Agroparks combine highly productive plant and animal production and processing in industrial mode with the input of high levels of knowledge and technology. The cycles of water, minerals and gases are skilfully closed within the cluster of different chains, and the use of fossil fuel is minimised, particularly by the processing of various flows of waste products and by-products in the chains. Non-agricultural functions such as energy production and waste and water management can also be integrated in the industrial process. An agropark may therefore be seen as the application of industrial ecology[4] in the agrosector. Agroparks are the outcome of a design process, in which a new balance is sought between agriculture as it functions in global networks, and the local environment of those same farms, especially in metropolises. An agropark is a systems innovation, i.e. not just the innovation of agricultural production itself but also of other relationships between the stakeholders concerned. In this regard the concept of sustainable development occupies centre-stage as a set of objectives that are simultaneously concerned with a reduction in environmental pollution, greater economic return and a better working and living environment for the people concerned.

Agroparks are currently the most far-reaching form of spatial concentration in the agricultural system[5]. They are the engine of the space pump in the countryside: an increasingly productive agricultural production and processing industry is concentrated in a limited productive area, and as a consequence there is much less space needed for agricultural use[6]. The focus in this publication is on agroparks but the final practice project under discussion in Chapter 6 will show that the agropark in most recent designs has become a part of an Intelligent Agrologistic Network, in which consolidation centres and rural transformation centres take on other important roles.

---

[4] The concept of industrial ecology was first used in industrial areas belonging to the chemical industry, whereby means were sought to use waste material from one production process as raw material for another. Frosch R.A. and N.E. Gallopoulos (1989). Strategies for manufacturing. *Scientific American* 261: 144-152.. The website of the *Journal of Industrial Ecology* defines industrial ecology as 'a rapidly growing field that systematically examines local, regional and global materials and energy uses and flows in products, processes, industrial sectors and economies. It focuses on the potential role of industry in reducing environmental burdens throughout the product life cycle from the extraction of raw materials, to the production of goods, to the use of those goods and to the management of the resulting wastes. Industrial ecology is ecological in that it (1) places human activity – industry in the very broadest sense – in the larger context of the biophysical environment from which we obtain resources and into which we place our wastes, and (2) looks to the natural world for models of highly efficient use of resources, energy and byproducts. By selectively applying these models, the environmental performance of industry can be improved. Industrial ecology sees corporate entities as key players in the protection of the environment, particularly where technological innovation is an avenue for environmental improvement. As repositories of technological expertise in our society, corporations provide crucial leverage in attacking environmental problems through product and process design' http://www.wiley.com/bw/aims.asp?ref=1088-1980&site=1, accessed on 29 July 2009.

[5] Although illegal, Dutch cannabis cultivation, usually in its tolerated form, which goes on in numerous attics and in only a few square metres per production location, can also be a very lucrative form of land independent agriculture.

[6] Intensive livestock farming no doubt needs large areas for arable farming in other parts of the world to produce fodder for livestock, but in these arable farming systems too, productivity is slowly rising and increasingly less space is needed per unit of product.

1.1 Scientific objective

The design process that I refer to as 'Expedition Agroparks' takes place at the interface between agricultural development and spatial planning. At the heart of it is the tension between the global network society and the local identity of rural areas and green areas outside cities.

I borrowed the term expedition from the description by Sloterdijk[7] of expeditions in the period of globalisation. An expedition is a methodological quest, set up as an enterprise, based around an ambition. The expedition in this book, the quest for new forms of agriculture in the metropolis from the perspective of sustainable development, began around 1992 in the spatial planning policy. From 1996 onwards it proceeded from the scientific institute Alterra, part of Wageningen-UR, that has the relationship between spatial planning and agriculture as one of its main focuses. Thereafter the agropark designs were further developed by working on system innovations for sustainable development with the focus on agriculture at TransForum.

Expedition Agroparks is the search for sustainable development and the position of agriculture and food provision within metropolises. In this publication this form of agriculture is referred to as metropolitan foodclusters. The book contains not only a report on this expedition but also a more in-depth theoretical explanation. Using various designs I shall give a scientific response to the question of whether modern agriculture can be part of the sustainable development of the metropolis. This is not only an ecological question about the relationship between agriculture and its physical environment. It is also a question about the economics of the new designs: are they profitable? Do they meet market demand? And it is also a question that looks at the needs of people in and around the system. Is it feasible? Does it provide an inhabitable landscape?

In addition to these questions aimed at the content of the designs, I will also provide a scientific assessment of the quality of the design process itself. Has this really been a successful innovation? Does this process do justice to the various stakeholders and has their knowledge been sufficiently taken into account? What are the learning experiences?

---

[7] Sloterdijk P. (2006). *Het kristalpaleis. Een filosofie van de globalisering.* Uitgeverij Boom/SUN, Amsterdam, the Netherlands: 105. 'The expedition is the routine form of a deliberate business process of searching and finding. Therefore the decisive movement of real globalisation is more than a spatial case of expansion; it belongs to the core process of the modern history of truth. Expansion would not be able to take place if unless it were defined beforehand, technically in relation to truth and so technically without further explanation, as a disclosure of what so far had been concealed' [translated from Dutch].
But although Sloterdijk limited the concept to the period of globalisation that, according to his definition, finished around 1945, I prefer to continue to apply the concept in the *Global Age*.

The scientific aim of this book is to find answers to the questions of whether agroparks contribute to sustainable development in metropolises, how an agropark is developed and how it must be designed.

1.2 Method

The scientific method used is Action Research[8]. In the language of designers the method of working can be interpreted as research by design or co-design[9]. It is engineering with regional designs as the end products, on which scientific research can be carried out in terms of feasibility and suitability studies but also in terms of process evaluations, which focus on generating more generic knowledge about the dilemma outlined above. That research is interdisciplinary, comprising both natural science (beta dimension) and social science aspects (gamma dimension), but must also take into account the aesthetics of designs and the cultural history of the areas, where work is done, with the historical development of agriculture and with the communication about the design to all stakeholders involved (humanities or alpha dimension).

The action in action research, or the design in research by design, is aimed at generating inventions and interventions, culminating in system innovations, which are necessary for finding to make a connection between agriculture in the spatial planning of metropolises and the new challenges of globalisation and network societies. Or in more real-life terms, to design and actually implement agroparks. Since it always involved practical situations of spatial planning, in which (interdisciplinary) scientific knowledge was constantly reiterated with the practical knowledge of many who were involved in the concrete projects, the practice is transdisciplinary[10].

Research by design, which is implemented on the basis of various designs, is centred around three theories. The first two concern the outcome of the design and implementation process, the third concerns the design process itself.

---

[8] Termeer C.J.A.M. and B. Kessener (2007). Revitalizing stagnated policy processes: Using the configuration approach for research and interventions. *Journal of Applied Behavioral Science* 43: 256-272.

[9] De Jonge J. (2009). *Landscape architecture between politics and science. An integrative perspective on landscape planning and design in the network society*, Wageningen University and Research Centre, Wageningen, the Netherlands, 233 pp.

[10] Termeer C. (2006). *Vitale verschillen. Over publiek leiderschap en maatschappelijke innovatie. Oratie, 7 december 2006*. Wageningen Universiteit en Researchcentrum, Wageningen, the Netherlands, 48 pp.

The first is the resource use efficiency theory devised by Wageningen production ecologist De Wit[11]. An agropark is primarily aimed at the production and processing of plant and animal products. De Wit's theory is that the efficiency of the agroproduction process in a chain increases with the increase in yield per hectare. In this publication I will investigate whether this also applies to the next innovative step in the development of agroproduction, namely agroparks.

The second theory concerns the three-dimensional landscape, extensively described by Dirx *et al.*[12] and Jacobs[13]. Starting from the principle of landscape ecology as a natural science, these authors develop a landscape theory, which encompasses both natural science aspects like soil, water, vegetation, etc. (matterscape) and social science aspects, like power relationships between people and groups in the landscape and the economic aspects thereof (powerscape). And last but not least, embracing the subjective aspects like aesthetics, history and communication, part of humanities (mindscape). From the perspective of the three-dimensional landscape theory, an agropark is regarded as a landscape, in which matterscape, powerscape and mindscape all play an important role and must be explicitly designed.

The third theory concerns the design process itself. What criteria must be satisfied by the design of a complex system innovation like an agropark, if it has any chance of working in today's society? How does the process evolve from design to implementation? Which parties must be involved in that design and how do they work together? When answering these questions, experience previously acquired in the participative plan-making processes in the so-called Landscape dialogue[14], was built on in the design of the seven agropark projects. In De Jonge's thesis[15] there is a theoretical basis to this, which is defined as Co-design.

---

[11] De Wit C.T., H.H. Huisman and R.R. Rabbinge (1987). Agriculture and its environment: Are there other ways? *Agricultural Systems* 23: 211-236..
De Wit C.T. (1992). Resource use efficiency in agriculture. *Agricultural Systems* 40: 125-151..
[12] Dirkx G.H.P., M. Jacobs, J.M. De Jonge, J.F. Jonkhof, J.A. Klijn, A. Schotman, P.J.A.M. Smeets, J.T.C.M. Sprangers, M. Van den Top, H. Wolfert and E. Vermeer (2001). Kubieke landschappen kennen geen grenzen. In: *Jaarboek Alterra 2000*, Alterra, Wageningen, the Netherlands.
[13] Jacobs M. (2004). Metropolitan matterscape, powerscape and mindscape. In: Tress G., B. Tress, W.B. Harms, P.J.A.M. Smeets and A. Van der Valk (eds.), *Planning metropolitan landscapes. Concepts, demands, approaches, Delta series*, Wageningen University, Wageningen, the Netherlands, pp. 26-39.
[14] Van Mansfeld M., M. Pleijte, J. De Jonge and H. Smit (2003). De regiodialoog als methode voor vernieuwende gebiedsontwikkeling. De casus Noord-Limburg. *Bestuurskunde* 12: 262-273.
[15] De Jonge J. (2009). *Landscape architecture between politics and science. An integrative perspective on landscape planning and design in the network society*, Wageningen University and Research Centre, Wageningen, the Netherlands, 233 pp.

In Chapter 5 working hypotheses are derived from these theories:

- An agropark realises lower costs, more added value and less environmental pollution per unit of product and surface area.
- An agropark can only be arrived at on the basis of an integral design of matterscape, powerscape and mindscape both on the global scale of Intelligent Agrologistics Networks and on the local scale of a landscape.
- An agropark is a knowledge-driven system innovation and makes a significant contribution to sustainable development.
- The design and implementation of system innovations like agroparks necessitates the participation of knowledge institutions, enterprises, NGOs and governmenal organisations (KENGi). It is a transdisciplinary process in which the explicit knowledge of research institutes and the tacit knowledge of the other partners are constantly evolving. KENGi brokers act as the facilitators of this transdisciplinary process.
- In all decision-making concerning the realisation of the integral agroparks design, involving matterscape, powerscape and mindscape aspects, arguments from the world of justice and trustfulness take precedence over the arguments from the world of truth.

These working hypotheses are further developed using the various agropark designs and culminate in an answer to the question of whether the development of agroparks really makes a contribution to sustainable development. Together they make contributions to the method of research by design used as a specific development of the Landscape Dialogue or Co-design method.

1.3 Reading guide

Since the urbanisation process is the driving force behind the emergence of metropolitan foodclusters, this publication begins by describing the context in Chapter 2 with a general description of urban society in the 21st century: the network society[16].

In Chapter 3, 'Delta Metropolises' the process of urbanisation is central, as seen in North West Europe but also in other deltas around the world. Delta metropolises are also the birthplace of key innovations in agriculture. Since agroproduction is central in this urban society, the perspective taken is more physical and material than most modern descriptions of this urban society in our time.

[16] For a more detailed description of the body of ideas involved here, see Asbeek Brusse W., H. van Dalen and B. Wissink (2002). *Stad en land in een nieuwe geografie. Maatschappelijke veranderingen en ruimtelijke dynamiek.* Wetenschappelijke Raad voor het Regeringsbeleid, The Hague, the Netherlands.

In Chapter 4 the emergence of industrial agriculture in the Northwest European Delta Metropolis is closely examined and its evolution into a system of metropolitan foodclusters is described. Various scientific exercises conducted in recent years at the Alterra research institute, part of Wageningen-UR, have focused on the future urban development of agriculture in the Netherlands and North West Europe.

Where spatial planning has moved cautiously in the last 12 years from blueprint planning to spatial development policy, the products that in the same context made the position of agriculture central, took the lead and went from agricultural main structure map images[17] via scenario planning[18] to learning networks[19].

In Chapter 5, 'Theoretical principles of Research by Design', I describe the basic principles (resource use efficiency, three-dimensional landscape, design process), which serve as a basis for the research by design and I derive from that the starting hypotheses which will be used to test the real-life designs of agroparks.

In Chapter 6, 'Research by Design on Agroparks', the new prospect of metropolitan foodclusters is examined using a number of examples of agroparks on which I have worked in recent years. I describe the examples in their regional context and as an example of research by design. In this chapter the actual action research takes place on the basis of these examples. Some examples are real-life designs that have been implemented. Other designs were important targets in a future 'utopian' scenario, around which it was possible to define ambitions in urban development policy, which then worked as a catalyst for new territorial design.

The hypotheses which are tested and developed using the various examples in this chapter, culminate at the end of the chapter in general conclusions which contain the answers to the scientific questions which form the basis of this publication.

Finally, in Chapter 7, I relate these conclusions back to the research by design. First to the theory. On the basis of the conclusions from the research by design, the validity of the resource use efficiency theory will be further expanded. On the basis of other conclusions the landscape theory will give shape to spatial development policy. I also work out a methodical feedback system by combining and anchoring the findings

[17] Van Eck W., A. Wintjes and G.J. Noij (1997). Landbouw op de kaart. In: *Jaarboek 1997 van het staring centrum*, Staring Centrum, Wageningen, the Netherlands, pp. 4-20.
[18] Dumont M.J., R. Groot, R. Schröder, P.J.A.M. Smeets and H. Smit (2003). *Nieuwe bruggen naar de toekomst. Weergave van een speurtocht naar nieuwe perspectieven voor het Gelders landelijk gebied.* Report 674, Alterra, Wageningen, the Netherlands.
[19] Veldkamp A., A.C. Van Altvorst, R. Eweg, E. Jacobsen, A. Van Kleef, H. Van Latesteijn, S. Mager, H. Mommaas, P.J.A.M. Smeets, L. Spaans and H. Van Trijp (2008). Triggering transitions towards sustainable development of Dutch agriculture: Transforum's approach. *Agronomy for Sustainable Development* 29: 87-96.

from the design process into the method of co-design. I conclude the chapter with two contributions to actual discussions in society that can be deduced from the publication. The first is a discussion about the role agroparks should play in the spatial development policy of the Netherlands, and the second concerns the knowledge infrastructure needed to realise the potential of the system innovation of agroparks.

2. The network society

2.1 The emergence of the network society

Over the last fifty years the world has been rapidly evolving, in terms of population growth, economy, communication and IT, into a society where knowledge and creativity are crucial characteristics for people and communities to trigger even more progress. Progress in the sense of less poverty, longer lives, better food and a better environment, for more and more people who no longer live under the daily threat of war and total chaos. That is a positive evolution, but we could do better. There are still too many people living below the absolute poverty line and there are still major food shortages. For yet more people, food quality is extremely poor. The prospect of a world without hunger and with good food is entirely possible but has not yet been achieved[20].

The revolution can be defined by two terms: globalisation, and the network society. The world as one increasingly connected system, in which billions of people can communicate directly with each other via mobile phones and the internet, the two most significant accomplishments of the IT revolution, which since the 1970s characterise the network society.

But it is not just people that are connected in the network society. Enterprises, from primitive farms to financial derivative trading companies and universities, also operate on a global scale and are anchored within the same network. It is also in this context that agriculture and the spatial planning thereof must be redefined. In order to be able to interpret the complexity of agriculture and spatial planning, I will describe this context in the first part of this book. When dealing with the development over time, I will use the work of two authors in this chapter: Manuel Castells[21], who in his trilogy 'The information age: economy, society and culture', described the emergence

[20] See Rabbinge R. (2000). World food production, food security and sustainable land use. In: El Obeid A.E., S.R. Johnson, J.H. H. and L.C. Smith (eds.) *Food security: new solutions for the twenty-first century. Symposium honoring the tenth anniversary of the world food prize.* Wiley & Sons, New York, USA, pp. 218-235; and Wetenschappelijke Raad voor het Regeringsbeleid (1994). *Duurzame risico's: Een blijvend gegeven*, Report 44, Wetenschappelijke Raad voor het Regeringsbeleid, The Hague, the Netherlands.
See also Rabbinge R.R. and P.S. Bindraban (2005). Poverty, agriculture and biodiversity. In: Riggs J.A. (ed.) *Conserving biodiversity.* The Aspen Institute, Washington, DC, USA, pp. 65-77.

[21] Castells M. (1996). *The information age: economy, society and culture. Volume 2: the power of identity*. Blackwell, Oxford, UK; Castells M. (2000a). *The information age: economy, society and culture. Volume 3: end of millennium*. Blackwell, Oxford, UK; Castells M. (2000b). *The information age: economy, society and culture. Volume 1: the rise of the network society*. Blackwell, Oxford, UK.

of the network society, and Peter Sloterdijk[22], who in his book 'Het kristalpaleis' (The Crystal Palace) deals with the community in the urbanised nuclei of this network society, but who primarily analyses the problem of the rift between the global and local level of scale.

2.1.1 Network society or 'Crystal Palace' as third development phase of mankind

The new world order of the Network Society has also been alluded to as the Global Age or 'Crystal Palace'. According to both authors above, it is the third development phase of human society. While the two authors talk in detail about the same phenomena in the present day, they put the emphasis on different aspects in history and differ in their interpretation of the start and end dates.

In the hunter-gatherer society, and in the agricultural society that lasted from 10,000 BC to 1750, nature dominated culture according to Castells' description, and social organisation was focused on the struggle to survive. According to Sloterdijk, in the last century of this agricultural society, i.e. in 1492, the era of globalisation began with the journeys of Columbus and his discovery of America. The earth turned out to be a world full of water, not flat but round and the expeditions by explorers, conquerors and merchants across the world's seas laid the foundations for the next phase.

This second phase was the industrial society, where commercial capitalism first emerged with early-industrial production. For example, in the Republic of the Seven United Netherlands, where forms of proto-industrialisation in shipbuilding, textile manufacture and agriculture were already present before 1750[23], and which developed on the back of the industrial revolution in England into the capitalistic global system.

Culture was able to dominate nature in the industrial society, and social organisation was focused on conquering nature, which in the eyes of the conquerors also included

---

[22] Sloterdijk P. (2006). *Het kristalpaleis. Een filosofie van de globalisering.* Uitgeverij Boom/SUN, Amsterdam, the Netherlands.

[23] Wallerstein I. (1980). *The modern world system ii. Mercantilism and the consolidation of the european world economy 1600-1750.* Academic Press, New York, NY, USA, writes thus about the Netherlands in the Golden Age: 'if it is to be asserted (...) that the Netherlands was the first country to achieve self-sustained growth, it is primarily because no other country showed such a coherent, cohesive and integrated agro-industrial production complex'.

See also Bieleman J.J. (1992). *Geschiedenis van de landbouw in Nederland, 1500-1950. Veranderingen en verscheidenheid.* Boom, Meppel, the Netherlands: 26. 'In the course of the 'long 16th century' an interregional agricultural economy took shape. The Flemish and later the Dutch cities functioned as a focus within this economy. The influence extended right into the North West European hinterland, encompassing even Denmark and the Baltic states.'

Taylor P.J. (2004) *World city network: a global urban analysis.* Routledge, London, UK: 8, places the first global trading system in the 12-13th century.

the primitive people and the land they inhabited. The class struggle determined the dynamics of society and the value of labour was the central concept in analyses made by economists at that time.

In addition to the somewhat longer use of wind- and water-mills, at the heart of the industrial society was the increasingly large-scale use of fossil fuels, the 'labour-saving device', which allowed for dramatic improvements in productivity[24]. In Holland, fossil fuel was available in the form of peat, and in England from 1750 onwards in the form of coal, and later worldwide in the form of coal, oil and natural gas.

In Castells' view, the third phase, that of the network and information society, began around 1970.[25] He postulates that from that point onwards knowledge became the essential source of productivity improvement and power. The information technology revolution is the driving force and it includes microelectronics, information technology, communication technology and gene technology. The defence of the subject against the logic of apparatus and markets determines the dynamics in society.

According to Castells in the network economy the network enterprise is the basis for the information society. Labour has been transformed: networking, unemployment and flexi-time are the norm. Only a small elite is globalising. Capital is global, uses the network and moves in instant time; labour is local, disaggregated, fragmented, diversified and under pressure to work according to clock time.

The culture of the network society is that of real virtuality. The mass media with unilateral communication from the industrial era has developed into multimedia with full communication. The reality itself is captured entirely in the virtual image which doesn't communicate the experience, but becomes the experience itself.

---

[24] Sloterdijk P. (2006). *Het kristalpaleis. Een filosofie van de globalisering.* Uitgeverij Boom/SUN, Amsterdam, the Netherlands: 249. 'Both the liberals and the Marxists of the nineteenth century also made serious attempts to interpret the phenomenon of the industrial society: but in neither system was the phenomenon of fossil fuel even mentioned, let alone considered. Because the ruling ideologies of the nineteenth and early twentieth century stubbornly based all their explanations of wealth on the doctrine-loaded concept of the value of labour, they remained chronically incapable of understanding that the industrially acquired and used coal is not a 'raw material' like any other, but the first big labour-saving agent. Thanks to this universal "natural labourer" (...) the principle of surplus took up residence in the hot-house of civilisation.'

[25] Castells M. (2000). *The information age: economy, society and culture. Volume 1: the rise of the network society.* Blackwell, Oxford, UK: 508. 'a new stage in which culture refers to culture, having superseded nature to the point that nature is artificially revived ("preserved") as a cultural form. (...) Because of the convergence of historical evolution and technological change we have entered a purely cultural pattern of social interaction and social organization. This is why information is the key ingredient of our social organization and why flows of messages and images between networks constitute the basic thread of our social structure.'

2.1.2 Network society and 'Crystal Palace' as a turning point in history

Castells also calls the beginning of the network society the beginning of the history of mankind, because for the first time, humanity has the level of knowledge and social organisation necessary to live in a social world[26].

In Sloterdijk's vision, the process of globalisation ends with the Bretton Woods agreement in 1944; then the Global Age begins. Inspired by the Crystal Palace, which in the 19th century was a symbol for what capitalism later had to offer, Sloterdijk uses the metaphor of the 'Crystal Palace' to describe the essence of the Global Age:

> '...whereby nothing less than the universal inclusion of the outer world in a fully worked-out inner space was at stake.'[27]
>
> 'The big comfort structure will probably continue to embrace countless new citizens for a long time to come, by making inhabitants of the semi-periphery fully fledged members, but it will also exclude former members. (...) The semi-periphery is in those places where "societies" still have a broad segment of traditional agricultural and cottage industries.'[28]
>
> 'The Crystal Palace [comprises] at the beginning of the twenty-first century a third of the species *homo sapiens* (...) [up to], a quarter or even less, [this] can be explained by the systemic impossibility to organise the incorporation of all members of the human race in an homogeneous welfare system under the current technical, energy-political and ecological conditions.'[29]

The 'Crystal Palace' is organised into 5 stages and offers its inhabitants the opportunity to take advantage of whatever is available there:[30]

- It offers a performance-free income, where the social safety net forms the firm basis.
- It gives political security to its inhabitants without them having to fight for it.
- It guarantees access to immunity provisions without a personal history of suffering.
- It enables the consumption of knowledge without any need for experience.
- It enables individuals to become famous by means of unvarnished self-publicising without any performance or published work.

---

[26] Castells M. (2000). *The information age: economy, society and culture. Volume 1: the rise of the network society.* Blackwell, Oxford, UK: 508-509. 'This is not to say that history has ended in a happy reconciliation of humankind with itself. It is in fact quite the opposite: history is just beginning, if by history we understand the moment when, after millenniums of prehistoric battle with nature, first to survive, then to conquer it, our species has reached the level of knowledge and social organization that will allow us to live in a predominantly social world.'

[27] Sloterdijk P. (2006). *Het kristalpaleis. Een filosofie van de globalisering.* Uitgeverij Boom/SUN, Amsterdam, the Netherlands: 191.

[28] *Ibid.*: 210.

[29] *Ibid.*: 211.

[30] *Ibid.*: 231.

In short, the world has become one global system and agroproduction is organised in that global system as well. Around the world people are trying to get access to the 'Crystal Palace'.

The big question is whether the inequality between inside and outside is systemic, in other words whether the 'Crystal Palace' can only exist for a minority at the cost of a majority who must live in poverty. As far as agroproduction is concerned, the prerequisite of the production-ecology is that agriculture productivity can be improved to such an extent that sustainably produced good food becomes available for everyone[31]. In this publication agroparks are presented as a contribution to that aim.

2.2 Time and space in the network society

The agricultural society lived according to nature's clock, according to the rhythms of the sun and moon, along with a few religious extras. The industrial society invented clock time, crucial for production in a capitalist system, in which the division of labour was the essential prerequisite.

In the network society 'timeless time' will dominate, not as the only form of time, but certainly as the predominant one. Timeless time is dispersed time due to the confusion of events and their simultaneous nature. Timeless time is characterised by the disappearance of rhythms, both biological and social. The timelessness of multimedia hypertext is making our culture both ephemeral and eternal. The denial of death obliges the sterilisation of society. Flexitime businesses work via just-in time delivery and working hours differ between companies, networks, jobs and employees. Time is managed as a resource, which differentiates to other companies, networks, processes and products.

Time and space are interconnected. Because society is changing, and dealing with time differently, its spatial planning is also changing. Harvey[32] states that:

> '...we can argue that objective conceptions of time and space are necessarily created through material practices and processes which serve to reproduce social life (...) It is a fundamental axiom of my enquiry that time and space can not be understood independently of social action.'

In his interpretation of spatial planning in the network society, Castells makes the distinction between spaces of place and spaces of flows:

---

[31] See for example, Wetenschappelijke Raad voor het Regeringsbeleid (1992). *Ground for choices; four perspectives for the rural areas in the european community.* Wetenschappelijke raad voor het Regeringsbeleid, The Hague, the Netherlands.

[32] Harvey D. (1989). *The condition of postmodernity: An enquiry into the origins of cultural change.* Blackwell, Oxford, UK: 204.

> 'A place is a locale whose form, function and meaning are self-contained within the boundaries of physical contiguity.'[33]

By definition, that also applies to a place as a carrier of natural processes. But because time and space can not be understood independently of social action space also changes with the transition to timeless time, to the network society. In the first two phases of human society, all places could be seen as contiguous spaces of place.

2.2.1 Spaces of flows and spaces of places

In the network society there emerge spaces of flows, in which processes that belong to timeless time are enacted. The space of flows is the material organisation of social practices that divide up time and operate via flows but that no longer need to possess any spatial contiguity. They are mutually bound in the network via flows. There are different flows: information, technology, organisational interaction, images, sounds and symbols.

According to Castells, the space of flows consists of 3 layers:
- The layer of the electronic network, which forms the basis of the information and communication flows in the network.
- The layer of hubs and centres, which organises the division of labour. The places of decision-making: the head offices, gated communities, hotels, airports, VIP rooms and the informal places for decision-making (golf courses, theatres, etc.). In addition to these there are the hierarchical production complexes, consisting of innovation environments, skilled manufacturing sites, assembly lines and market-oriented factories.
- Finally, the layer of the management elite. Managers are cosmopolitan and so move between the hubs and the centres.

The space of flows is not the only spatial logic in the network society, but it is the dominant one. Spaces of flows are materially linked to spaces of places but according to their own logic and often in complete dissonance with the place where they actually flourish. That applies to hubs, centres and the infrastructural connections between them. But every space of flows is also a contiguous space of place. They are simultaneously the places where people live, the managers as well as the autochthonal inhabitants, but also the people of the fourth world – those who have searched for happiness in the global network society, didn't find it and were rejected by the network society. A space of place is therefore a colourful mix, a palette of historically developed residue from the agricultural and industrial societies, with the accompanying natural

[33] Castells M. (2000). *The information age: economy, society and culture. Volume 1: the rise of the network society.* Blackwell, Oxford, UK: 453.

clock rhythms and clock times, with the non-places in which the fourth world resides and with the material carriers of the layers that go to make up the space of flows[34].

The local space of the spaces of places is not the antonym of the global space of flows. The global society is not merely the sum total of the local spaces without borders, just as a plane can be seen as a collection of points. This misconception is evident in a word like glocalisation, which insinuates that people will attribute new meaning to their local environment as a reaction to the globalisation process. Both spaces imply fundamentally different experiences of time and distance[35]. Processes like biological reproduction, the raising and education of children, the handing down of culture, in short learning how to live, cannot be compressed into timeless time and treated as a component of the global society in the space of flows. They cannot be traded on the world market but are autonomous and are still most closely related to the housing function of spaces of places, which is why Sloterdijk calls housing the mother of asymmetry between the global and the local.

In this contrast between the global space of flows and the local spaces of places, such as described by Sloterdijk and Castells[36], we find the philosophical core of the public debate about contemporary agriculture. Food production, which is still only a part of agroproduction, has withdrawn from the spaces of place, from the incompressible, inhabitable expanses, from the local world. More specifically, it has disappeared from the residential environment, has even become antagonistic towards it, and has globalised, into part of the space of flows. There is an economic and ecological rationale behind this process, which is indigestible for most people. As consumers they buy the products of this agroproduction on a daily basis and profit handsomely from the increasingly low prices resulting from improved productivity, but as residents they often regard a farm as a nuisance, in terms of the smells, light, heavy traffic, fine dust, etc. This nuisance element is reinforced by veterinary disasters or food scandals played out in the media, even though they bear little or no relation to the industrial nature of

---

[34] *Ibid.*: 497. 'The space of flows (...) dissolves time by disordering the sequence of events and making them simultaneous, thus installing society in eternal ephemerality. The multiple space of places, scattered, fragmented and disconnected, displays diverse temporalities from the most primitive domination of natural rhythms to the strictest tyranny of clock time.'

[35] Sloterdijk P. (2006). *Het kristalpaleis. Een filosofie van de globalisering.* Uitgeverij Boom/SUN, Amsterdam, the Netherlands: 276-282.

[36] Castells M. (1996). *The information age: economy, society and culture. Volume 2: the power of identity.* Blackwell, Oxford, UK: 124, has a similar description of the tension between global and local: 'What is distinctive of new social structure, the network society, is that most dominant processes, concentrating power, wealth, and information, are organised in the space of flows. Most human experience, and meaning are still locally based. The disjunction between the two spatial logics is a fundamental mechanism of domination in our societies, because it shifts the core economic, symbolic, and political processes away from the realm where social meaning can be constructed and political control can be exercised.'

the production process. Many citizens also have difficulty with the industrial nature of modern agriculture, because they have memories of farming, at least up until the latter half of the last century, as the backbone of village life in which many urban dwellers still have all or some of their roots. This is reinforced by the many adverts for modern agricultural products, which increasingly refer back to this romantic and Arcadian notion of farming as it used to be, in order to conceal the industrial provenance of agricultural products in the supermarket.

2.2.2 Communicative self-steering in delayed time

From their local environment, their space of place, people resist the development of modern farming. In this respect too there have been quite a few changes in the last few decades. Like Sloterdijk and Castells, the cultural philosopher Arnold Cornelis[37] describes in his work the tension between the social control system on the one hand and forms of what he calls 'communicative self-steering' on the other. Key words in the social control system are hierarchy, planning, control and implementation. It is an achievement of the industrial society, but there is no feedback between those who control and those who are controlled[38].

In Cornelis' view this had everything to do with an understanding of the concept of time as merely the fourth dimension and as such comparable with the three dimensions of space[39]. He defines a shift in human thought, which is linked to a different approach to time. In contrast to the timeless time of Castells, he propagates the slowing down, the communicative self-steering, which people use to think about possibilities, about the control, which they can use for self-preservation in an increasingly fast-moving world.

The concept of communicative self-steering is the essence of the perspective to which Cornelis refers. Only by thinking in delayed time can people develop a programme which they can use to operate independently in the world, which operates in timeless time. Like upbringing, biological reproduction and the handing down of culture, it is another paradigm that is characteristic of the incompressible, inhabitable expanses that Sloterdijk defines as the centre of the world in the 'Crystal Palace'. The development of

[37] Cornelis A. (1999). *De vertraagde tijd*. Stichting Essence, Amsterdam, the Netherlands, 174 pp.
[38] *Ibid.*: 8. 'A retrospective look at the twentieth century reveals a society that learned how to conduct trade in an organised and technological way, in a social control system of obedience and political hierarchy, but without acknowledgement. Power was totalitarian and people were obedient. The twentieth century produced the silent man without self-steering and without communication.'
[39] *Ibid.*: 44. 'I call the time a hidden programme, the events of the future are not concealed in space but in time (...) Without the concept of a "hidden programme" we cannot conceive of the dimension of time. If we don't have this concept, because development is reduced to a fourth dimension of space, dominated by repetition, such as in the thinking of the twentieth century, then there is no time for human consciousness.'

this capacity for communicative self-steering is a lot less predetermined. Learning is no longer controlled from an external perspective only; there is also internal control, on the basis of imagination, internalisation, enthusiasm and communication with others. In a situation of rising levels of education and an increasing openness and dynamic, people are increasingly taking responsibility themselves, and wondering whether the institutional norm employed is consistent with the values to which they are aspiring.

Only democratic systems have the capacity to translate the communicative self-steering, which is in full swing in the locally inhabited expanses of spaces of places, via representative representation, into the institutions that have the capacity to regulate the space of flows[40].

But this democratic system then runs the risk of becoming paralysed between space of flows and space of places, between universal space and local space. While the administration tries to introduce quality to the local environment of increasingly vocal inhabitants, it is increasingly dominated by the laws of the network society, which is actually separate from space and time and which suffuses culture with real virtuality. Sloterdijk formulates it thus:

> 'The characteristic of the established globality is the state of forced neighbourliness with innumerable people co-existing by chance. This state is best described by the topological term "density" (...). Inhibition becomes second nature due to chronic residence in dense environments.'[41]
> 'In the post-historic dense situations (...) every impulse is destroyed by its reactions, often before it has even been able to evolve. Long before the first spade enters the ground, everything that wants to move forward, that challenges the boundaries, that wants to build, is reflected in protests, objections, counter-proposals, swan songs; measures are overtaken by counter-measures – most restructuring proposals could be realised with a twentieth of the energy that is used to reformulate, water down and temporarily postpone (...) Governments are currently groups of people who specialise in making it look as though the state of a country can be improved within the inhibiting context.'[42]

[40] *Ibid.*: 11. 'Man was a spatial object about which decisions were taken, from the centre of political power. But at the turn of the century (...) we see power being decentralised. In itself that is an inevitable and logical development, from the policy standpoint (...) As a politician you can try to get out of it, but then you are left with a country lagging behind (...) Because this concerns a cultural knowledge change in the human spirit. The more a social system is decentralised, the more intelligent that system becomes. Because the number of decision points for control multiplies, because of self-steering and communication between people.'

[41] Sloterdijk P. (2006). *Het kristalpaleis. Een filosofie van de globalisering.* Uitgeverij Boom/SUN, Amsterdam, the Netherlands: 192.

[42] *Ibid.*: 207.

This inhibiting context, which, in the case of the agricultural modernisation, is further reinforced by resistance to the development of modern agriculture in Western society, is the hurdle that must be cleared with the implementation and therefore also the design of the agroparks. Many arguments calling for the continuing industrialisation of agriculture, with agroparks as the next step, come from the space of flows. The consumer, who supports the development of modern agriculture by eagerly purchasing the products thereof, also operates on the global scale in his modern kitchen. Resistance arises as soon as the development of modern agriculture materialises in companies with actual production, processing and logistics, that are built 'in the consumer's backyard.' Then the consumers emerge from their kitchens and become citizens in the drawing room, look outside and see their space of place under threat and ask themselves whether or not they should support the NGOs, which want to resist the building of mega-stalls, agricultural development areas or distribution centres.

2.3 'The environmental movement' as a synthesis of spaces of flows and spaces of places

Is it possible to bridge this rift between, on the one hand, communicative self-steering citizens, who operate from spaces of places and have developed a formidable inhibiting context there, and on the other, the modern consumers, who not only personify and preserve the global space of flows in their work but also enjoy it daily in an unknown culinary diversity?

Castells[43] sees another approach to dealing with time in the current era than the dominant form of clock-time and timeless time. As an opponent of timeless time, the 'environmental movement'[44] appeals to glacial time, in which the long-term evolutionary relationship between culture as a product of human civilisation and nature is considered.

> 'I propose the hypothesis that there is a direct correspondence between the themes put forward by the environmental movement and the fundamental dimensions of (...) the network society (...): science and technology as the basic means and goals of economy and society; the transformation of space; the transformation of time; and the domination of cultural identity by abstract, global flows of wealth, power, and information constructing real virtuality through media networks.'[45]

---

[43] Castells M. (2000). *The information age: economy, society and culture. Volume 1: the rise of the network society*. Blackwell, Oxford, UK.

[44] Castells uses *green culture* as an umbrella term for what is referred to in the Netherlands as the environmental movement, including the Green political parties.

[45] Castells M. (1996). *The information age: economy, society and culture. Volume 2: the power of identity*. Blackwell, Oxford, UK: 122.

'...the environmental movement is precisely characterized by the project of introducing a "glacial time" perspective in our temporality, in terms of both consciousness and policy. Ecological thinking considers interaction between all forms of matter in an evolutionary perspective. The idea of limiting the use of resources to renewable resources (...) is predicated precisely on the notion that alteration of basic balances in the planet (...) may over time undo a delicate ecological equilibrium, with catastrophic consequences. (...) In very direct personal terms, glacial time means to measure our life by the life of our children. (...) To propose sustainable development as intergenerational solidarity brings together healthy selfishness and systematic thinking in an evolutionary perspective.'[46]

The introduction of a new socio-biological identity can be discerned behind this plea for sustainable development. This is not ahistorical and takes full account of all kinds of cultural authenticity. But the environmental movement takes the world as a whole as a starting point, because only on that level are there ultimately relevant and fairly absolute limits. However, this global thinking is at odds with the nation state, because this forces the unity of people within territories, which often have little relevance in an ecological sense. The global thinking is also incompatible with Sloterdijk's 'Crystal Palace', which is also defined to a large extent on the basis of these nation state boundaries:

'...ecologists are at the same time, localists and globalists: globalists in the management of time, localists in the defence of space. Evolutionary thinking and policy require a global perspective. People's harmony with their environment starts in their local community (...). Environmentalism supersedes the opposition between the culture of real virtuality (...) and the expression of fundamentalist cultural or religious identities. It is the only global identity put forward on behalf of all human beings, regardless of their specific social, historical, or gender attachments, or their religious faith. However since most people do not live their lives cosmologically, the critical matter for the influence of new ecological culture is its ability to weave threads of singular cultures into a human hypertext, made out of historical diversity and biological commonality.'[47]

This characterisation of the environmental movement gives direction to designs for modern agriculture. They are made starting with the parameters and new possibilities emerging from a world that is increasingly functioning as one global system. They must also be completely embedded in the local setting in which they function, and must not make any concessions to precisely those characteristics which cause offence

---

[46] *Ibid.*: 125-126.
[47] *Ibid.*: 127.

to the people living in their neighbourhood. It sounds idealistic – and it is. But there is no other way.

Therefore, according to Sloterdijk's definition of expeditions, quoted in Chapter 1, Expedition Agroparks can be outlined beforehand, 'technically in relation to truth and so technically without further explanation, as a disclosure of what has so far been concealed'. At any rate, in the broad sense of time dimension, it was postulated above that the task of working on sustainable development must follow on naturally from the above-cited goal of the environmental movement.

The aim behind the design of agroparks is to put the position of agroproduction simultaneously in its global perspective, fully functioning in the network society and under the laws of clock-time and timeless time, and in the perspective of sustainable development (glacial time), and to allow it to flourish in a local setting in which businesses and citizens and their representatives in governments and NGOs must be seen to concur. That and nothing less is the task of the design.

It is precisely in the metropolitan environment that the places where modern agriculture is practised are too big and too present not to be affected by their environment. That also applies to the areas from which agriculture is retreating. It is not difficult to find a new role within the influence of a metropolis. But those areas where space is plentiful are becoming marginalised.

In Netherlands, at any rate, this is the domain of spatial planning. In order that the expedition can flourish in this domain, it is also necessary to explore the context of this policy terrain.

2.4 Spatial development policy

While it will become clear in Chapter 6 that the place of action for the first expeditions was in fact a bigger space than the entire surface area of the Netherlands, the following examination of the aspect of spatial planning will focus primarily on the Netherlands. This is because spatial planning is still organised in multifarious ways in the different nation states that make up the North West European territory. While spatial planning in the Netherlands is largely decentralised, in North Rhine-Westphalia the regional level of the *Länder* and the *Regierungsbezirke* dominates and in Belgium spatial planning is practically non-existent.

Using the concept of Negative Space, De Geyter *et al.* (Figure 2) shows the difference that spatial planning makes to domestic policy.

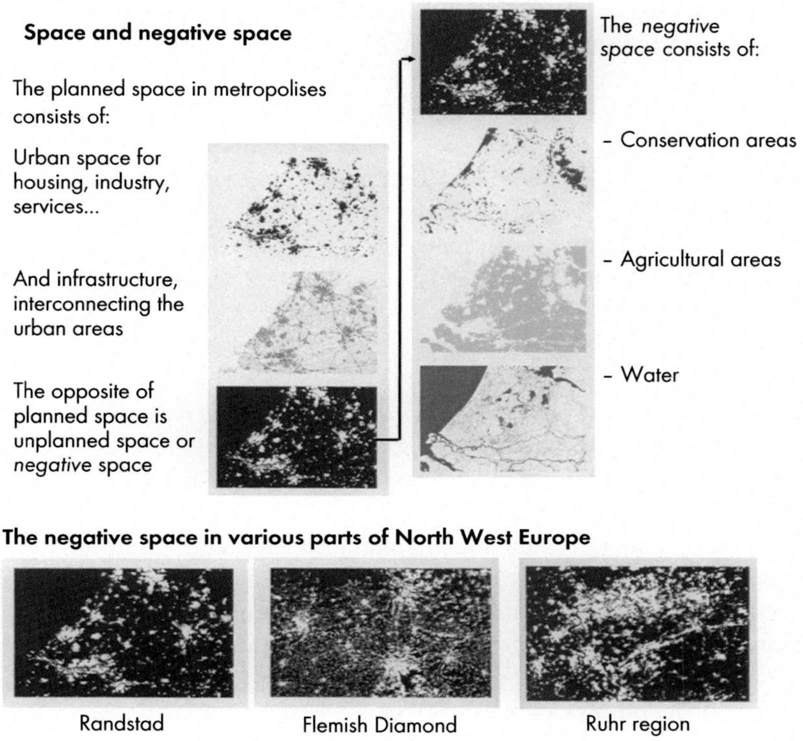

Figure 2. Space and negative space in North West Europe.
Figure taken from various images and definitions in De Geyter et al. (2002).

What is particularly striking here is the difference between the Randstad and Flanders in the presence of green open space in an urbanised area.

The spatial planning in the Netherlands in the 1990s also included the above-described 'inhibiting context'. To use spatial planning jargon, this was called permissive planning, in which moreover various ministries implemented their own spatial planning policy and also the synchronisation between different government levels (state, provinces, municipalities) was far from optimal. In 1998, with the aim of modernising spatial planning, the Wetenschappelijke Raad voor het Regeringsbeleid (WRR) (Scientific Council for Government Policy) formulated a new concept for spatial planning under the name of spatial development policy[48]. In this the Council tried to stick to the principles of spatial policy which had come under tremendous pressure up to that point:

- concentration of urbanisation: compact urbanisation in order to preserve the spatial quality of the green space;
- spatial coherence: holding together of urban functions, which appears increasingly difficult and so causes congestion;
- spatial differentiation, which is disappearing fast as a result of stealthy urbanisation;
- spatial hierarchy, which is declining fast as a result of the dispersal of high-value urban functions; and
- spatial justice, which is being undermined by extensive suburban development[49].

The WRR proposals basically boil down to coupling planning with spatial investment and, in parallel, with an extension of the spatial planning competencies of the national government by the implementation of a rough 'spatial main structure', in which areas and connections, important for domestic policy, can be established. In these 'national projects' national policy is a guide for regional or local planning. In the other areas (development areas, where the state creates a leitmotif, and basic areas, where the state limits itself to defining the basic qualities to be respected), lower authorities will acquire more possibilities to implement their own planning.

In the Fifth National Policy Document on Spatial Planning the spatial planning policy-makers undertook a last attempt to implement strong town and country planning with the help of a rigid system of red and green contours[50]. But the attempt ran aground

---

48 Wetenschappelijke Raad voor het Regeringsbeleid (1998). *Ruimtelijke ontwikkelingspolitiek.* Wetenschappelijke raad voor het Regeringsbeleid, The Hague, the Netherlands. Instead of the term 'spatial development policy', many authors used the term 'development planning', but this term completely overlooks one of the aims that the WWR explicitly wanted to achieve, namely lifting spatial planning out of the sectoral and political realm, for which in the Netherlands not only the Ministry of Housing, Spatial Planning and Environment, but also the Ministry of Transport, Water and Public Works, the Ministry of Agriculture, Nature and Food Quality and the Ministry of Economic Affairs should be jointly responsible. See for example, Dammers E., F. Verwest, B. Staffhorst and W. Verschoor (2004). *Ontwikkelingsplanologie. Lessen uit en voor de praktijk.* Ruimtelijk Planbureau, NAi Uitgevers, Rotterdam, the Netherlands.
For an explicit plea for the use of the term 'spatial development policy', see Rabbinge R. (2006). Ruimtelijke ontwikkelingspolitiek. In: Aarts N., R. During and P. Van der Jagt (eds.) *Te koop en andere ideeën over de inrichting van Nederland.* Wageningen UR, Wageningen, the Netherlands, pp. 195-200.

49 *Ibid.*

50 Ministerie van Volkshuisvesting Ruimtelijke ordening en Milieubeheer (2001). *Ruimte maken, ruimte delen. Vijfde nota over de ruimtelijke ordening 2000/2020 vastgesteld door de ministerraad op 15 december 2000, Den Haag.* SDU, The Hague, the Netherlands.

before the policy could be established, because the government which had drafted this policy document fell from power before it was brought into voting.[51]

The subsequent government adopted most of the WRR proposals in the Nota Ruimte[52] (the Spatial Planning Document that followed the Fifth document). But it is still too early to say whether spatial development policy has led to more successful spatial planning.

The main author of the WRR report sounded a fairly sombre note in a mid-term review:

> '...there is a policy increasingly taking shape, that denies the actual developments, that neither uses the power and potential, nor is able to maintain the spatial basic principles and that doesn't combine the pleas of the WRR for planning and resources (...). Extensive lip service to the regional spatial policy has, though, been proved and the provincial authorities have been made responsible for spatial policy. The spatial main structure pleaded for by the WRR was not given, no work was done on the spatial basic quality that should be used when testing local plans, regional development areas are not being identified and there is no mention of national projects. In short, under the pretext of delegating responsibilities, no spatial policy is being carried out which would maintain the basic principles of spatial policy.'[53]

As far as a national main structure is concerned, we are operating in a policy-free period. There is no explicitly established national spatial plan with established priorities. It is not certain that a broad vision on the spatial main structure of the Netherlands will emerge again in the coming years. Not only does the national government seem unwilling to make a contribution in this area, but its role here is diminishing; on the one hand because of the increasing dominance of European policy, which despite the lack of an explicit spatial policy still has major spatial consequences[54], and on the other hand because cities are beginning to have a better understanding of their role

---

[51] For a description and critique of the policy in the *Vijfde Nota*, refer to Asbeek Brusse W., H. van Dalen and B. Wissink (2002). *Stad en land in een nieuwe geografie. Maatschappelijke veranderingen en ruimtelijke dynamiek.*, Wetenschappelijke Raad voor het Regeringsbeleid, The Hague, the Netherlands: 164 ff.

[52] Ministerie van Volkshuisvesting Ruimtelijke Ordening en Milieu (2004). *Nota ruimte, ruimte voor ontwikkeling*. SDU Uitgevers, The Hague, the Netherlands.

[53] Rabbinge R. (2006). Ruimtelijke ontwikkelingspolitiek. In: Aarts N., R. During and P. Van der Jagt (eds.) *Te koop en andere ideeën over de inrichting van Nederland*. Wageningen UR, Wageningen, the Netherlands, pp. 195-200.

[54] For an overview of the spatial consequences of European policy on spatial planning, see Van Ravesteyn N. and D. Evers (2004). *Unseen Europe: a survey of eu politics and its impact on spatial development in the Netherlands*. NAi Publishers, The Hague, the Netherlands, 157 pp.

in the world-encompassing metropolitan network of the global age and are becoming increasingly autonomous in capitalising on this position[55].

That is an advantage in some ways for 'Expedition Agroparks', because the inhibiting context is effectively absent from this policy, which can exert huge top-down influence on the development possibilities in the rural area. From a regional perspective, the national government is not even really necessary for the implementation of broadly supported local and regional development projects. This way, however, the Netherlands does run the huge risk of losing characteristic qualities that define the identity of the country, and that motivated the WRR to outline the basic principles for spatial development policy. But that is a consequence of relinquishing the decision to implement spatial development policy.

But the fact that there was no explicit vision of the spatial main structure in the most recent Nota Ruimte does not mean that it doesn't implicitly exist. The principles of spatial planning, as defined on the previous page, still stand. By repeatedly using the collective intelligence of participants and external experts within the process of local or regional landscape design via peer reviews and scenarios, this vision can justifiably be included in these plans on the global level of the network society.

That creates space to work on the bottom-up implementation of spatial planning. The main structure can be left implicit, as tacit knowledge among the participants in a local or regional planning process, by taking the locality which Sloterdijk talks about together with the capacity for communicative self-steering of participating citizens as a starting point and, as the WRR urges, proceed from there in a thematic way, and bring about supported change processes from the bottom up. That will result in a much more direct dialogue between citizens and their municipality or provincial council.

An agropark is a design for agriculture in the network society, which is constructed from the perspective of two other disciplines, in which the Netherlands excels on an international level: urban planning and spatial planning. The concept of agriculture in the network society comprises the three areas of urban planning, spatial planning and agriculture, and shows that the traditional antithesis of city and countryside has ceased to exist, because the core activity of what used to be rural – agriculture – is now an urban production process. This means a far-reaching redefinition of agriculture as

---

55 See for example, Hemel Z. (2008). Middelpunt zoekende krachten. *Stedebouw & Ruimtelijke Ordening* 89: 28-34. 'Partly because of the chronic indifference towards big city development (...) and the preoccupation with distributive justice in The Hague, Amsterdam is increasingly focusing on the international playing field. It is looking to collaborate with other European cities, is active in Brussels, and is entering into relations with New York, Seoul, San Francisco and London. (...) And with some success. Amsterdam is seen as an international centre, and is increasingly regarded by the world as a creative core within the Northwest European lowlands.'

an economic sector, as a category in policy, which traditionally characterised its own domain as 'non-urban' and only applied spatial planning at the level of parcels, lots and land reallotment schemes. It is this transcending of all these traditional borders that makes the work on the system innovation of agroparks an interesting way of really perpetrating spatial development policy.

3. Delta metropolises

3.1 The historical development of cities: centralisation and dispersal

Urbanisation, the process whereby people come to live in an ever-decreasing area, became possible from the moment man discovered agriculture and started creating food surpluses[56]. Following the industrial revolution and the accompanying rise in agricultural productivity, the urbanisation process persisted at an even greater pace and, according to United Nations' statistics, there are now more people living in cities than in the countryside the world over[57].

3.1.1 Central cities and trading cities

Many scholars have reconstructed the emergence of the first cities. Mumford[58] describes an exchange between city rulers and priests on the one hand, and farmers on the other side as the basis for one type: the central city. The rulers and priests offered protection against external enemies, took care of the laying of irrigation systems and other infrastructure, created a marketplace, performed rituals and acted as intermediaries with higher authorities. The farmers relinquished part of their production to feed the urban population. In the idealised version of this set-up a 'social contract' was voluntarily entered into by equal parties. The historical reality is that there was usually an unequal relationship between external, often nomadic conquerors and settled farmers, from whom the surplus contribution was forcibly extracted. Ibn Khaldun[59] describes how hardened nomads from the mountains, the desert or the steppes looked on farming communities of the river valleys and coastal areas, and their cities of abundance, with a mixture of disdain and envy. They moved in and took power, only to be softened up by the life of luxury and fall victim to new usurpers over the course of time.

There is also a fundamentally different type, the trading city. McNeill[60] describes its emergence in the case of the ancient Greek cities. Greece consists chiefly of mountains and has many rocky coasts. Fertile valleys and plains for growing grain are scarce. Wine and olive oil, on the other hand, can be produced here in abundance. When grain as a basic food item became indispensable, the Greeks were forced to go elsewhere

[56] Diamond J. (2000). *Zwaarden, paarden en ziektekiemen.* Uitgeverij Het Spectrum, Utrecht, the Netherlands.
[57] UNFPH (2008). State of world population. Unleashing the potential of urban growth. Available at: http://www.unfpa.org/swp/2007/english/introduction.html.
[58] Mumford L. (1961/1989). *The city in history.* Harcourt, London, UK: 30-38.
[59] Ibn Khaldun (1967). *The muqaddimah.* Princeton University Press, Princeton, NJ, USA: 141-142.
[60] McNeill W. (1963/1991). *The rise of the west.* The University of Chicago Press, Chicago, IL, USA: 201.

to acquire it in exchange for their surplus wine and oil. This was the start of trading between Greece and the northern coasts of the Black Sea, which gradually included more products in the trade flows, and from other parts of the former Mediterranean world as well. As a consequence, a great number of coastal settlements developed into flourishing trading cities.

From the beginning the two types of city led to different urbanisation patterns. Central cities in agricultural areas tended to increasingly extend their radius of action, mimicking the territorial expansion drift of their rulers. Previously independent cities in newly conquered areas were sucked dry, as it were, by the central city. Trading cities don't compete to conquer their direct environment, but to conquer distant markets. Trading cities were therefore able to exist side by side and continue developing. The average size of trading cities is much smaller than that of agricultural central cities. Trading cities are often clustered, and at smaller distances from one another; agricultural central cities are located in an expanse of countryside where the next central city is a long way off.

3.1.2 Emergence of great empires – urban hybrids

Historical developments have, sometimes alternately, sometimes simultaneously, stimulated one or another type of city and urbanisation pattern, adding new layers of complexity each time. Capital cities like Rome accelerated more in their growth and reached an unprecedented size. With the continuing expansion, distances became so big that the role of original centres of independent agricultural regions drawn within the empire could no longer be completely absorbed by the capital city. So they continued to play an administrative and economic role, though now as a satellite of the capital. Hereditary rulers of these regional centres, who went all the way back to earlier conquerors, were replaced by bureaucrats appointed from the capital city.

With the emergence of the great empires, original trading cities were able to expand their market, but capital cities and regional administration centres also attracted trade. Industry developed in both, albeit with a different orientation: trading cities were more supply-oriented, towards the processing possibilities of imported raw materials; central cities were more professionally oriented, towards a growing market of rich city dwellers. This resulted in hybrids with a wider range of functions: administrative-religious centre, as well as agricultural market town, trading city, and industrial city. Such cities evolved primarily in the deltas of major rivers, where great expanses of land, with increasingly fertile soil and a permanent supply of water, facilitated large-scale highly productive farming, and where the location on the coast and river facilitated transport by water, both overseas and inland.

The level of urbanisation in the early Middle Ages in Europe greatly regressed. Thereafter, there was a repeat of the emergence and development of both city and urban types. The distribution of the kingdom of Charlemagne among his three sons played a role here. The western and eastern part were maintained as major entities with an overwhelmingly agrarian economy, most noticeably in the west where the foundations had already been laid for the later unparalleled position of Paris. The middle kingdom of Lotharius, which stretched in a band from Italy to the North Sea, soon fragmented, with a much stronger representation of trade and industry as their economic basis. In the famous drawing by the French spatial planning institute DATAR[61] (see Figure 3), the kingdom of Lotharius is immediately recognisable in the 'blue banana,' as an area of urban sprawl, while the major cities are situated on the periphery. Looked at like this, the overall pattern of European urbanisation has not changed much in twelve hundred years.

3.1.3 Dutch trading cities

It is tempting to compare the early flourishing of the trading function in West Holland, as part of the European middle zone, with the situation in ancient Greece. Here agriculture was practised on reclaimed peat that initially lay high enough and allowed sufficient natural draining for the successful cultivation of grain. The draining process oxidised the peat, which lowered the surface level and the land became so wet that

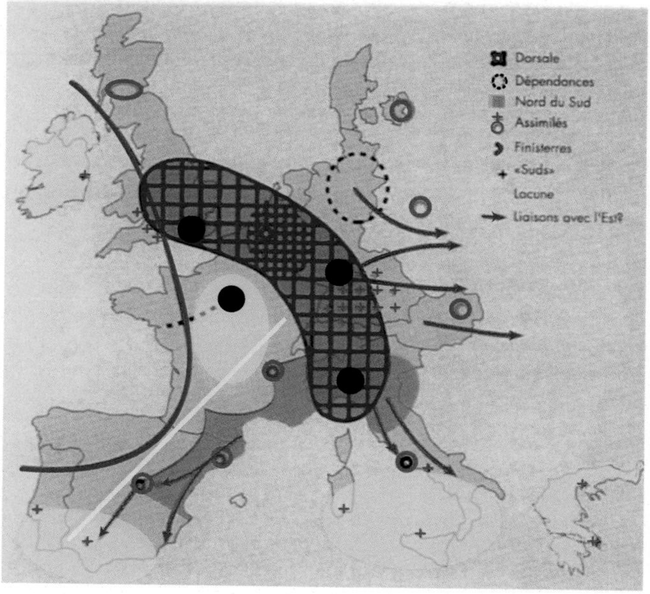

Figure 3. In 1989 the 'Blue Banana' covered 18% of the territory of the EU-12 and Switzerland and Austria, but 47% of the conurbations with more than 200,000 inhabitants (RECLUS, 1989).

[61] RECLUS (1989). *Les villes europeénnes.* DATAR. RECLUS, Montpellier, France.

it was no longer possible to grow grain. What remained was dairy farming with a surplus yield of dairy products[62]. These were then exchanged in a similar way to the wine and oil of the Greeks, and what the Black Sea was for the Greeks as a grain store, so the Baltic Sea became for the Dutch, after initial imports from England and Northern France. Once there, the Dutch brought back other products from that area such as wood or fur, while taking back products that inhabitants there could not make themselves because of the harsh climate. These did not have to come from the Netherlands itself, but could be sourced elsewhere: for example, wine from France or spices from the Lebanon or later Portugal. When, during the course of the eighty year war (1568-1648), the Portuguese harbours closed to the Dutch, the latter had no choice but to sail to the Far East themselves, whereupon trade made a leap from the inter-regional to the intercontinental level[63].

3.1.4 The colonial era

Overseas intercontinental trading heralded the era of Europe as a global coloniser. First Spain and Portugal evolved into centres of large overseas empires[64], then the Netherlands and England[65], followed in the nineteenth century by France and Germany[66]. Another sort of colonial empire developed in parallel, this time as connected land masses from the north and east of Europe deep into Asia: Sweden, Prussia, Russia, Austria[67]. The last two had to compete with the Turkish empire which had already begun a similar expansion starting from Istanbul. The capital cities of the colonial empires started to amass surplus in the same way as old agricultural central cities, on an unprecedented scale. These colonial capital cities were often former agrarian central cities, where centralised growth had rapidly accelerated: Madrid, Stockholm, Vienna, St. Petersburg, Berlin, Paris. They were less frequently cities with a trading history like Lisbon or Amsterdam. London was a blend of both.

---

[62] Slicher van Bath B. (1960/1980). *De agrarische geschiedenis van West-Europa 500-1850*. Uitgeverij Het Spectrum, Utrecht, the Netherlands.
Bieleman J.J. (1992). *Geschiedenis van de landbouw in Nederland, 1500-1950. Veranderingen en verscheidenheid*. Boom, Meppel, the Netherlands, 423 pp.
[63] De Nijs T. and E. Beukers (eds.) (2002). *Geschiedenis van Holland (4 dln)*. Verloren, Hilversum. the Netherlands.
Lesger C. (2001). *Handel in Amsterdam ten tijde van de opstand*. Verloren, Hilversum, the Netherlands.
[64] Thomas H. (2003). *Rivers of gold*. Phoenix, Londen, UK.
[65] Israel J.I. (1996). *De republiek 1477-1806*. Van Wijnen, Franeker, the Netherlands, 1368 pp.
James L. (1994). *The rise and fall of the british empire*. Little, Brown and Company, London, UK.
[66] Wesseling H.L. (2003). *Europa's koloniale eeuw*. Uitgeverij Bert Bakker, Amsterdam, the Netherlands.
[67] Imber C. (2002). *The Ottoman empire*. Palgrave (Macmillan), Basingstoke, UK.
Darwin J. (2007). *After Tamerlane*. Allen Lane (Penguin Books), London, UK.

Amsterdam, as a trading city that became a colonial capital city, acquired a dual character. On a regional level the city remained part of a cluster of more or less equal adjacent cities (also a result of the federal structure of the Republic); in its role on the world stage, Amsterdam increasingly distinguished itself from these surroundings. In the fifteenth and early sixteenth century, the biggest city in the Netherlands was Dordrecht, not Amsterdam. By the seventeenth century Amsterdam had not only become the biggest city, but it was double the size of Rotterdam, number two on the list.

3.1.5 Effects of infrastructure

The question of whether developments promoted centralisation or spread is still a matter of debate in the expansion of the infrastructure. Initially, during the Middle Ages and in the first two centuries of the colonial period, transport by water was predominant. This had a centralising effect, at least for narrow waterways, along the coast and along a few navigable rivers, which then became trading outlets. But this also stimulated dispersal because of the strong indentation of the European coast, where in many places the sea forced its way deep into the continental land mass. There were many hundreds of locations along the coast and rivers which were ideal for a trading city.

3.1.6 The industrial revolution and the railway

The industrial revolution was the next significant influence on the European urbanisation pattern, in particular the second stage dominated by coal and steel. The first stage was still concerned primarily with the mechanisation of the textile industry, in settlement patterns that had existed much longer. The second stage heralded the arrival of a new sort of industrial city in areas where coal and iron ore were found side by side in the ground: Central and Northern England, Scotland, North Sweden, North West Spain, South East Germany and South Poland, the half moon of the western Rhine Valley and the Ruhr area over North East France and southern Belgium to the French North West, in whose periphery the Netherlands' South Limburg lay.

This dispersal of coal and ore sources also had a decentralising effect on the European urbanisation pattern. On a regional scale the pattern was also decentralised: clusters of more or less equal medium-sized cities in close proximity, though these were not primary trading cities but industrial cities (although it was sometimes existing trading cities which thus acquired an industrial dimension)[68].

[68] McNeill W. (1963/1991). *The rise of the west*. The University of Chicago Press, Chicago, IL, USA.

The railway infrastructure also had a dispersing effect in this period. For the first time it was possible to travel across country at great speed and at very little cost[69]. There were also centralising powers at work on one important point: the financing of industrial development. Capital accumulated from the colonial period in the capital cities of the global empires was invested in this area. As such these capital cities were able to add a new dimension to their economic power, as financial directors of the industrial revolution. This effect was felt most in London, and least in Amsterdam, because the industrial revolution was slower to develop in the Netherlands. In Amsterdam, accumulated capital was often invested in English companies and infrastructure, rather than being used for enterprises in the Netherlands[70]. As such the head start that Amsterdam had gained on the other Dutch cities in the 17th century was lost, and Rotterdam was able to catch up in a big way because it became the predominant seaport for industrialisation in the German hinterland.

3.1.7 The arrival of the automobile

In the third stage of the industrial revolution, with its focus on electrical engineering and chemistry, and its shift from the mindset of English mechanical engineering to the science-based technology of the Germans and the French, the pattern of urbanisation did not change much. The centralisation of management roles in financial centres continued, as did the dispersal of production activities made possible by the gradual expansion of the railway infrastructure. The automobile created a new boost in the 20th century. It was much more of a decentralising force than the railways. Before the Second World War, the freight truck was of primary importance, facilitating as the train had done, the efficient and rapid transport of large quantities of goods, but this time across a much more finely woven network, which because of the relatively cheap cost of building new roads, became even more tightly interwoven. Even areas with no rail connection were now able to be take part in the industrial revolution. After the war, cars for individuals dominated, with major consequences for the pattern of urbanisation, although this time not on the European scale, but at the regional level. From the end of the 19th century suburbanisation was already taking place because of the railways, or, as in London, the underground lines, but mass suburbanisation only really occurred from the 1960s onwards in Europe with the advent of mass motoring.

69 For the Netherlands: Van der Woud A. (2007). *Een nieuwe wereld*. Uitgeverij Bert Bakker, Amsterdam, the Netherlands.

70 Israel J.I. (1996). *De republiek 1477-1806*. Van Wijnen, Franeker, the Netherlands, 1368 pp.

3.2 Polycentric mega city regions

3.2.1 Increasingly complex mix of centralising and dispersal forces

In the network society, from the 1970s onwards, centralising and decentralising forces occurred increasingly simultaneously in complex mutual relationships. As for which of the two dominated, this differed by domain (state, economic, cultural), by economic sector (industry, agriculture, service industry, consumer services), by mode of transport (ship, plane, train, car, telecommunication), and by geographical location. The complexity was further enhanced by the fact that, because of increasingly fast and cheap transport and even more importantly the development of telecommunications, functional centralisation and physical concentration were no longer inseparable. In the global network that has since emerged, functional centralisation and physical dispersal go well together. Only where face-to-face contacts are essential, such as with financial, technological or cultural innovation, does physical concentration remain an inevitable prerequisite of functional centralisation and the ensuing advancement[71].

3.2.2 Developments by domain and by sector

In the domain of the state, the end of the great colonial empires, followed later by the collapse of the Soviet Union, led to an explosive growth in the number of independent countries, each with its own capital city. This had a great decentralising effect on the global level. Within countries the pattern of dispersal depends too on the administrative structure. Countries with a strongly centralised government, as seen in many developing countries, but also in developed countries like France, also have a centralised urbanisation pattern, with a disproportionately big capital city. Countries with a decentralised, often federal structure, demonstrate decentralised patterns of urbanisation, as for example in Germany.

In the economic domain, the consumer-oriented manufacturing industry is heavily dispersed. For example, the countless sweatshops in low-income countries, first in Southern Europe and Japan, then Hong Kong, Korea, Taiwan, Singapore, then in a number of Southeast Asian countries especially in China, directly followed by India, and soon no doubt Africa. In agriculture there is still great dispersal due to the hundreds of millions of small farmers who provide food for their own needs and sell some surplus to their immediate environment. But powerful centralisation is also taking place around modern agricultural forms, in globally operating food concerns. In logistics the basis is also one of extreme dispersal but current centralising trends are certainly as strong as in agriculture. Consumer services are still not very centralised in proportion and that is likely to remain so by its very nature. The business service sector,

---

[71] Sassen S. (2001). *The global city. New York, London, Tokyo.* Princeton University, Princeton, NJ, USA, 447 pp.

focused on the international market and within that the financial sector primarily, continues to be strongly centralised. To the trio of London, New York and Tokyo, will probably be added cities in China and India as these giants develop their economies further, and there are probably also opportunities for a city in the Middle East, or South America or Africa, but dozens of financial centres in the second order, such as Amsterdam, Frankfurt or Paris, will undoubtedly be forced to take a back seat.

Finally, the cultural domain: the pattern of urbanisation here seems to be similar to that of the business services and financial sector. Here too there are strong centralising trends, undoubtedly boosted by the global media and information and telecommunications networks, and problems encountered by smaller cities trying to maintain a flourishing cultural life[72].

3.2.3 Cities as spaces of flows

Cities and metropolises in the network society of the global age are processes, spaces of flows, centres of command and control, of innovation, of management of network enterprises and of markets for products and innovations. According to Taylor[73] cities are the places where the globalisation of the world economy is actually taking place. He points to the antithesis of nation states and cities, with far-reaching consequences, not only for the economic development process itself but also for the development of cities and even for social science, of which geography is a part.

The essence of cities is their capacity for import replacement:

> 'Import replacement is a city process because it requires special places to be generated and sustained. Only cities can embody a critical mass of people and ideas with the skills and flexibility to create the necessary new production. Such production does not have to be only of finished goods; the various producer goods and services that are required can be part of the vibrant new economy. (....) For import replacement to create a net gain in economic wealth there has to be a network of cities.'[74]

This statement is crucial. A city as a space of flows can never stand alone, it can only function in relation to other cities:

> 'Cities cannot be studied in isolation. Each human settlement is connected to other settlements in many different ways and through many different actors. These connections include flows of information, capital, goods and persons, which travel along such infrastructures as roads, railways, waterways, airlines and increasingly telecommunications. While actors such as companies,

[72] Florida R. (2002). *The rise of the creative class.* Basic Books, New York, NY, USA.
[73] Taylor P.J. (2004). *World city network: a global urban analysis.* Routledge, London, UK, 241 pp.
[74] *Ibid.*: 43.

> institutes, households or individual persons maintain these connections, on a more abstract level it is also possible to distinguish relationships between cities (...). These intercity relationships can be considered the aggregates of all the multifarious types of flows between the many urban actors.'[75]

That makes the identification of cities difficult because they are always part of a bigger whole. Nevertheless, there are many classifications of metropolises and new ones are being added all the time. But starting from the existence of the bigger network in which cities function, every classification is relative.

The traditional lists, where the starting point was provincial boundaries and number of residents, are no longer adequate. Since Castells defined the network society, research has been taking place on the basis of this new paradigm, whereby attention is no longer focused on the databases which are territorially limited on the basis of nation states and their subdivisions. Now spaces of flows take centre stage: The city is seen as a process and what is important is the intensity of the processes in the global network which make the city a place of residence. This new focus not only results in attention being paid to other kinds of data, where boundaries are also a constant problem. But the city itself is becoming a collection of networks. Hall and Pain define the 'polyopolis' as follows:

> 'A new phenomenon is emerging in the most highly urbanized parts of the world: the polycentric mega-city region. (...) It is a new form (...). It is no exaggeration to say that this is the emerging urban form at the start of the 21st century. (...) A key figure of these regions is that in different degrees they are polycentric (...). They are becoming more so over time, as an increasing share of population and employment locates outside the largest central city or cities, and as other smaller cities and towns become increasingly networked with each other.'[76]

Within this *mega-city region* the economic process revolves around clusters, defined by Porter (1998) as:

> '...geographic concentrations of interconnected companies, specialized suppliers, service providers, firms in related industries, and associated institutions (...) in particular fields that compete but also co-operate.'[77]

---

[75] Meijers E. (2007). *Synergy in polycentric urban regions. Complementarity, organising capacity and critical mass.* Delft University of Technology, Delft, the Netherlands: 3.
[76] Hall P. and K. Pain (2006). *The polycentric metropolis. Learning from mega-city regions in europe.* Earthscan / James & James, London, UK: 3.
[77] Porter M.M.E. (1998). Clusters and the new economics of competition. *Harvard business review* 76: 77, quoted in Hall P. and K. Pain (2006). *The polycentric metropolis. Learning from mega-city regions in Europe.* Earthscan / James & James, London, UK, 256 pp.

Within these clusters, focus is centred mainly on the advanced producer services (APS):

> '...the informational mode of development (...) is the emergence of the (...) advanced producer services (...): a cluster of activities that provide specialized services, embodying professional knowledge and processing specialized information, to other service sectors.'[78]

The advanced producer services (APS) on which the classifications of metropolises used here are based include accountancy firms, banks and financing companies, insurance groups, the legal profession, management consultants and the advertising world. Given the great emphasis on knowledge in the network society and the increasing entanglement between traditional knowledge institutes and the other KENGi parties (see Chapter 1), it is only logical to add the private and public research & development institutes to the APS list. In addition to the advertising agencies, other components of what until recently was known as the creative industry, can also be regarded as part of the advanced production services[79].

3.2.4 Metropolises

In his Metropolitan World Atlas, Van Susteren defines the metropolitan area as:

> 'Regions where global relationships dominate over local ones and which are characterized spatially by a high concentration of global connections and a high concentration of people.'[80]

Van Susteren (2005)[81] classifies the 101 most important metropolitan areas on the basis of 6 listings: namely (1) the 50 biggest population concentrations, and five 'flow characteristics': (2) the 25 biggest ports, (3) the 25 biggest airports in terms of passenger numbers, (4) freight and (5) flight movements and (6) the 30 most important telecom hubs. This classification produces 89 areas. In order to reach the figure of 101 metropolitan areas and so to enable comparison between every individual area with 100 others, Van Susteren added 12 metropolises on the basis of their political, cultural or religious status. Although this last choice in particular is arbitrary and Van Susteren has to make combinations throughout Western Europe in order to include places like the Randstad, the Flemish Diamond and the Rhine-Ruhr area in the ranking, Van

---

78 Hall P. and K. Pain (2006). *The polycentric metropolis. Learning from mega-city regions in Europe.* Earthscan / James & James, London, UK: 4.

79 Innovatieplatform (2005). Creativiteit. De gewichtloze brandstof van de economie. Available at: http://www.innovatieplatform.nl/assets/binaries/documenten/2005/creatieve_industrie/rapportcreatieveindustrie2.pdf.

80 Van Susteren A.W.C. (2005). *Metropolitan world atlas.* O10 Publishers, Rotterdam, the Netherlands: 7.

81 *Ibid.*: 10. The added metropolitan areas are Melbourne, Barcelona, Berlin, Montreal, Lisbon, Geneva, Athens, Vancouver, Oslo, Jerusalem-Tel Aviv (MCR), Tangier and Baghdad.

Susteren's Metropolitan World Atlas is a refreshing overview, particularly because of the consistency with which data from each metropolitan area is presented.

Hall and Pain[82] and Taylor[83] have also drawn up classifications of the most important metropolises. As a result of their criticism of earlier classification systems such as that of Sassen[84], which measure attribute data from the metropolises and not flow data, they try to gain insight on the basis of the actual flows that take place in the network and an analysis thereof. To this end Taylor uses information from the top 100 APS companies he drew up himself[85], and their distribution across head and subsidiary offices around the world. He ranks these offices on the basis of the position that office occupies in the company and then scores the cities on the total weighting of APS offices that they accommodate. Hall and Pain have attempted to go a step further in their analysis and have tried to depict the real information flows and contacts (the flows) via an online survey of the employees of the APS offices. However, the response to these surveys was too low for a reliable analysis.[86] Hall and Pain present the Loughborough classification of cities, in which the ranking is achieved on the basis of a combined weighting of the significance of cities in, respectively, the world of accountancy, advertising, banking and the legal profession[87].

Both Hall and Pain and Taylor[88] are aware of the data problem that an analysis of spaces of flows creates if the available data is primarily from spaces of places. This data problem is especially prevalent as regards the definitions of network cities and/or polycentric metropolises. As a result, both the above-mentioned classifications include the cities of Amsterdam, The Hague, Rotterdam and Utrecht, but these are combined in the analysis of Hall and Pain[89] into the polycentric metropolis Randstad. The same applies to all other polycentric metropolises.

---

[82] Hall P. and K. Pain (2006). *The polycentric metropolis. Learning from mega-city regions in Europe.* Earthscan / James & James, London, UK, 256 pp.
[83] Taylor P.J. (2004). *World city network: A global urban analysis.* Routledge, London, UK, 241 pp.
[84] Sassen S. (2001). *The global city. New York, London, Tokyo.* Princeton University, Princeton, NJ, USA, 447 pp.
[85] Taylor P.J. (2004). *World city network: A global urban analysis.* Routledge, London, UK: 66.
[86] Hall P. and K. Pain (2006). *The polycentric metropolis. Learning from mega-city regions in Europe.* Earthscan / James & James, London, UK: 86.
[87] Beaverstock J.V., P.J. Taylor and R.G. Smith (1999). A roster of world cities. *Cities* 16: 445-458..
[88] Taylor P.J. (2004). *World city network: A global urban analysis.* Routledge, London, UK, 241 pp.; Hall P. and K. Pain (2006). *The polycentric metropolis. Learning from mega-city regions in Europe.* Earthscan / James & James, London, UK, 256 pp.
[89] *Ibid.*

3.2.5 Polycentric mega-city regions

The polycentric nature of these metropolises makes them scale-less in a spatial sense precisely because of their essential characteristic as spaces of flows. The global age is one network; in the metropolitan areas it takes the form of a polycentric mega-region[90], which consists of yet more networks:

> '...the eight (...) mega-city regions [of North West Europe] together make up a much bigger (though discontinuous) mass of no less than 72 million people, and while not contiguous they are sufficiently close and highly linked as to form a super MCR: Europolis, fully comparable with the largest such regions in Eastern Asia.'[91]

The following important characteristics of these polycentric mega-city regions (MCR) are highlighted in an analysis by Hall and Pain[92]:

- Measured primarily by the settlement of APSs, each MCR has one primary city[93], which ensures the connection with the global network from that MCR. The communication flows are at their most intense within the primary city, and there is also further clustering of APSs within the primary city in a limited number of locations.
- Although electronic communication in many forms is dramatically increasing, face-to-face contacts remain essential and mobility is rising in proportion. A good infrastructure is therefore vital for the functioning of the MCR and most particularly for the primary city within. The priorities in infrastructure development then reconfirm the dominant position of the primary city.
- With the exception of the hardware for the APSs, the quality of the city-buzz is critical for the business environment. As a result the APSs have made explicit demands on the authorities concerning transport, educational facilities, housing and urban planning. Conversely, an address in the primary city is important for the competitive edge of the individual enterprises.
- The delimitation of the MCR is of no relevance to the APSs. They do business within the local or regional MCR network, or outside with neighbouring regions or in the global network – in essence, within the universe. Hall & Pain do in fact confirm the asymmetry between universal and local space that Sloterdijk also refers to (see Chapter 2).

---

[90] *Ibid.* define eight polycentric mega-regions (PMR) in North West Europe (but could also have included in this area the ECRs of Central and North England, Hamburg and Copenhagen): Dublin and surroundings, South East England (London), Randstad (including Amersfoort, Breda and the hub of Arnhem Nijmegen), the Flemish Diamond, Rhine-Ruhr, Rhine Main, central Paris and Zurich. In Figure 2 these PMR are included, but with Van Susteren's codes.

[91] *Ibid.*: 13.

[92] *Ibid.*: 120.

[93] Dublin, London in South East England, Amsterdam in the Randstad, Brussels in the Flemish Diamond, Düsseldorf in Rhine-Ruhr, Frankfurt in Rhine Main, Paris and Zürich.

3.2.6 The world outside Europe

Aviation and telecommunications belong to the developments currently embracing the whole world. As a result of this the processes also going on outside Europe have already been included in the description of the centralising and dispersing trends accompanying these developments. That begs the question: is the way in which centralisation and dispersal occurs outside Europe also applicable in general to that within Europe? And does it have the same result? In other words, is the whole world reflecting the same historically evolved distinction between, on the one hand, the swarming of not very big cities at relatively small distances from each other, and on the other hand much bigger solitary cities? A distinction that is still very recognizable in Europe, as for example the earlier picture of the 'blue banana' shows, despite the increase in complexity and the ensuing nuances and intermediary forms

The comparison with the rest of the world is distorted by scale differences. There is already a scale difference with that part of the world closest to Europe, namely North America. The average North American city is twice the size of the average European city, which gives rise to the tendency to compare a trading and industrial city with a central city in Europe merely on the basis of a similarity in size. The patterns do seem to be substantially similar, when corrected for the scale difference by expressing all city sizes, on both continents, in relation to the average[94]. Looking back at the American development, the dispersal forces there seem to have been stronger, and the centralising forces weaker, than in Europe. Given the federal structure of the US, that confirms the previously established relationship with administrative development.

When compared with Asia, the difference in scale is even bigger. The average city there is five to ten times bigger than that in Europe. There too it is common practice to standardise by expressing city size as a fraction or multiple of the average. In this case there is also a lot more similarity with the European, and so too the American, patterns. It should however be noted that centralisation has played a bigger role here than in North America, which again confirms the previously established relationship with the administrative structure. In most key Asian countries (with the exception of federal India, where the city size distribution is considerably more even), government is very centralised and occurs in combination with a serious lack of basic infrastructure and facilities in the area of education, health care and knowledge development, right across the country. Given the limited development budgets (and sometimes too the lack of interest in the rest of the country), only the capital city manages to do anything about this, reinforcing the pressure on businesses and institutions to settle there. Only in the capital city is there an international airport with a reasonable number of flight destinations, only here is there a more or less guaranteed supply of electricity

[94] Van Susteren A.W.C. (2005). *Metropolitan world atlas.* O10 Publishers, Rotterdam, the Netherlands, may serve as a good basis for such an exercise.

and drinking water, good telecommunications systems, hospitals complying with international standards, good schools and universities, and modern and well-stocked shopping centres, etc.[95].

3.2.7 Hundredseven metropolises

Table 1 gives an overview of the three classifications: 100 metropolitan areas from the Metropolitan World Atlas[96] supplemented with those areas from Taylor's[97] analysis, and those from the Loughborough classification. To the first classification are added cities from the second and third that scored higher in both these last classifications than Van Susteren's cities[98]. The result is a total of 106 cities. An average ranking is calculated from the three league tables and from the ranking on population size. This determines the position of the area in the table. By reason of the above-mentioned uncertainties about the data, this table should also be seen as a relative ranking. Where Van Susteren clustered cities in the Metropolitan World Atlas which are mentioned separately in both other indices, this is indicated with the abbreviation 'MCR' (mega-city region) or 'ECR' (European city region).

3.3 Spatial planning of metropolises in the network society

Taylor concurs with Jacobs[99] in his description of the spatial planning principles of the metropolis and its environment in the global age. Every settlement that develops its potential for import replacement, becomes a city. The core of the city comprises five forces of expansion which are unleashed in its economic landscape:

'...

- new market force: a rapid increase in the size of the city market for new imports through the city network;
- new employment force: a rapid increase in the quantity and the diversity of city jobs associated with import replacements;
- relocation force: a more rapid turnover in city industries as older enterprises move out of the city
- new technology force: an increased use of new city technologies; and
- new capital force: a generation of new city capital.'[100]

---

95 See for example, the description of Jakarta by Buijs S.C. (1990). De stedebouwkundige ontwikkeling van Jakarta. In: Rijksplanologische Dienst (ed.) *Ruimtelijke verkenningen 1990*. Ministerie VROM, The Hague, the Netherlands.

96 Monterey (California, USA) has been omitted from this list because of unclear classification.

97 Taylor P.J. (2004). *World city network: A global urban analysis*. Routledge, London, UK, 241 pp.

98 These areas are Zürich, Prague, Caracas, Warsaw, Budapest and Munich.

99 Jacobs J. (1984). *Cities and the wealth of nations*. Vintage, New York, NY, USA, in: Taylor P.J. (2004). *World city network: A global urban analysis*. Routledge, London, UK, 241 pp.

100 *Ibid*.: 45.

Table 1. Metropolis ranking: R1: Ranking according to van Susteren (2005), R2: Ranking according to Taylor (2003), R3: Ranking according to Beaverstock (1999). RT: Ranking on the basis of the average ranking of population, R1, R2 and R3.

	Population	R1	R2	R3	RT
Tokyo Yokohama (MCR)	33,190,000	1	5	1	1
New York	21,767,000	2	2	1	1
London	13,945,000	5	1	1	3
Los Angeles	17,263,000	3	9	5	4
Paris	10,600,000	8	4	1	5
Hong Kong	9,180,000	7	3	5	6
Chicago	9,549,000	16	7	5	7
São Paulo	17,720,000	19	16	15	8
San Francisco – Oakland (MCR)	7,154,000	6	17	11	9
Mexico City	19,620,000	35	18	15	10
Randstad Holland (EMR)	6,600,000	3	12	21	11
Singapore	4,163,000	10	6	5	12
Taipei	7,260,000	13	20	21	13
Seoul Incheon	22,877,000	28	41	19	14
Madrid	5,300,000	16	11	15	15
Frankfurt (EMR Rhine Main)	2,600,000	9	14	5	16
Washington – Baltimore	7,910,000	16	37	21	17
Sydney	3,997,000	23	13	11	17
Buenos Aires	13,390,000	33	23	41	19
Milan	4,050,000	42	8	5	19
Jakarta	13,330,000	55	22	21	21
Toronto	5,470,000	45	10	11	21
Moscow	13,100,000	55	34	19	23
Shanghai	13,580,000	42	31	41	24
Bangkok	7,250,000	30	28	36	24
Antwerp – Brussels (EMR Central Belgium)	3,725,000	37	15	15	24
Miami	5,289,000	11	25	41	27
Mumbai	20,043,000	55	21	56	28
Dallas – Ft. Worth	5,785,000	11	61	21	29
Beijing	13,160,000	55	36	36	30
Rhine Ruhr (EMR)	11,100,000	55	50	21	31
Zürich (EMR N. Switzerland)	3,500,000	55	19	11	32
Manila	14,140,000	55	46	41	33
Atlanta	3,500,000	15	33	41	34
Istanbul	10,430,000	55	35	41	35
Houston	5,176,000	25	62	21	36
Kuala Lumpur	3,025,000	28	26	41	37

Table 1. Continued.

	Population	R1	R2	R3	RT
Melbourne	3,367,000	55	24	21	38
Johannesburg	5,530,000	55	43	21	39
Kobe – Osaka – Kyoto (MCR)	16,390,000	14	125	21	40
Santiago de Chile	6,061,000	55	57	21	41
New Delhi	13,730,000	55	52	56	42
Boston	5,815,000	55	60	21	43
Stockholm	1,684,000	40	27	36	44
Cairo	14,000,000	55	59	62	45
Barcelona	3,766,000	55	32	41	46
Copenhagen	1,524,000	21	44	41	47
Rio de Janeiro	10,810,000	52	69	56	48
Prague	1,193,000	55	29	21	48
Rome	3,900,000	52	53	36	50
Hamburg	2,593,000	40	48	41	51
Berlin	4,101,000	55	51	41	52
Bogota	6,990,000	55	55	62	53
Montreal	3,216,000	55	47	41	54
Caracas	1,823,000	55	58	21	55
Warsaw	1,615,000	55	40	36	55
Seattle	2,712,000	19	68	62	57
Budapest	1,825,000	55	45	41	57
Minneapolis – St. Paul	2,389,000	31	77	41	59
Calcutta	13,940,000	55	87	74	60
Philadelphia	6,010,000	55	76	56	60
Lisbon	3,000,000	55	42	62	60
Lima	7,420,000	55	80	62	63
Karachi	11,020,000	55	84	74	64
Detroit	5,415,000	37	85	62	64
Auckland	1,290,000	50	38	56	66
Munich	1,194,000	55	49	41	67
Kaohsiung	1,494,457	22		74	68
Geneva	399,000	55	67	21	69
Athens	3,188,000	55	56	71	70
Denver	1,984,000	25	73	74	71
Bangalore	5,687,000	55	82	71	72
Vancouver	2,118,000	55	65	62	73
Dubai	1,171,000	55	54	62	74
Chennai	6,700,000	55	102	74	75
Lagos	10,030,000	55	122	74	76

Table 1. Continued.

	Population	R1	R2	R3	RT
Orlando	1,157,000	42		74	77
Le Havre	815,089	36		74	77
St. Louis	2,078,000	50	81	74	79
Oslo	780,000	55	66	62	80
Perth	1,340,000	45	79	74	81
Taichung	983,684	45		74	82
Jerusalem – Tel Aviv (MCR)	1,040,000	55	91	56	83
Memphis	972,000	55		74	84
Louisville	864,000	55		74	85
Dhaka	8,610,000	55	150	74	86
Phoenix	2,907,000	27	139	74	87
Tanger	497,147	55		74	88
Anchorage	339,286	55		74	89
St. Petersburg	5,410,000	55	142	71	90
Lahore	5,920,000	55	155	74	91
Pittsburg	1,753,000	55	119	74	92
Indianapolis	1,219,000	55	114	74	93
Charlotte	759,000	55	108	74	94
Tehran	10,740,000	55	201	74	95
Cincinnati	1,503,000	55	147	74	97
Durban	2,117,650	48	164	74	98
Nagoya	8,837,000	31	233	74	99
Las Vegas	1,314,000	33	196	74	100
Hyderabad	6,390,000	55	227	74	101
Tianjin	9,920,000	55	256	74	102
Sacramento	1,393,000	37	215	74	103
Busan	3,650,000	23	264	74	104
New Orleans	1,009,000	52	208	74	105
Kinshasa	5,750,000	55	274	74	106
Baghdad	5,400,000	55	307	74	107

As mentioned earlier, this city development takes place in networks of cities, and within these networks cities can fluctuate between phases of strong and weak economic growth, because they compensate for each other. Where urban networks lose the capacity to operate in a non-synchronised fashion, they also lose their place in the core of the global age and fall back into the periphery. Borrowing this model of centre, semi-periphery and periphery from Wallerstein[101], Taylor describes a basic spatial model for cities in the global age and the relationship with their periphery which also identifies five types of peripheral spaces, connected to the five above-mentioned forces of expansion:

> '...the five types of peripheral spaces are supply regions created by city markets, abandoned regions created by city jobs, cleared regions created by city technologies, transplant regions created by exported city industries, and cityless regions distorted by city capital. (...) These five types of space actually divide into two groups: three types of disrupted places (abandoned regions, cleared regions and warped cityless regions), and two types of connected regions (supply regions and transplant regions) (...) Finally, there are peripheral spaces that are the opposite end of the world-economy's wealth creation spectrum: bypassed places, regions that have lost their cities and city ties or have never historically developed them, This is the world of subsistence economies, the rural backwaters of the third world.'[102]

The whole model can be made dynamic with this split into peripheral spaces. Where Wallerstein uses the concept of the semi-periphery to characterise the dynamic transitions in time between regions that are first central and later peripheral and vice versa, Taylor does this with cities. New central cities arise from the connected regions, and then in particular from the supply regions:

> 'Of course, they can arise only through their strong connections with the core's city network. (...) Supply regions from which Hong Kong, Seoul and Singapore have (...) grown.'[103]

3.3.1 *Inhabitants of the network cities*

As a physical hub of the network society, the metropolises are places where, to use Sloterdijk's phrase, 'souls with no place' can be found: the managers and knowledge workers in and around the advanced production services, who move from head office to branch office to airport to hotel and to meeting place, managing the global age. The demand for services, provided by cheap labour, is high among these managers. There is lots of room for 'transition places' here because there is a lot of traffic, the economy is concentrated and metropolises are historically places with a story and so attractive to tourists: traffic (stations, streets, shopping centres) as well as short-stay options (holiday homes, industrial terrains, asylum centres), all of which have to be built, run and maintained.

---

[101] Wallerstein I. (1979). *The capitalist world-economy.* Cambridge Universty Press, Cambridge, UK, 305 pp.

[102] Taylor P.J. (2004). *World city network: A global urban analysis.* Routledge, London, UK: 46.

[103] *Ibid.*: 48.

As contiguous spaces of place these are also the places where many people live. Because when the 'souls with no place' are not working, they want to enjoy good living standards. The quality of the space of place becomes a critical factor in attracting these people:

> 'Where young skilled people and senior business decision-makers want to live, and to work, thus becomes a crucial determinant of the geography of the regional knowledge economy.'[104]

3.3.2 The fourth world

But these same cities also function as the great promise of happiness for all those people outside the 'Crystal Palace', who come in great numbers, legally or illegally, to seek their fortune. Some succeed, many fail and are excluded by the network society[105]. Castells defines this as an anti-society: the fourth world. These are the people, the city nomads with nothing, who populate the backstreets and slums of the metropolises and the people who live in deterritorialised groups in the rural backwaters of the third world:

> 'The emergence of a powerful, competitive Pacific economy, and the new processes of industrialization and market expansion in various areas of the world, regardless of recurrent crises and systemic instability, broadened the scope and scale of the global economy, establishing a multicultural foundation of economic interdependence. Networks of capital, labour, information and markets linked up, through valuable functions, people, and localities around the world, while switching off from their networks those populations and territories deprived of the value and interests for the dynamics of global capitalism. There followed the social exclusion and economic irrelevance of segments of societies, of areas of cities, of regions, and of entire countries, constituting what I call the "Fourth World". The desperate attempt by some of these social groups and territories to link up with the global economy, to escape marginality led to what I call the "perverse connection", when organized crime around the world took advantage of their plight to foster the development of a global criminal economy.'[106]

The above description shows that the network cities can be interpreted spatially from the way in which they function in the network, and that the arising functional

[104] Hall P. and K. Pain (2006). *The polycentric metropolis. Learning from mega-city regions in Europe.* Earthscan / James & James, London, UK: 113.
[105] Sloterdijk P. (2006). *Het kristalpaleis. Een filosofie van de globalisering.* Uitgeverij Boom/SUN, Amsterdam, the Netherlands: 163 ff.
[106] Castells M. (2000). *The information age: economy, society and culture. Volume 3: end of millennium.* Blackwell, Oxford, UK: 368.

relationships with their hinterland are also very dominant in the design of the hinterland, sometimes as functionally linked regions and often as dislocated regions.

3.4 Delta metropolises[107]

Behind the previously discussed interpretations of cities as the cores of the network society, lies the assumption that the physical significance of the space is less important than the significance it acquires from the social processes that take place there. That certainly holds true if the focus is on the processes that take place largely in the virtual world, like financial traffic, ICT, the media and entertainment industry, tourism and recreation. But since this publication is about agroproduction, the physical quality of the space deserves more attention, simply because it matters in agroproduction: in the traditional sense as soil quality, hydrology and climate, but also on the basis of recent relevant parameters as logistical accessibility or availability of waste and by-products.

From now on, therefore, the focus will be on the special form and physical environment of those cities in which the agroproduction highlighted here has emerged by focussing attention on a common functional feature of the metropolises that have ruled the global economy since the Middle Ages[108]: Venice, Genoa, Antwerp, Amsterdam, London and New York but also Shanghai and Hong Kong are all port cities, situated at the mouth of a river, which functionally links them to a huge hinterland.

Delta metropolises, as the name suggests, are located in water-rich, low-lying and fertile river deltas. Deltas are formed by the deposit of sediment at the place where a river flowing into the sea loses speed and with it the capacity to transport sediment.

---

[107] In the period between 2002 and 2005 a team of researchers from Alterra and RIZA worked on a publication about Delta Metropolises, but due to a lack of funding this work was not completed. This section uses unpublished components of this collaboration. Other publications arising from this collaboration are:
Smeets P.J.A.M., W.B. Harms, M.J.M. Van Mansfeld, A.W.C. Van Susteren and M.G.N. Van Steekelenburg (2004). Metropolitan delta landscapes. In: Tress G., B. Tress, W.B. Harms, P.J.A.M. Smeets and A. Van der Valk (eds.) *Planning metropolitan landscapes. Concepts, demands, approaches, Delta series.* Wageningen University, Wageningen, the Netherlands, pp. 103-114.
Oosterberg W. and C. Van Drimmelen (2006). *Rode delta's.* Ministerie van Verkeer en Waterstaat, The Hague, the Netherlands.

[108] Taylor P.J. (ed.) (2003). *European cities in the world network, The european metropolis 1920-2000.* Eramus University, Rotterdam, the Netherlands. Summary of descriptions from different authors in the list Venice, Genoa, Antwerp, Amsterdam, London and New York. See also:
Wallerstein I. (1974). *The modern world system. Capitalist agriculture and the origins of the European world-economy in the sixteenth century.* Academic Press, New York, NY, USA; Wallerstein I. (1980). *The modern world system ii. Mercantilism and the consolidation of the european world economy 1600-1750.* Academic Press, New York, NY, USA, 370 pp.; Wallerstein I. (1989). *The modern world system iii. The second era of great expansion of the capitalistic world-economy, 1730-1840s.* Academic Press, New York, NY, USA, 372 pp.

Many rivers flow to a so-called tectonic basin, where natural subsidence occurs. When the sedimentation occurs faster than the subsidence, the delta becomes not only higher but wider, in the direction of the sea. Areas with more recent mountain formation and a tropical climate drop a lot of sediment. If a river in one of these areas also flows out over a continental plate (a shallow sea), this can create a large delta in a relatively short period of time. That is why South East Asia, with its combination of young mountain ranges and continental plates in various places, has many delta metropolises. Other good examples are the Rhine and Po delta in Europe and of course the Nile delta in Egypt.

Delta metropolises are polycentric mega-city regions with a special character. Not only are they of historical interest, but they offer especially favourable conditions for the development of the high productive agriculture that this publication focuses on.

3.4.1 Delta metropolises historically developed in Europe and China

McNeil[109] refers to the difference in development that took place at the beginning of globalisation in two places in the world, namely China and Europe:

> '...the special pattern of European transport, [which], before the age of the railways, rendered port cities on or near the mouth of major river systems unusually important (...) The transport system in the most north-westerly part of the European peninsula concentrated supplies and economic functions in port cities – most of which were situated "on the edge" of the continent, and weren't capital cities of important nations.'

The caravan was the dominant transport system in the rest of the world, and in many places until well into the 19th century. Trade was conducted by camel, mule and llama, and even on people's backs, when these animals were not available because of disease. These caravan systems not only had accompanying physical facilities but also their own laws, which regulated water and food provisions and safety during the long journeys.

In Europe and China the big river deltas provided the possibility of another system: transport across water, by sea on sailing ship and on rivers by barge, enabled much bigger loads to be transported than the caravan system. Delta metropolises arose with the transfer of these much greater loads. It was also from this point on that a significant part of the globalisation written about by Sloterdijk, took place. In the period between 1400 and 1850 the dominant world powers were the sea powers of Venice, the Republic of the Seven United Netherlands and Great Britain, each operating from the Delta

---

[109] McNeill W.H. (1996). *De excentriciteit van het wiel en andere wereldhistorische essays*. Uitgeverij Bert Bakker, Amsterdam, the Netherlands: 37.

of one or more low-lying rivers[110]. In time Great Britain functioned as an island from several deltas interconnected by canals. With the passage of time the originally dominant position of the London port has been completely superseded by other ports, and London no longer appears in the top 25 list of the Metropolitan World Atlas. In the European colonies this spatial configuration was, where possible, copied in areas later called New York and New Orleans.

3.4.2 Delta metropolises as trading cities

In this way, delta metropolises became the intersections of global trade, because they linked their Hinterland, their supply regions in the river basin, via the river with other areas overseas.

As well as fulfilling a transport function, delta metropolises also became concentrations of industrial activity. The supply of nutrient-rich water and sediment, and the presence of water and land, means not only that the biodiversity of deltas is naturally very great by nature, but it also offers the possibility of food production. At the same time, there is usually a big market outlet in the immediate surroundings. Delta metropolises are therefore often characterised by an intensive aquaculture, agriculture and horticulture industry. These are sometimes so intensive that the products have become an important export item. The combination of highly productive agriculture also facilitated the emergence of a high-quality food and textile industry. The establishment of the oil industry in the vicinity of delta metropolises is not only linked to the presence of ports; it is also because the substratum of the delta regions is often rich in oil and gas.

The presence of fishing grounds, fertile land and waterways for transport has always drawn people to delta areas. However, there are also deltas that are sparsely populated or where no cities grew up, such as those of the Ebro and Danube. Residences were traditionally concentrated on the higher banks of the rivers, which were dry and sturdy, or just outside the delta. But as urbanisation accelerated, the basin areas were also reclaimed and then built on. A delta metropolis therefore usually consists of several cities or a conurbation, which together form a polynuclear urban network.

This historical development still has consequences for culture and society in delta metropolises. For example, there are only a few delta metropolises that are an administrative unit, some are even frontier regions (Mekong, Ganges-Brahmaputra, Rhine-Maas-Schelde) The borders of nation states seldom coincide with the borders of a delta. Or else deltas are incorporated in their entirety within a certain state

---

110 Wallerstein I. (1974). *The modern world system. Capitalist agriculture and the origins of the European world-economy in the sixteenth century.* Academic Press, New York, NY, USA; Wallerstein I. (1980). *The modern world system ii. Mercantilism and the consolidation of the European world economy 1600-1750.* Academic Press, New York, NY, USA, 370 pp.

or the river or centre of the delta forms the border between nation states, an easy defence line as it historically was. The administrative authorities of deltas often settle elsewhere and usually don't recognise the delta metropolis phenomenon, showing little interest in their mega-identity. Furthermore, this identity is – almost by definition – multicultural, because as world ports, delta metropolises are nearly always places of heavy immigration flows and have strong relationships with the global network.

3.4.3 Problems in Delta metropolises[111]

The nature of deltas is under threat. Many deltas are dyked, causing the clay sediment within the dykes to become bedded down and further reinforcing the subsidence in the polders. Between the dykes, where the river flows, the sediment is then deposited even more quickly, causing the river bed to rise. This increases the risk of flooding and the water balance is almost constantly having to be reviewed. The flood risk is increasing because of deforestation of the basin areas and expected climate change. The delta metropolises will then be caught between the increasing drainage of rivers and rising sea levels. The flooding of the delta causes a reduction in the size of the area of nature reserves and wetlands, in turn causing a decrease in the natural resilience to natural disasters (high flood waters, tsunamis, hurricanes, etc.). In addition, when dams are erected upstream for irrigation and flow management, deltas then suffer from periods of drought. This also has major consequences for safety. The condition of dykes can deteriorate and coastal erosion may occur. Nature and food production (fish and rice crops) also suffer as a result.

In most delta areas there are extensive areas of wetlands. They are consequently also important for preserving global biodiversity. The wetlands constitute a key link for fish that grow up in the sea and spawn in rivers and streams. Furthermore, all major delta areas in Asia, Europe and North America are invaluable stepping stones on the migratory path of birds between the Arctic and tropical regions. Permanent large wetland areas are necessary to provide the birds with enough food. The natural wetlands (including mangroves) and their diversity are threatened primarily by urban sprawl and intensive food production such as shrimp farming. In Asia water contamination caused by, for example, the intensification of farming, is a growing problem. But in Europe and North America rivers are now being revitalised and critical fish species like salmon are being reintroduced.

[111] Makaske B. (2008). De kwetsbaarheid van delta's. Zeven plagen in een geologisch perspectief. *Geografie* 17: 50-55.

3.4.4 Delta metropolises in the world

There is no single list of delta metropolises in the world. In their report on Red Deltas, Oosterberg and Van Drimmelen[112] give a list of metropolises relevant from the perspective of flood management, but this summary is incomplete.

Of the 107 metropolises mentioned in Section 2.2, Table 2 indicates those cities that can be characterised as the core of a delta metropolis.

3.5 The Northwest European delta metropolis

The Northwest European delta metropolis (Figure 4) covers the region between Lille, Amsterdam and Cologne and was charted in the Benelux Structure Sketch.[113] The pattern of urbanisation of this delta metropolis is described chronologically. The transport infrastructure along which the proceeding urbanisation is taking place is the network in which corridors of housing and industrial activity are developing.[114] The bigger network of these corridors was first charted in the European Commission report entitled Europe 2000+[115] (Figure 5).

Insight into the evolution of the polycentric metropolis of North West Europe can also be gained from the German concept of *Groß-* or *Agglomerationsräume*, which is defined as the area from whence the centre of a major city can be reached within one hour of travel. When applied to cities with more than 500,000 inhabitants, and replacing an hour's travel with a 50 km distance, it is evident that the *Agglomerationsräume* in the area between Lille, Amsterdam and Cologne overlap one another. There is no bigger collection of cities to be found in Europe (Figure 6).

The pressure on space is high in the delta metropolis. But, together with the high cost of labour, this is the key driving force behind the intensification of land use and the accompanying productivity benefits. When seen like this, the Northwest European delta metropolis is not a threat but a prerequisite and a future prospect for the continued qualitative and quantitative growth of agriculture in the network society.

---

[112] Oosterberg W. and C. Van Drimmelen (2006). *Rode delta's.* Ministerie van Verkeer en Waterstaat, Den Haag, the Netherlands. Discuss the flooding problem of the Thames Delta, the Seine, the Elbe, the Rhine and the cities of New Orleans, Tokyo, Wuhan (China), Dhaka (Bangladesh) and Venice.

[113] Van den Broeck J., M. Barendrecht, P. De Boe, F. D'hondt, P. Govaerts, P. Janssens, R. Kragt, M. Van Ginderen and W. Zonneveld (1996). *Ruimte voor samenwerking; tweede Benelux structuurschets,* Brussel, Belgium, 185 pp.

[114] Verkennis A. and T. Groenewegen (1997). Ontwikkelingen in de regio Randstad-Rijn/Ruhr. In: Zonneveld, W. and F. Evers (eds.) *Van delta naar Europees achterland.* NIROV-Europlan, The Hague, the Netherlands.

[115] European Commission (1995). *Europe 2000+.* Office for Official Publications of the European Commision, Luxembourg, Luxembourg.

Table 2. Delta metropolises in the world.

All population data from van Susteren (2005), unless otherwise stated.

Delta	Central city	Estimated population in core metropolis	The whole delta comprises	Estimated population of whole Delta metropolis
Yangtze	Shanghai	13,580,000		93,336,700[1]
Pearl River	Hong Kong	9,180,000		19,500,000[2]
				37,187,000[1]
Rhine-Maas	Amsterdam	713,000	Randstad, Antwerp-Brussels, Rhine-Ruhr	21,425,000
	Brussels	294,000		
	Cologne (2002)	729,000	City rings, Central Belgium, Rhine-Ruhr area	28.075.712[3] (2000)
Hudson	New York (2003)	8,086,000		21,767,000
Thames	London (2001)	2,766,000	Greater London	13,945,000
			South East England	18,984,298[3]
Brahmaputra, Ganges	Calcutta (2000)			13,940,000
Nile	Cairo (2000)			14,000,000
Rio de la Plata	Buenos Aires (2000)			13,390,000
Po	Milan (2001)	3,790,000		4,050,000
Elbe	Hamburg	1,726,000		2,593,000
Brahmaputra, Ganges	Dhaka (2000)			8,610,000
Mississippi	New Orleans (2000)			1.009.000

[1] Shi *et al.* (2002).

[2] Rohlen (2000).

[3] Hall and Pain (2006: 21).

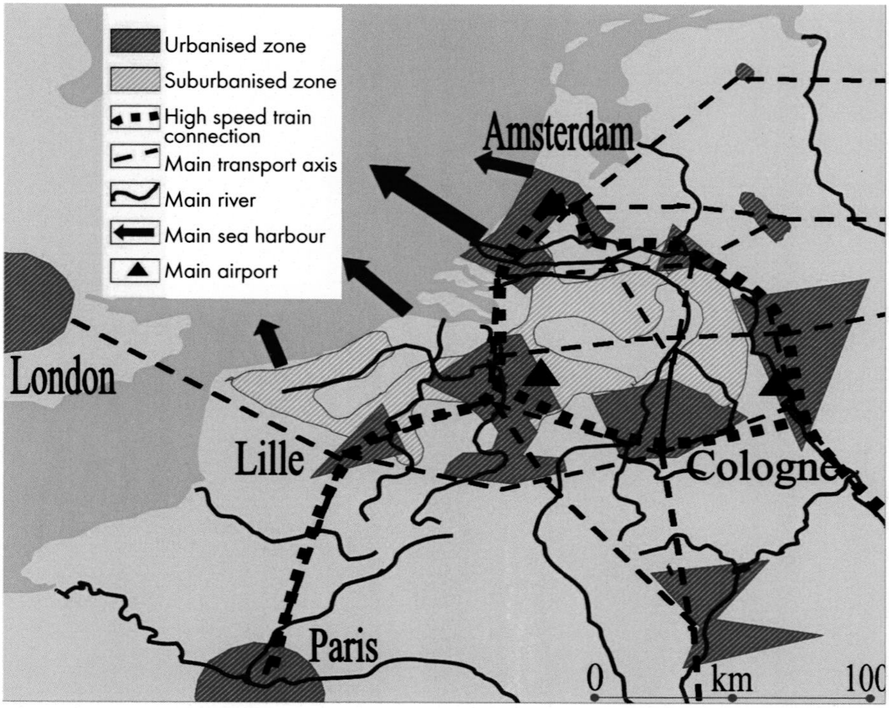

Figure 4. The Northwest European delta metropolis (Van den Broeck et al., 1996).

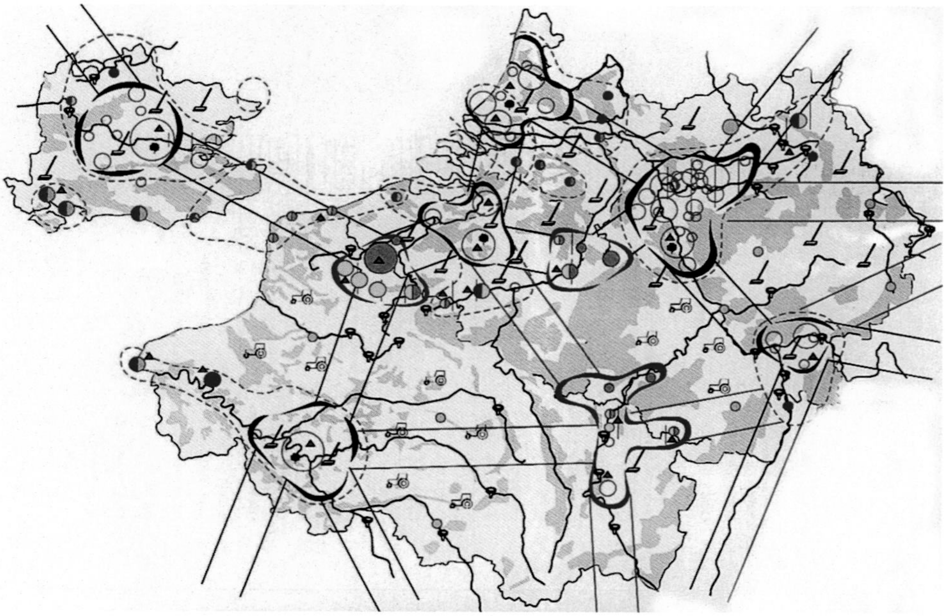

Figure 5. The mega-corridors in North West Europe according to Europe 2000+ (European Commission, 1995: 181).

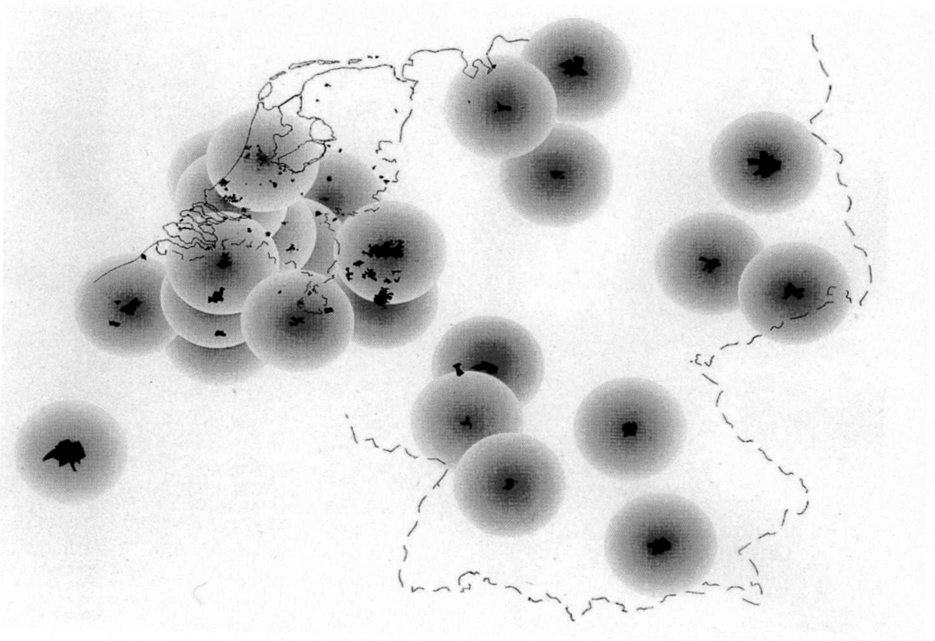

Figure 6. Overlapping Agglomerationsräume *in North West Europe together form the Northwest European delta metropolis.*

In the preceding chapters, it has been argued that there are multiple reasons why the further development of agricultural productivity with a heavy accent on modern industrial agriculture and metropolitan foodclusters is urgently required. In the first place, because the world's population will continue to grow dramatically in the upcoming decades and because all this growth will take place in cities. Secondly, because people in cities will develop new consumption patterns. Thirdly, because people are abandoning the countryside, leaving fewer and fewer people who want to produce food in the traditional way.

Seen in this light, agriculture and food provision are extremely important aspects of sustainable development. And sustainable development will not be possible without radical modernisation of agricultural production.

Conversely, traditional forms of farming still take up a large part of the increasingly scarce space in and around metropolises. The future of metropolitan foodclusters and agroparks in the metropolis itself is as a space pump, freeing up land for other urban functions due to the rise in productivity in agriculture.

In the following chapter I will look in more detail at the development of agriculture up to the current day in North West Europe.

4. Agriculture in the network society

4.1 Mega-trends in the urbanised network society

The long lines in history, sketched in Chapter 2 as the driving force behind the emergence of the network society and in Chapter 3 as the urbanisation process, can also be summed up as a number of 'mega-trends'. The mega-trends are described here insofar as they are relevant and apply as parameters for the special development of agriculture that is the central tenet of this book, i.e. agriculture in the network society.

Mega-trend 1. Not all the world and not all the world's citizens are part of the Global Age. World politics will be dominated by the conflicts that emerge in this field of tension.

In other words, there is a danger that the 'Crystal Palace' will close, causing an even bigger contrast to emerge between inside and outside or in geographical terms between centre on the one hand and semi-periphery and periphery on the other. Food production, and more particularly food distribution, play a central role here. The development of it, together with other sectors in society, which are directly and indirectly closely connected to agroproduction (energy, water, waste), largely determines whether the luxury position of the 'Crystal Palace' will become accessible to even more people.

How should we deal with this rift between inside and outside? According to the standards and values of the 'Crystal Palace' itself, or by resorting to the historical approach from the era of globalisation? In the words of Castells:

> 'The territorial unevenness of production will result in an extraordinary geography of differential value making that will sharply contrast countries, regions, and metropolitan areas.'[116]
>
> 'New forms of warfare (...) will be used by individuals, organizations, and states, strong in their convictions, weak in their military means, but able to access new technologies of destruction.'[117]

[116] Castells M. (2000). *The information age: economy, society and culture. Volume 3: end of millennium.* Blackwell, Oxford, UK: 385. The contrast is not territorial but can be found on all spatial scales, in which the 'crystal palace' manifests itself. On the global scale between rich and poor continents and countries but also within cities between the gated communities of rich citizens and the ghettos of the poor.

[117] *Ibid.*: 387.

> 'Geopolitics will also be increasingly dominated by a fundamental contradiction between the multilateralism of decision-making and the unilateralism of military implementation of these decisions.'[118]

But it's not just about military conflicts: Castells talks about the exclusion of the excluders by the excluded and the advent of globally organised criminal networks:

> 'Because the whole world (...) will increasingly be intertwined (...) under the logic of the network society, opting out by people and countries will not be a peaceful withdrawal.'[119]
>
> 'The global criminal economy will be a fundamental feature of the twenty-first century and its economic, political, and cultural influence will penetrate all spheres of life.'[120]

Mega-trend 2. The network society is made up of multilateral institutions, which are themselves networks

While the systematic expansion of the borders of the 'Crystal Palace' requires strong leadership, the diminishing power of the nation state is reducing this capacity to lead. At the height of globalisation, the nation state, an invention of the 19th century, will not simply disappear but the capacity of individual nation states to intervene in the global network society will continue to decline.

> 'Nation-states will survive, but not so their sovereignty. They will band together in multilateral networks, with a variable geometry of commitments, responsibilities, alliances and subordinations (....) The global economy will be governed by a set of multilateral institutions, networked among themselves.'[121]

Both Castells and Sloterdijk point to the European Union as an example of such a multilateral institution[122]. The example is so striking because it also shows the weakness of the administrative power of networks in the Global Age.

---

[118] *Ibid.*: 388. Sloterdijk P. (2006). *Het kristalpaleis. Een filosofie van de globalisering.* Uitgeverij Boom/SUN, Amsterdam, the Netherlands: 259, calls the recent American military action an explicit example of the impossibility of solving problems in the *Global Age* with tools from the era of globalisation: 'Bush's America is abandoning post-historicity and re-enters history'.

[119] Castells M. (2000). *The information age: economy, society and culture. Volume 3: end of millennium.* Blackwell, Oxford, UK: 386.

[120] *Ibid.*: 385.

[121] *Ibid.*: 386.

[122] As stated by Sloterdijk P. (2006). *Het kristalpaleis. Een filosofie van de globalisering.* Uitgeverij Boom/SUN, Amsterdam, the Netherlands: 179, 'Europe (...) consumed its first strike-power during the opening up of the planet and burned its surplus energy in two great wars (...). The present-day Europe (...) has become exemplary in another way, because it entails an almost mature concept of post-imperialistic policy.'

This withdrawal of the nation state was described in Chapter 2 with reference to spatial planning. At a time when processes like urbanisation and agricultural modernisation are under discussion and threaten to develop into an even sharper division between the 'haves' and 'have-nots', the strong control of the nation state in the background is disappearing and there are effectively no alternatives in place.

Mega-trend 3. Individuals are becoming disengaged and resorting to their primary networks

> 'People (...) will be increasingly distant from the halls of power, and disaffected from the crumbling institutions of civil society. They will be individualized in their work and lives, constructing their own meaning. (...) When subjected to collective threats, they will build communal heavens (...) The twenty-first century will not be a dark age (...). Rather it may well be characterized by informed bewilderment.'[123]

Therefore, the crucial question is whether there will be reasons in the Global Age for spatial planning that transcends individual housing, other than collective threats, which up until now have not resulted in much more than extremely defensive reactions, such as the control of oil supplies and terrorism. There will be others in the coming decades. They will arise from the other mega-trends, such as the problem of climate change, the rise of other superpowers, the increasingly dire absolute shortage of water and the partially related rise in food shortages.

At the same time new network technology provides an opportunity to connect with other individuals around the world.

This trend seems to be the counterpart of communicative self-steering, on which Cornelis pinned his hopes. It is a warning that the self-steering capacity of citizens in the network society, which is anticipated by the spatial development policy, cannot necessarily be counted on.

All in all, these three trends combine to form a sombre outlook for the future. It is reminiscent of Marcuse, who concludes his bleak work 'De eendimensionale mens'[124] [The One-dimensional man] with the statement: *Nur um Willen der Hoffnungslosen ist uns die Hoffnung gegeben* [Only for the sake of the hopeless, ours is given hope]. Sloterdijk turns this statement on its head:

[123] Castells M. (2000). *The information age: economy, society and culture. Volume 3: end of millennium.* Blackwell, Oxford, UK: 388.
[124] Marcuse H. (1970). *De een-dimensionale mens: studies over de ideologie van de hoog-industriële samenleving.* Paul Brand, Bussum, the Netherlands, 279 pp.

> 'Hope, therefore, is not a principle but a result. There are two things that keep hope alive (...): Firstly the fact that people occasionally have new ideas, which bring changes in life in the transition from model to application, both on the micro level and on a large scale. (...) Secondly, the fact that, normally speaking, where there is sufficient density, a practicable test is filtered out of the torrent of ideas begging to be implemented, which is the best offer for most people if not all. The reason behind the density works as a series of filters that ensure the elimination of unilateral offensives and directly harmful innovations.'[125]

New ideas will emerge from the communicative self-steering that Cornelis talks about. New ideas, that are developed into a practical test, supported by the collective intelligence, that can be found in the metropolitan density of the 'Crystal Palace'.

Expedition Agroparks is a report on the work performed on such a new idea and on the practical test, the long journey through the filters and the inhibiting contexts, that have to be put aside before any result can be achieved in the 'Crystal Palace'.

In contrast to the above three bleak trends that define the regime, in which new ideas must come to fruition, there are four promising trends, which form the basis for the work on agroparks. They are inextricably linked and I will focus more on their significance for agriculture. My approach here largely corresponds to that of Rabbinge[126].

Mega-trend 4. Increased productivity

The increase in productivity in terms of labour and land, which has been an ongoing trend since the beginning of globalisation in the 15th century[127] and has accelerated since the start of the network society, will continue and expand[128]. There is also

[125] Sloterdijk P. (2006). *Het kristalpaleis. Een filosofie van de globalisering.* Uitgeverij Boom/SUN, Amsterdam, the Netherlands: 192-193.
[126] Rabbinge R. (2000). The future role of agriculture. In: Boekestein A., P. Diederen, W. Jongen, R. Rabbinge and H. Rutten (eds.) Towards an agenda for agricultural research in Europe. Wageningen Pers, Wageningen, the Netherlands, pp. 161-168.
[127] Castells M. (2000). *The information age: economy, society and culture. Volume 1: the rise of the network society.* Blackwell, Oxford, UK. But see also Penning de Vries F.W.T., R. Rabbinge and J.J.R. Groot (1997). Potential and attainable food production and food security in different regions. *Philosophical transactions of the Royal Society of London. Series B, Biological sciences* 352: 917-928.
[128] Castells M. (2000). *The information age: economy, society and culture. Volume 3: end of millennium.* Blackwell, Oxford, UK: 384-385. 'The information technology revolution will accentuate its transformative potential (...). The twenty-first century will be marked by the full flowering of the genetic revolution (...) To prevent the evil effects of biological revolution we need not only responsible governments, but a responsible, educated society.'

likely to be a broadening of the range of energy sources, on the one hand because climate change is limiting the use of fossil fuel, and on the other hand because a characteristic of the continued productivity development is that more can be done with the constant supply of non-fossil fuel available on the planet[129]. This is a hopeful trend but its continuation requires far-reaching interventions in spatial planning and energy policy. This trend applies explicitly to agriculture: the continuous development in productivity, in terms of land and labour, and since the 1970s of the use of fossil fuels. This means that fewer and fewer people can do more and more with increasingly less land. According to the WRR[130], as far as land productivity is concerned, this trend has been on-going since the 15th century and is independent of social processes like demographic or economic growth or shrinkage. The driving forces behind the continued productivity development are not only the development of technology and the cost of labour and land. Recent decades have also seen environmental demands, which have led to industrialisation of the production process, in which the use of energy and raw materials per unit of product can be reduced to a minimum. Boone *et al.*[131] summarise the environmental performance of Dutch agriculture and horticulture as follows:

> 'In the nineties agriculture and horticulture have been able to achieve a sizeable reduction in energy consumption, nitrogen and phosphate surpluses, ammonia emissions, net pollution from heavy metals and the use of pesticides (...) In the years 1995-2003 energy consumption by agriculture and horticulture fell by about 10%. The emissions of methane, nitrous oxide and carbon dioxide (each responsible for a third of the contribution of agriculture and horticulture to the climate problem) were reduced in the period 1990-2002 by about 15% and have been virtually stable ever since. The surpluses of nitrogen and phosphate, the emission of ammonia and the net pollution

[129] In an essay written in 1995 by employers at the Ministry of Housing, Spatial Planning and the Environment, which had a short-term goal of generating a common vision on the 'natural' collaboration between the various parts of that ministry, an interesting statement was defended, namely that the solution to environmental problems is a question of sufficient energy supplies, with which these problems can be solved. And because the world will have to make the transition from fossil to renewal energy sources, and these latter depend on the amount of space available for the generation thereof, the amount of space in the world was postulated as the ultimately determining variable. The policy translation of this reasoning was that the environment and energy problems which occupy the ministry can ultimately be solved with good spatial planning. The unity of policy has not materialised since 1995 (Ministerie van Volkshuisvesting Ruimtelijke Ordening en Milieu (1995). *Milieu, ruimte en wonen; tijd voor duurzaamheid.* VROM, The Hague, the Netherlands, 91 pp.).

[130] Wetenschappelijke Raad voor het Regeringsbeleid (1992). *Ground for choices; four perspectives for the rural areas in the European community.* Wetenschappelijke raad voor het Regeringsbeleid, The Hague, the Netherlands.

[131] Boone K., K. de Bont, K.J. van Calker, A. van der Knijff and H. Leneman (2007). *Duurzame landbouw in beeld. Resultaten van de Nederlandse land- en tuinbouw op het gebied van people, planet en profit.* Report 2.07.09, LEI, The Hague, the Netherlands..

> by heavy metals were all nearly halved in the period 1990-2002. There has been a dramatic fall in the supply of animal manure, partly as a result of the reduction in the number of animals, as well as fertiliser. In addition, low-emission spreading of animal manure has played an important role in the reduction of ammonia emissions. (...) The use of pesticides (expressed in active ingredient) also halved in the period between 1990 and 2003.'

It is important to realise here that production as well as production value increased in many cases. So the EE index (defined as the primary fuel consumption per unit of output compared to the reference year 1980) of the most energy-intensive agrosector in the Netherlands, i.e. glasshouse horticulture, fell between 1980 and 2005 to 46%[132].

Theoretically the increased efficiency means that there is an increasingly favourable ratio between costs and yield, known as the resource use efficiency theory[133], which I will come back to in Chapter 5.

Mega-trend 5. Industrialisation and rationalisation

Directly related to the increased productivity trend and in the last two centuries one of the driving forces itself, is the mega-trend of industrialisation. The principles from the industrial revolution are also being used to their full in agricultural production. In the first instance this means the use of the 'natural labourer fossil fuel', in the production of fertilisers and pesticides and in the replacement of human and animal labour by machines. Industrialisation in agriculture is also the uniformisation and optimisation of land and water management and land consolidation and then the manipulation of climatic conditions by means of advanced stables and greenhouses and the replacement of monotonous manual labour by automated systems such as manure spreading, feed machines, milking- and picking-robots.

The transition from industrial to information society also takes place in agroproduction. With the input of yet more knowledge the agro-eco systems are adapted to the demands of producers and consumers: as a result of the ICT revolution, biotechnology and genetic modification are developing rapidly.

---

[132] Wetzels W., A.W.N. van Dril and B.W. Daniëls (2007). *Kenschets van de nederlandse glastuinbouw.* Report ECN-E--07-095, Energiecentrum Nederland, Petten, the Netherlands: 14.

[133] De Wit C.C.T. (1992) Resource use efficiency in agriculture. *Agricultural Systems* 40: 125-151; De Wit C.C.T., H.H. Huisman and R.R. Rabbinge (1987) Agriculture and its environment: are there other ways? *Agricultural Systems* 23: 211-236. See also Rabbinge R., H.C. van Latesteijn and P.J.A.M. Smeets (1996). Planning consequences of long term land-use scenario's in the European union. In: Jongman R.H.G. (ed.) *Ecological and landscape consequences of land use change in Europe.* European Centre for Nature Conservation, Tilburg, the Netherlands.

Globalisation is reflected in worldwide sourcing and sales. The role of agrologistics is increasing proportionately.

Mega-trend 6. Chain and network formation

This trend too is partly a result of the previous trend. Industrialisation enables control of the quality and quantity of intermediary products in the agricultural production process to such an extent that extensive specialisation can occur in parts of this process, via independent production units, which are interconnected in clock time:

- Manure and fertiliser management: development from the deep litter stable system, whereby minerals from heather and other common pastures were concentrated for use on arable land, to industrial production of inorganic fertilisers and compost and to redistribution of manure from specialised livestock farms to arable businesses.
- Production of raw material: whereas in the old system the production process from 'seed to cutlet' took place entirely within the farm itself, these processes have also been unravelled into increasingly smaller components and taken over by specialist companies. For example, in intensive pig farming there are specialist companies that take care of the production of sperm; grandparental animals; breeding sows; piglets and final fattening; slaughter and secondary processing of meat products.
- The processing and marketing of agricultural products, which are exceptionally consumed by the primary producer himself or directly sold to the consumer. Not only is there more and more industrial processing of agricultural half-products like milk, pig carcasses, etc., but an extensive distribution and trading system has also arisen, in which the large retail companies dominate.
- Agrologistics, which handles the transportation between all these specialised chain components. In the Netherlands one in three lorries transports agroproducts and a quarter of all logistical movements are related to agrologistics[134].

In this context the transition from industrial to network society translates into the emergence of agroparks. That is the essence of this publication, which will be discussed in detail in Chapter 6.

These three trends of productivity increase, industrialisation and chain formation are highlighted in Figure 7 which shows the situation of the Dutch agrosector in 2002. This figure also clearly shows the urban industrial nature of Northwest European agriculture. Within the Dutch part of the delta metropolis (South and Central Netherlands), the relative share of the high productive industrial plant, dairy and meat producers are incidentally much greater than this figure demonstrates.

[134] Brouwer F.M., C.J.A.M. De Bont, H. Leneman and H.A.B. Van der Meulen (2004). *Duurzame landbouw in beeld.* Landbouweconomisch Instituut, The Hague, the Netherlands: 35.

- Total economy: € 410,000 mln
- Agrocomplex total: € 40,000 mln (10%)
- Agrocomplex on basis of domestic raw materials: € 23,000 mln (5.5%)
- Primary production: 9,000 mln (2.2%)
- Employment: 10%
- Use of space: 19,470 km²
- Spatial productivity: (€ mln/km²)

Glass:	27
ILF:	11
OGH:	1.1
Dairy:	0.17
Arable:	0.17

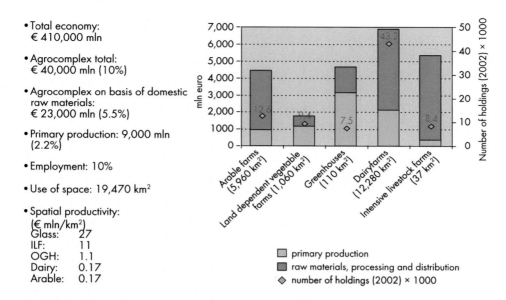

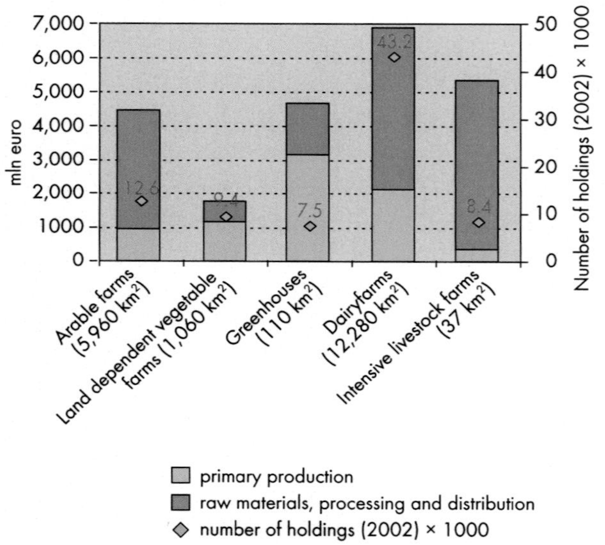

Figure 7. Overview of the significance of the Dutch agrosector in the economy and in space. Situation in 2002.

The figure is drawn up on the basis of data from the Ministerie van Landbouw Natuurbeheer en Visserij (2004).

Figure 7 also reveals some other important features of industrial agriculture: taken as a whole, the share of the primary sector is no more than 20% of the total turnover; but in glasshouse horticulture more than 2/3 of the turnover is earned by primary producers. There are extreme differences in production per ha: in glass horticulture this amounts to €270,000, in intensive livestock farming €110,000 and in arable and dairy farming €1,700.

It should also be noted that the production of concentrates, which occurs in dairy farming and in intensive livestock farming largely on the basis of imported arable products, is included in the costs but not in the surface areas. The output per ha of dairy farming is therefore lower and that of intensive livestock farming much lower.

Figure 7 also points to a third delimitation problem. The question is whether all advanced producer services, which in terms of origin and actual focus are largely related to this agrosector, should be included or not. The figures from the Wageningen UR Agricultural Economics Research Institute, shown in Figure 7, at least disregard part of this – such as the contribution by the financial economic services sector, government and knowledge institutions to the agrocomplex. That is a serious underestimation of the exceptional position that the agrologistic complex in North West Europe already occupies in relation to the dominant trend, in which the manufacturing industry is migrating to low-wage countries. True enough the number of primary producers (agricultural enterprises) is declining at a fast rate, but if the number of employees in the service sector serving the agrosector is also added to this sector, then the total employment does seem to have been falling for a longer period in relative terms. However, that is because employment in other parts of the economy has been increasing more dramatically. In absolute terms employment remains fairly constant. Table 3 shows the global picture.

Pluri-activity as a mega-trend?

As an addition to the three above-mentioned trends, Rabbinge calls the fourth trend that of monofunctionality to multifunctionality or pluri-activity. I don't subscribe to this trend. It's true that in recent decades there has been a visible move towards multifunctionality in the more urbanised, tourist and marginal areas of Europe, but this should be seen from the perspective of agriculture as more of a greatly supported slowing-down in the decline in the number of enterprises in the primary sector than a long-term trend towards more pluri-activity agriculture. In densely populated metropolises, of course, multifunctional land use is the obvious solution for the extreme pressure on space, but this would be better resolved by a collaboration of specialists in conservation, recreation, tourism, water management, etc., than by a new kind of universal farmer taking on all these professions, including the accompanying complex systems of legislation and permits. The expansion towards pluri-activity of agricultural enterprises is more likely to be the beginning of the end than a viable longer term future. The role of enterprises with a kind of broadbased operation therefore seems extremely limited in terms of the overall importance of agriculture for the national economy, as Table 4 shows. Van Eck *et al.* also subscribe to this need for professionalisation and specialisation.

> 'The financing of both development directions [i.e. experiential farming and agricultural management of nature and landscape] still requires a lot of attention, especially if the difference in operational management with production-oriented agriculture increases, and the revenue flows from production agriculture declines further. (...) It is likely that professionalisation of the agricultural management of nature and landscape and experiential farming will lead to specialisation and the link with production-oriented agriculture will be broken.'[135]

It is true that, like other sectors in the economy, agriculture will have to take account of the possible negative impact of its own activities on its surroundings. These surroundings are big and comprise many aspects because agriculture is still the biggest land user. Water management and management of nature and landscape are

[135] Van Eck W., A. Van den Ham, A.J. Reinard, R. Leopold and K.R. De Poel (2002). *Ruimte voor landbouw; uitwerking van vier ontwikkelingsrichtingen.* Report 530, Alterra, Wageningen, the Netherlands.

Table 3. Relative and absolute change in employment in the Dutch agrosector between 1995 and 2005.

	1995		2000		2004		1995-2004	
	absolute (×1000 work years)	% of total	absolute (×1000 work years)	% of total	absolute (×1000 work years)	% of total	absolute decrease	relative decrease
Agriculture and horticulture	228	4.0%	227	3.5%	209	3.2%	-8.3%	-19.57%
Food and luxury foods	144	2.5%	136	2.1%	124	1.9%	-13.89%	-24.44%
Total agrofood	372	6.6%	363	5.7%	333	5.2%	-10.5%	-21.45%
Transport, storage and communication	355	6.3%	403	6.3%	407	6.3%	14.65%	0.60%
30% assigned to Agrofood	107		121		122			
Total hardware agrofood	479	8.4%	484	7.5%	455	7.1%	-4.89%	-16.55%
Financial and business services	956	16.9%	1293	20.1%	1,276	19.8%	33.47%	17.11%
% Hardware assigned to agrofood	81		97		90			
Government	704	12.4%	731	11.4%	806	12.5%	14.49%	0.46%
% Hardware assigned to agrofood	59		55		57			
Total hard-, org en software agrofood	619	10.9%	636	9.9%	602	9.3%	-2.72%	-14.65%
Total	5,663	100.0%	6,423	100.0%	6,454	100.0%	13.97%	

Source: http://www3.lei.wur.nl/ltc/Classificatie.aspx, Table 12b. Totale arbeidsvolume werkzame personen, naar bedrijfstakken en sectoren. Accessed on 19 February 2007.

Table 4. Significance of pluri activity agriculture 2003.

	Number	Turnover (mln €)	Income (mln €)
Agricultural management of nature and landscape	9,580	30	12
Recreation and care	2,730	36	9
Storage	3,840	10	4
Wind energy	430	43	17
Total	16,580	119	42 = 1.5%

Figures based on Berkhout and Van Bruchem (2004).

therefore playing an increasingly important role in the operational management of land-dependent farming, and livestock holdings are having to implement ever more complex environmental management policies. But, with the exception of management of nature and landscape, these activities do not generate additional income, and are therefore not a basis for a future perspective in pluri activity.

Mega-trend 7. Diversification of the product package

This trend is also known as the development from food to fashion. The list is referred to by the author as the '12ff of RR' and is shown in Figure 8. The figure clearly demonstrates that agroproduction is fast diversifying and exiting traditional markets for food (and even sooner those of energy supply, fibre and building material). In the new markets being penetrated, a number of important features of traditional agriculture no longer apply, in particular the inelasticity of demand, as is evident from the inserts in Figure 8, borrowed from The Economist (2004)[136].

Furthermore, the higher up on the '12ff' ladder the main product is, the bigger is the quantity of waste and by-products in general in the production process, which can then be used again lower on the ladder. In this way, manure and waste from public parks can be fermented and turned into raw materials for bioenergy, and a large part of Dutch pig feed comes from leftovers and by-products of human food production.

A significant share of the move from food to fashion and pharma can be seen in the glasshouse horticulture industry, but this is not the only sector. Intensive livestock farming produces important raw materials for pharmaceutical products, such as

[136] Economist (2004). Rags and riches. Survey on fashion. *The Economist* 370.

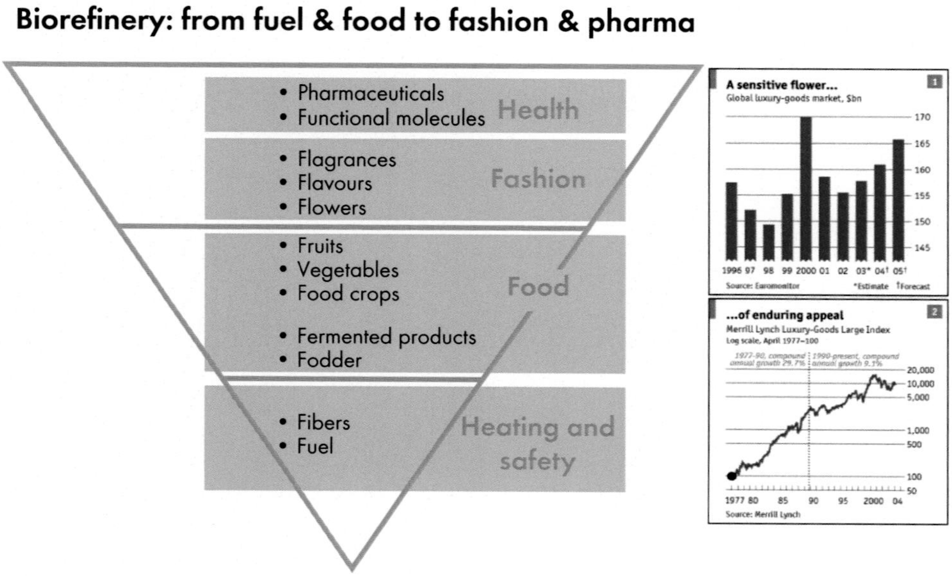

Figure 8. Most important product groups of agricultural production from bottom to top with increasing profit (in €) per ha as identified by Rabbinge. Two figures from The Economist (2004) have been added, which show at a glance the total absolute scope of the market for luxury goods between 1996 and 2005 and the relative growth of this market between 1977 and 2004.

gelatine, and with its contributions to functional foods and health products, the dairy industry is also moving to the top of the '12ff' list.

An extreme form of innovation in the area of pharmaceuticals is the development of Dutch home-grown cannabis cultivation, which has occurred in the Netherlands in the last 15 years. Based in fact on glasshouse horticultural techniques and seed production, this cultivation has not only undergone extreme improvements with a doubling of the quantity of active ingredient[137]. With the development of do-it-yourself growth sets, the (semi)illegally cultivated Dutch cannabis has largely suppressed the import of cannabis and has itself become an export product. The CBS (Central Bureau of Statistics) estimated that in 2001 the total trading value of cannabis from the Netherlands was € 1,410 m, of which the quantity of marihuana was € 1,190 m. The export value of cannabis in 2001 was € 810 m[138].

---

137 Pijlman F.T.A. (2005). Strong increase in total delta-thc in cannabis preparations sold in Dutch coffee shops. *Addiction Biology* 10: 171-180..

138 CBS (2006). *De Nederlandse economie 2003*. Centraal Bureau voor de Statistiek, Voorburg, the Netherlands.

4.1.1 Agroparks as innovation

The basis of the idea behind agroparks stems from mega-trend 4: the possibility of continuing productivity development and, linked to that, mega-trends 5 (industrialisation), 6 (chain and network formation) and 7 (diversification of the product package). Only by realising that perspective will it be possible to avert mega-trend 1 (exclusion of people from participation in the global age) by at least making food available for an increasing number of people, which accompanies the wealth of the 'Crystal Palace', without jeopardising other important features of that 'Crystal Palace' (sufficient good water, sufficient energy and good care of the environment). But just when the implementation of this possibility demands strong management, the capacity of an important steering party in this, the nation state is diminishing, as mega-trend 3 states. The network society as a whole needs collaborating institutions, and in the cores of the network society, in the metropolises, an inhibiting context is emerging, which opposes innovation and which comprises disengaged citizens who are seemingly less and less concerned about collective interests (mega-trend 4).

4.2 Agriculture in the network society

4.2.1 Development of planning and research into agriculture in the network society

In 1992 the Scientific Council for Government Policy (WRR) published the report 'Ground for choices'[139]. This report is the starting point for the thought process on the development of agriculture, which is further developed in this dissertation. On the basis of four political scenarios, the report explores the boundaries of the playing field within which European agriculture of the 12 member states of the European Union at that time could develop, but it focused on the future of land-dependent agriculture. The report centred on the mega-trend mentioned in Section 4.1 of continuing productivity increase in the development of agriculture and the most significant result of its scenario calculations was that, depending on the chosen scenario, 30-70% of the area of productive land for cultivation in the then member states of the European Union would be superfluous by 2020, if this increase in productivity continued as it has done in the preceding centuries.

[139] Wetenschappelijke Raad voor het Regeringsbeleid (1992). *Ground for choices; four perspectives for the rural areas in the European community.* Wetenschappelijke raad voor het Regeringsbeleid, The hague, the Netherlands.

The most important conclusion was that, given the differences between the chosen political scenarios, in which free trade, employment, nature and the environment occupy centre stage respectively, there are some important decisions to be made for European Union policy. Fifteen years down the line the second scenario, with employment in agriculture as the most important aim, has been the guiding principle of European politics to date. In that scenario the number of hectares of agricultural land that was expected to be freed up was the lowest, i.e. 40 million (from 120 million in 1992 to 80 million in 2020).

It is difficult to determine to what extent this exploration of the situation in 2020 is supported by actual development. Nevertheless, it wasn't the aim of the WRR study to make predictions. But the policy of the European Union in the area of land-dependent agriculture was also not an explicit choice, but rather the increasingly complex continuation of the original Common Agricultural Policy, whose key pillars, the price and export subsidies, will not be scrapped before 2013.

In addition, since the beginning of the 1990s, there has been a policy, alongside this one mainly in support of big enterprises, to assist smaller enterprises and businesses in vulnerable areas so that they can continue making their supposed contribution to the management of nature and landscape.

The Landelijke Gebieden en Europa (Rural Areas and Europe)[140] project, which was implemented by the Rijksplanologische Dienst (National Physical Planning Agency) between 1992 and 1997, is a continuation of the WRR report. The project was undertaken as preparation for the planning cycle of the Fifth National Policy Document on Spatial Planning and concentrated on developments in the Netherlands and North West Europe. As a consequence a few ideas which had particular relevance for the spatial interpretation of the situation in North West Europe were added to the focus on productivity development of land-dependent agriculture in the WRR report. In addition to the development of land-dependent agriculture, which was central to the WRR report, the relationship between agriculture and other dominant processes in land use, such as urbanisation and the development of tourism, was also included.

[140] Projectgroep Landelijke Gebieden & Europa (1995). *De toekomst van het landelijk gebied. Discussienota Eurokompas '95.* Ministerie van Volkshuisvesting, Ruimtelijke Ordening en Natuurbeheer, The Hague, the Netherlands; Projectgroep Landelijke Gebieden & Europa (1997). *Landelijke gebieden in Europa: Eindrapport.* Ministerie van Volkshuisvesting, Ruimtelijke Ordening en Natuurbeheer, The Hague, the Netherlands.

Furthermore, the focus was extended from land-dependent agriculture[141] to industrial agriculture[142], to marginalisation[143] and to pluri activity[144].

The Rural Areas and Europe project was relevant for this publication because it was also the prelude to Expedition Agroparks. This is evident from the following selection of conclusions from the final report:

> 'National and European policy [must] strive for an open spatial policy; not the autonomous development in agriculture but the interplay between agriculture, urbanisation and ecologicalisation is taken as the starting point for the policy for rural areas (...).
> The criteria for developing a good competitive position for the intensive sectors, with (...) an increase in spatial quality, must also be taken from spatial planning. (...)
> The combination of intensive agriculture and ecologicalisation offers a future for ecotechnology in the form of modern agriculture: integrated management of diseases and pests, substrate cultivation, manure processing, industrial dairy farming and so on.'[145]

Some of the recommendations of the Rural Areas and Europe project can be found many years later in the most recent National Policy Document on Spatial Planning[146], in particular where it defines five Greenports as a means of implementing the policy for the intensive sectors.

From 1997 onwards, work that was started in the context of the Rural Areas and Europe project has been continued in projects of the former Staring Centre, now

---

[141] Van Eck W., B. Van der Ploeg, K.R. De Poel and B.W. Zaalmink (1996). *Koeien en koersen; ruimtelijke kwaliteit van melkveehouderijsystemen in 2025.* Report 431, Staring Centrum, Wageningen, the Netherlands.

[142] Bolsius E.C.A. (1993). *Pigs in space.* Ministry of Housing, Spatial Planning and the Envirionment, The Hague, the Netherlands, 63 pp.; Alleblas J.T.W. (1996). *Vier kassengebieden in Europa; visie op ruimtelijke kwaliteit.* Landbouw-Economisch Instituut, The Hague, the Netherlands.

[143] Baldock D., G. Beaufoy, F. Brouwer and F. Godeschalk (1996). *Farming at the margins; abandonment or redeployment of agricultural land in Europe.* Institute for European Environmental Policy; Agricultural Economics Research Institute, London, UK; Bethe F. and E.C.A. Bolsius (1995). *Marginalisation of agricultural land in the Netherlands, Denmark and Germany.* National Physical Planning Agency, The Hague, the Netherlands / Kopenhagen, Denmark / Bonn, Germany.

[144] Van de Klundert A.F., A.G.J. Dietvorst and J.v. Os (1994). *Back to the future; nieuwe functies voor landelijke gebieden in Europa.* Report 354, Instituut voor Onderzoek van het Landelijk Gebied (SC-DLO), Wageningen, the Netherlands.

[145] Projectgroep Landelijke Gebieden & Europa (1997). *Landelijke gebieden in europa: Eindrapport,* Ministerie van Volkshuisvesting, Ruimtelijke Ordening en Natuurbeheer, The Hague, the Netherlands: 12-13.

[146] Ministerie van Volkshuisvesting Ruimtelijke Ordening en Milieu (2004). *Nota ruimte, ruimte voor ontwikkeling.* SDU Uitgevers, The Hague, the Netherlands.

part of Alterra, part of Wageningen-UR, with the publication of future visions and scenario studies on the future spatial development of agriculture in the Netherlands and Europe. This work was initially analytical, exploratory and predictive[147]. Only later was it translated into active designs, in the form of national[148] and regional[149] policy recommendations and in the form of landscape designs and designs for business systems.

In the years since 1997 this work has resulted in a vision of a new symbiosis between urban and rural areas. In an internal collaboration between Wageningen-UR and TransForum, an organisation which encourages the necessary sustainable development of Dutch agriculture by linking it to its metropolitan environment, the name 'metropolitan agriculture'[150] was devised for this body of ideas but this soon became an ineffective melting pot since various forms of urban agriculture got connected to it. This is why in this English translation the term 'metropolitan agriculture' is replaced by agriculture in the network society (referring to the activity) and metropolitan foodclusters (referring to its physical appearance).

4.2.2 Agriculture in the network society[151]

The network society enables a system of agroproduction, which, by means of the new and intelligent connections (between producers, sectors, raw materials, energy and waste flows; between stakeholders and between their value systems), is capable

---

[147] For example, Bethe F.H. (1997). *Rural areas and Europe. Processes in rural land use and the effects on nature and landscape.* Ministry of Housing, Spatial Planning and the Environment, National Spatial Planning Agency, The Hague, the Netherlands, pp. 72.
Van Eck W, A. Van den Ham, A.J. Reinard, R. Leopold and K.R. De Poel (2002b). *Ruimte voor landbouw; uitwerking van vier ontwikkelingsrichtingen.* Report 530. Alterra, Wageningen, the Netherlands, pp. 78.
Van Eck W., A. Wintjes and G.J. Noij (1997) Landbouw op de kaart. In: *Jaarboek 1997 van het Staring Centrum.* Staring Centrum, Wageningen, the Netherlands, pp. 4-20.
Van Eck W., R. Groot, K. Hulsteijn, P.J.A.M. Smeets and M.G.N. Van Steekelenburg (2002). *Voorbeelden van agribusinessparken*, Report 594, Alterra, Wageningen, the Netherlands.
[148] For example Denktank varkenshouderij (1998). *Mythen en sagen in de varkenshouderij.* Wageningen Universiteit & Research, Wageningen, the Netherlands, pp. 46.
Raad voor het Landelijk Gebied (2005) *Plankgas voor glas? Advies over duurzame ontwikkeling van de glastuinbouw in Nederland.* Raad voor het Landelijk Gebied, The Hague, the Netherlands, pp. 55.
VROMraad (2004) *'Meerwerk' advies over de landbouw en het landelijk gebied in ruimtelijk perspectief.* VROM, The Hague, the Netherlands.
[149] For example, Goedman J., D. Langendijk, E. Opdam, S. Reinhard, I.d. Vries and M. Wijermans (2002). *Zee en land meervoudig benut; beknopt projectverslag.* Alterra, WageningenVan Mansfeld M., A. Wintjes, J. De Jonge, M. Pleijte and P.J.A.M. Smeets (2003). *Regiodialoog: Naar een systeeminnovatie in de praktijk*, 808, Alterra, Innonet, WISI, Wageningen.
[150] As an alternative for the English term Metropolitan Agriculture, former Minister Verburg of Agriculture recently suggested the term 'Duurzame Delta Landbouw' [Sustainable Delta Agriculture].
[151] This text is a reworking of a text on metropolitan agriculture produced by TransForum.

of meeting the changing and competing demands of the urbanised society on a sustainable basis.

This agriculture is first and foremost part of the space of flows. Trade in raw materials and products for this agriculture are arranged on a global basis. The logistical networks come together in and directly around the metropolises, which are the cores of the network society, and they make optimum use of the power of the metropolitan environment (logistical hubs, strong networks, trendsetting consumers, large and qualitatively diverse food requirements, organisational capacity, the flow of knowledge, concentrated purchasing power).

The Delta metropolis has traditionally offered a number of location advantages, which are interconnected and mutually reinforcing:

- good physical conditions for agriculture;
- a good infrastructure and logistics, where transport by water plays an important role;
- a high level of knowledge among entrepreneurs and management in the agro-industry, and logistics, and among people and advanced production services around them;
- direct sales opportunities to and feedback from many (critical) consumers;
- an abundance of cheap labour;
- a large supply of waste and by-products, which can be used again elsewhere in the complex;
- possibility to grow from monofunctional chains into network complexes.

The resulting metropolitan foodclusters in the Delta Metropolis effectively satisfies the growing need for adequate, safe, healthy, diverse and responsibly produced food in the metropolises, but they are also important economic engines in the same metropolis.

However, in terms of spaces of places, as a spatial ambition, the transformation to metropolitan foodclusters is much more difficult. The fact that food production took place outside the city was not a problem at first. As long as cities remained small, the distance to the food source was minimal and the food producing area didn't have to be very expansive to satisfy the needs of the whole urban area. However, with the emergence of much bigger cities the distances are increasing, as is the quantity of land required, while the growth of the city itself is swallowing up much more agricultural land. That is happening at a pace that is determined by population growth as well as an increase in wealth. The greater the wealth, the greater the need for more space per head of population. The expected doubling of the urban population in the coming decades, and the expected increase of income per head as well, will lead to a four-fold increase in urban spatial possession. That may lead to a substantial reduction in productive agricultural land in traditionally fertile areas where urbanisation is already high. If this were to halve while the urban population doubled, the productivity of the remaining agricultural land would need to be four times higher in order to be able to feed the urban population.

Obviously not all food needs to come from the immediate surroundings. There is an extensive trade in food in the network society which ensures that products that cannot be grown locally are supplied from elsewhere, for instance out-of-season products. But, in many cases, this external supply also comes from areas that are being rapidly urbanised. All in all, it is true to say that the ratio between the total number of city-dwellers that need to be fed and the total area of available agricultural land is rapidly decreasing. Extension of the area of productivity is not an option. Fertile and ecologically robust areas have been used for agricultural land since time immemorial. Expansion is only possible now in marginal areas where the extra production to be achieved is out of all proportion to the damage that will be caused to the fragile ecology.

Increased productivity is therefore the only solution, helped by a change in diet that is also the result of urbanisation and an increase in wealth. Wealthy city-dwellers have much lower energy requirements than poor rural dwellers who perform heavy manual labour. They therefore have less need for basic foods (rice, corn, wheat, potatoes, etc.) that provide mostly calories, and instead want food that is rich in proteins, minerals and vitamins. Above all they want food that is tasty, easy to prepare, healthy, varied and fashionable. In the first instance, therefore, we are talking about the shift to higher meat consumption, but also fish, and dairy products, and fruit and vegetables. It is precisely this kind of food that lends itself perfectly to intensive production, with relatively limited use of land, that can be slotted nicely into the urban environment[152]. If a large part of agricultural areas of productivity can be converted from basic food production to this more high-value food, then the deterioration of the ratio between the number of city-dwellers to be fed and the ground available for food production needs not ultimately be a problem. Quite the opposite in fact, as there will be enough land remaining to grow the cattle feed necessary for the higher meat consumption, and even to accommodate a drastic rise in the cultivation of industrial and energy crops. A central criterion is a sufficiently rapid and widespread growth in wealth. Only then does a shift in diet occur and only then do city-dwellers have the income to pay for this higher priced high-value food.

---

[152] In this context for meat production a distinction must be made between the extensive land-dependent production of beef, goat and sheep meat, for which larger areas of grassland are needed and the intensive production of calves, pigs, chicken and fish, with a much more favourable food conversion. In recent years there has been a strong tendency among consumers, stimulated by health concerns, to switch from red (beef and pork) meat to white meat.

4.3 Spatial planning of agriculture in the network society

From the perspective of agriculture in the network society the classic antithesis between city and countryside is disappearing. In the metropolitan region there are different densities of urbanisation but rural areas no longer exist.

> '...in the countryside there has been a transformation from the old "working village" to the "residential village". The countryside has transformed from a production space to a multifunctional consumption space with a range of managers, users and stakeholders.'[153]

In order to be able to carry out a good analysis of the living situation in the countryside in the Netherlands, the Sociaal Cultureel Planbureau[154] (Social Cultural Planning Agency) used a negative definition in terms of urbanisation: areas with fewer than 1000 inhabitants per km² are defined as countryside. Compared to the definition used by the OECD, for example, to differentiate between urban and non-urban around the world, population density must first be increased from 150 to 1000 inhabitants per km² in the definition of countryside. The map in Figure 9 represents the consequences of this redefinition[155]. While in the SCP definition only the black areas are defined as urban, according to the OECD definition only the white areas and some of the red areas are still defined as rural.

The countryside, in which agriculture was not only the dominant factor in spatial terms but also primarily from an economic perspective, has disappeared in the Northwest European Delta Metropolis. Between 1995 and 2002 the share in added value of the primary production in the slightly urbanised areas fell from 2.8% to 1.9% and in the more rural areas from 7.3 to 5.1%[156].

The largest part of the Dutch countryside has become the outer space of the Northwest European metropolis and the agriculture within has changed with it. In the cores of the metropolis but also in the smaller cities and even in the villages, the lives of inhabitants have become essentially urban. The non-built environment in the metropolitan outer region fulfils roles that are connected to the housing function of those inhabitants: water regulation, nature, landscape and other recreational activities. Primary production is increasingly bound to preconditions from that function: dairy farms have to keep cows outside because they belong to the landscape, industrial

---

[153] Steenbekkers A., C. Simon and V. Veldheer (2006). *Thuis op het platteland. De leefsituatie van platteland en stad vergeleken.* Sociaal en Cultureel Planbureau, Den Haag, the Netherlands: 14-15.
[154] *Ibid.*: 19.
[155] European Commission (1996). *Prospects for the development of the central and capital cities and regions.* Office for Official Publications of the European Commision, Luxembourg, Luxembourg.
[156] Steenbekkers A., C. Simon and V. Veldheer (2006). *Thuis op het platteland. De leefsituatie van platteland en stad vergeleken.* Sociaal en Cultureel Planbureau, The Hague, the Netherlands, 72 pp. Area definition based on share of rural and urban postcode areas respectively in COROP regions.

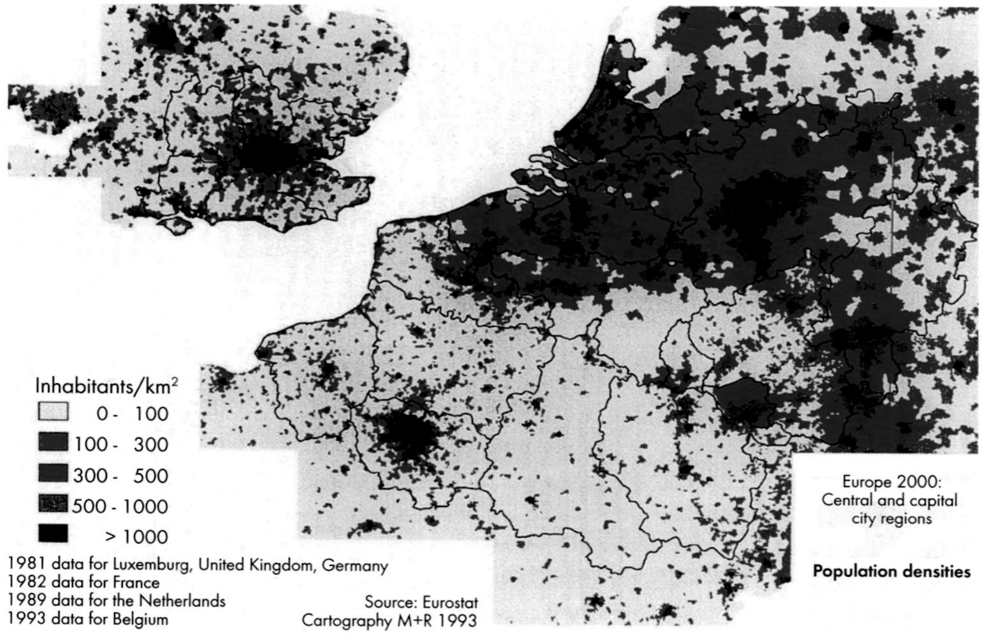

Figure 9. Population density in North Western Europe.
European Commission (1996).

buildings must fit in with the landscape and while street lamps burn, glass horticulture has to screen off its 'light pollution'.

The dominant function of the non-built environment in the metropolitan outer region is still that of agricultural production. To use Taylor's terms[157] (see quote in Section 3.3), these areas are the 'supply regions, created by city markets' and 'transplant regions, created by city industries'.

Outside the metropolis too, a significant portion of the countryside (with less than 150 inhabitants per km²) is used for these production functions. It makes them vulnerable because they depend on the same metropolitan foodclusters, of which only the primary production part takes place in the countryside.

In the countryside outside metropolises there are also many cleared regions, where a labour-extensive form of primary production has taken the place of previously labour-intensive forms of agriculture; or other economic activities, that have been

[157] Taylor P.J. (2004). *World city network: a global urban analysis.* Routledge, London, UK, 241 pp.

marginalised because of technological innovation. In the Netherlands, the river forelands are an example of a cleared region.

While the outer area (within the metropolis) or the countryside (outside) no longer plays an important role in the agriculture of the network society because of physical unsuitability, the other definitions from Taylor hold true: these are abandoned regions from whence the still remaining population retreats or cityless regions, where traditionally only marginal agriculture occurs.

4.3.1 Agricultural main structure?

The agricultural spatial main structure on a national level has been the subject of debate for years, but has never actually got beyond the passionate desires of various politicians. The Council for the Rural Areas rejects the idea in its opinion on this theme.

> 'The currently available legal and management set of instruments on the national, provincial and local level offers sufficient opportunities to protect and qualitatively steer the rural areas in our densely populated country with its diversity and differences in landscapes and strongly interwoven functions. There is no added value in having more instruments in the form of an agricultural spatial main structure at the national level in this situation.'[158]

In its opinion, the Council completely ignores the experiences from the spatial policy that was established between 1992 and 2004 by the national authorities for the rural areas in the Fourth National Policy Document on Spatial Planning (Extra)[159] in the Koersbepaling Landelijke Gebieden (Setting the Direction for Rural Areas) and that was only formally renewed with the establishment of the latest National Policy Document on Spatial Planning and the accompanying Structural Scheme for Green Space 2 in 2004.

Setting the Direction for Rural Areas was actually an attempt to formulate a generic national policy for rural areas, with the aim of spatially integrating changes in the rural area under the influence from agriculture. The dominant long term development direction was described for the whole of the Netherlands, and divided up into water catchment areas for this purpose: yellow direction: accent on agricultural production

---

[158] Raad voor het Landelijk Gebied (2007). *Samen of apart? Advies over de wenselijkheid van een agrarische hoofdstructuur op rijksniveau.* Report RLG 07/7, Raad voor het Landelijk Gebied, The Hague, the Netherlands: 4.

[159] Ministerie van Volkshuisvesting Ruimtelijke Ordening en Milieu (1992). *Vierde nota over de ruimtelijke ordening extra.* SDU Uitgevers, The Hague, the Netherlands. For a more detailed description of the body of ideas behind the 'directions' approach, see Kamphuis H.W. (1991). De vierde nota extra. Koersbepaling landelijke gebieden. *Landschap* 8: 47-58, and Kamphuis H.W., P.L. Dauvelier, J. Groen, H.C. Jacobs and G.J. Wijchers (1991). *Platteland op weg naar 2015*, Rijksplanologische Dienst, The Hague, the Netherlands.

function; green direction: accent on nature development; blue direction: accent on economic integration of various functions; and brown direction: accent on land-dependent agriculture in a mosaic of functions. The accompanying map image covered the entire Netherlands, with no differentiation between urban and rural areas.

The 'directions' policy completely failed[160]. So completely that it hasn't even been abolished[161]. In a self-assessment, included in the Fifth National Policy Document on Spatial Planning[162], the following conclusion was drawn:

> 'The directions policy was not well connected to the sector instrumentation and the financial tools from the Structural Scheme for Green Space (SGR). The policy had none of its own instruments. Although most provinces paid attention to these sorts of issues in their regional plans, they were unable in practice to handle the direction colours. The policy was much too passively designed to enable management of investments in agriculture.'

The Scientific Council for Government Policy also shared this opinion:

> 'Moreover, the continuation of this work in the planning framework of the Ministry of Agriculture, the Structural Scheme for Green Space, and the innumerable policy concepts that arose from this, is very limited.'[163]

In the light of these previous experiences with an attempt to obtain an agricultural spatial main structure, and given the previously described inactivity in which the spatial policy finds itself since the most recent Policy Document on Spatial Planning (see previously discussed comments on this issue in Section 2.4), there is no reason to be optimistic in this context. However, from the perspective of spatial development policy, it is possible to derive two parameters from the experiences of the directions policy, which should suffice for a national spatial policy:

- The national policy must restrict itself to the 'pearls' of the Dutch agrosector, which are relevant on the national level insofar as they justify the intervention of the state.
- These state priorities must be directly translated as political priorities and not as planning development into the policy of other departments and other government layers. If this does not happen then the policy will continue to be limited, just like the Direction Policy of the Fourth National Policy Document on Spatial Planning, to hollow phrases and colours on maps.

---

160 Driessen P.P.J. (1995). *Koersen tussen rijk en provincie: evaluatie van de doorwerking van het koersenbeleid voor het landelijk gebied naar het provinciaal ruimtelijk beleid.* Utrecht University, Utrecht, the Netherlands, 201 pp.
161 Van Duinhoven G. (2004). De vergeelde koersen. *Landwerk* 5 (2): 26-28.
162 Ministerie van Volkshuisvesting Ruimtelijke ordening en Milieubeheer (2001). *Ruimte maken, ruimte delen: Vijfde nota over de ruimtelijke ordening 2000/2020 vastgesteld door de ministerraad op 15 december 2000, Den Haag.* SDU, The Hague, the Netherlands.
163 Wetenschappelijke Raad voor het Regeringsbeleid (1998). *Ruimtelijke ontwikkelingspolitiek.* Wetenschappelijke raad voor het Regeringsbeleid, The Hague, the Netherlands: 29.

However, the aforementioned trends, and especially the trend described in Section 4.1 of industrialisation in agriculture and the emergence of metropolitan foodclusters, lead to yet another conclusion: an important part of the agricultural main structure to be formed is no longer situated in the rural areas. The core activities of agriculture in the network society, such as glasshouse horticulture, intensive livestock farming (which increasingly includes dairy farming) and other industrial primary production processes such as mushroom cultivation, and fish breeding have, like the other parts of agrologistics, such as processing, storage and distribution, become part of the urban system, of the 'planned space' to use De Geyter's terms (see Section 2.4). The most important parts of this agriculture and certainly those parts that are the most characteristic of Dutch agroproduction, have become part of the urban space. Translated into an agricultural main structure, they would only be a policy ambition of the national government, aimed at the optimal arrangement of spaces of place. However, the dominant decision making in the various chains and networks takes place, as in other dominant sectors in the economy, in the space of flows, where these nation state authorities have hardly any influence and which they can facilitate, but not manage, by means of spatial policy.

4.3.2 Greenports

In the most recent National Policy Document on Spatial Planning[164] a cautious start with this management is being made in relation to the Dutch horticulture and vegetable cultivation under glass by allocating five Greenports:

> 'In addition to two mainports and the brainport, the Netherlands has five Greenport areas where knowledge intensive horticulture and agribusiness is concentrated. From an international economic perspective the national government believes it is important for the horticultural function in these locations to retain its international importance and to be reinforced[165]. (….). The national policy is aimed at managing the spatial development of Greenports so that their function as a Greenport continues in the long term and/or is reinforced. Important areas of focus to this end are: the location in relation to the mainports, the physical accessibility and the restructuring task as a consequence of objectives in the area of environment, water, energy and spatial planning. The Greenports are: the South Holland glass district (Westland and Oostland), Aalsmeer and surroundings, Venlo, the Bulb region, and Boskoop.'[166]

---

[164] Ministerie van Volkshuisvesting Ruimtelijke Ordening en Milieu (2004). *Nota ruimte, ruimte voor ontwikkeling*. SDU Uitgevers, The Hague, the Netherlands.
[165] *Ibid.*: Summary.
[166] *Ibid.*: Chapter 3.4.6.1.

But the National Policy Document on Spatial Planning has already diluted the ambitions of Greenports:

> 'A long-term guarantee of the Greenport function does not automatically lead to guaranteed expansion space on the spot. Some greenports are situated such that expansion in that area is actually impossible, or will lead to serious damage to the surrounding landscape and/or water system. In such a case, expansion space will have to be found – if it is so desired – in existing or yet to be developed agricultural areas elsewhere in the Netherlands. Restructuring and good physical access to existing greenports are often all that is needed to maintain the strong international competitive position. This is primarily a job for the provinces.'[167]

At least four of the five allocated Greenports are situated such that any real expansion is not feasible: Parts of the Westland are currently being restructured for the extension of The Hague. The big expansion location of Oostland in the Zuidplaspolder is very controversial due to the low level of this polder. The orchard area of Boskoop lies in the middle of the Green Heart and the Bulb region is also more frequently discussed in the context of potential city expansion in the North Wing of the Randstad than in terms of expansion of the bulb crop. But finding 'expansion space on existing or yet to be developed agricultural areas elsewhere in the Netherlands' is a matter for the provinces. Alternative locations for expansion are few and far between in the province of South Holland.

So it seems that this subsequent attempt by spatial policy to regulate the development in agriculture differently from the 'directions' policy, is too selective (only horticulture and vegetable crops under glass) and is much too concerned with perpetuating the existing situation.

A much more ambitious effect could be achieved by taking into account the selectivity, which should follow the national spatial policy according to the principles of spatial development policy. Because of all the highly productive agricultural forms, only glasshouse horticulture has actually broken free of the typical occupational pattern of the Dutch countryside. Continuous glasshouse areas have emerged many hundreds, sometimes thousands, of hectares in size. The intensive livestock farming is still made up chiefly of family businesses based on original farmhouse parcels in rural areas. Bulb cultivation and ornamental plant cultivation also have a pattern of spatially distributed business residences, in family-run businesses.

When reconstructing the intensive livestock farming areas in the South and East of the Netherlands, an opportunity was missed to adapt this historical pattern to

[167] *Ibid.*: Chapter 3.4.6.1.

modern circumstances. The family business have remained the starting point, as has the occupation pattern with free-standing business parcels that belong to the entirely land-based system of yesteryear. In order to offer enough total capacity, far more agricultural development areas (where further growth is possible) had to be allocated than were needed in a more concentrated settlement pattern.

In the meantime, more and more businesses (often managed by a cooperative group of families)[168] are emerging that are much bigger than the original one-family business and applications are being submitted to build new stables which do not fit within the development areas, and are not wanted by those living in the area and local authorities[169]. Such businesses may then be a size too big for the open landscape in which they are established, and a size too small for an efficient operation. For example, they are not nearly big enough for slaughtering and processing to be carried out on the premises, which means that there is still no end in sight to the transportation of animals, with the accompanying costs, the suffering of the animals, and the risk of spreading diseases. If an optimal operation size had been the starting point, with the unavoidable conclusion that spreading settlements in the countryside (in its traditional form) was no longer possible, then it would have been possible to set up agricultural production and processing areas that would have been given the urban treatment in their positioning and design.

In addition to the already mentioned vertical integration possibilities (production and processing), horizontal integration would also be possible. That offers the opportunity for new, manufactured products with higher added value, and for the exchange of waste and by-products, energy and water which leads to lower costs, less environmental pollution and better working conditions overall. It seems as though the discussion now starting on mega-stables will lead to a repeat of history: fixation on the one-family business, underestimation of the increase in scale, needless damage to the open landscape, missed efficiency yields, and missed opportunities to drastically reduce environmental pollution.

The intensification, now of dairy farming as well, offers a golden opportunity to completely spatially integrate agriculture and urban development, which have been closely knit on a functional level in the Netherlands for a long time already, into a pattern in which recreation, nature and landscape also occupy an optimal place. There is a

---

168 Rienks W.A., W. van Eck, B.S. Elbersen, K. Hulsteijn, W.J.H. Meulenkamp and K.R. De Poel (2003). *Melkveehouderij op schaal. Nieuwe concepten voor grootschalige melkveehouderij*. Report 03.2.051, InnovatieNetwerk Groene Ruimte en Agrocluster, The Hague, the Netherlands.

169 This causes the problem of the mega-stalls. See College van Rijksadviseurs (2007). *Advies megastallen.* College van Rijksadviseurs, The Hague, the Netherlands, and Gies E., J. Van Os, T. Hermans and R. Olde Loohuis (2007). *Megastallen in beeld*. Report 1581, Alterra, Wageningen, the Netherlands.

need for development in the form of agroparks: multifunctional development areas for non-land-dependent agriculture, where various production forms (glass horticulture, pigs, chickens, calves, aquaculture, mushrooms, etc.) are brought together, along with processing companies and trade and distribution businesses, and with all kinds of support services (storage facilities and supply of packaging material; construction and maintenance of greenhouses, stables, installations, and such; knowledge development, research and certification; financial services and project development; branding and marketing, etc.) The free-standing pig or chicken enterprise, that is too big for its environment, but too small to be economically viable in the future relationships, will eventually have to disappear from the Dutch landscape. And that will ultimately be the case for non-land-dependent dairy farming too.

At the heart of this book are several designs for agroparks which have been made in recent years. Before beginning with the systematic analysis of these designs, I will first elaborate on the theoretical framework necessary for this in the next chapter.

5. Theoretical starting points in research by design

5.1 Research by design

In order to systematically describe as a research by design the learning experiences arising from the work on seven system innovation projects on agroparks, and to convert them into scientific theory, I apply the method proposed by Action Research, based on the work of Catherine Termeer[170]:

> 'First of all action research is engaged in action on real life issues with those who experience these issues directly (...). It demands an integral involvement by the researchers in an intent to change organizational rules and an intent by the public managers to take action. Besides action, action research can be considered as learning and researching. New knowledge is created through an interactive process in which actors reflect on their actions and underlying assumptions (...). Moreover, this learning must not remain restricted to the actors concerned, but is also aimed at making the experiences meaningful to others (...). Finally, we mention the characteristic of intensive collaboration between researchers and organisational members. The actors who take action, in our cases the public managers, also actively participate in the research.'

Figure 10 is the starting point for this purpose. In this approach the agropark designs are considered as applications of emergent theoretical concepts. The design and implementation thereof are action-oriented interventions and the methodical reflection, the description of the implicit meaning, the explicit meaning and the publication are carried out via this book. I describe the theoretical development on the basis of seven different designs, whereby the implicit assumption is that the development of the different theories has also taken place over time along these design steps. The reality is more chaotic, and is structured by a number of iterations. Not just because some designs ran in parallel but also because the insight sometimes only came later.

5.1.1 Three theories

Using this method of action research three theoretical concepts are further developed step by step on the basis of the seven agropark designs. The first two look at the result of the design and implementation process.

[170] Termeer C.J.A.M. (2008). Barriers for new modes of horizontal governance. A sensemaking perspective, Twelfth Annual Conference of the international Research Society for Public Management Queensland University of Technology, Brisbane, Australia.

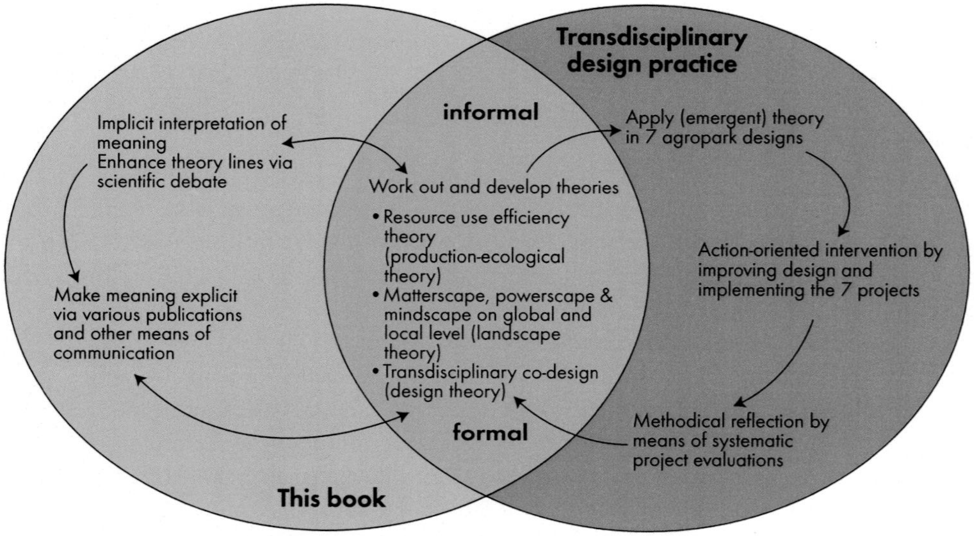

Figure 10. Research by design based on the Action Research method, as applied in this publication.

The efficient resource use theory formulated by Wageningen production ecologist De Wit[171], examines the productivity development in agriculture, both in ecological and economic terms.

The theory on the three-dimensional landscape, described in detail by Dirx *et al.*[172] and Jacobs[173], lifts landscape ecology out of the domain of natural sciences into landscape theory which covers both the area of natural sciences and the areas of social sciences and experience sciences.

The third theory, on co-design, concerns the design process itself[174]. It explores the position of the designer between the theory and the policy and social reality in which designs have to be realised, the role of different stakeholders in the process, and the way in which collaboration between these parties can be accomplished.

---

[171] De Wit C.T. (1992). Resource use efficiency in agriculture. *Agricultural Systems* 40: 125-151.
[172] Dirkx G.H.P., M. Jacobs, J.M. De Jonge, J.F. Jonkhof, J.A. Klijn, A. Schotman, P.J.A.M. Smeets, J.T.C.M. Sprangers, M. Van den Top, H. Wolfert and E. Vermeer (2001). Kubieke landschappen kennen geen grenzen. In: *Jaarboek Alterra 2000.* Alterra, Wageningen, the Netherlands.
[173] Jacobs M. (2004). Metropolitan matterscape, powerscape and mindscape. In: Tress G., B. Tress, W.B. Harms, P.J.A.M. Smeets and A. Van der Valk (eds.) *Planning metropolitan landscapes. Concepts, demands, approaches.* Delta series, Wageningen University, Wageningen, The Netherlands: 26-39.
[174] De Jonge J. (2009). *Landscape architecture between politics and science. An integrative perspective on landscape planning and design in the network society.* Wageningen University and Research Centre, Wageningen, the Netherlands, 233 pp.

At the beginning of the research by design one or more working hypotheses are formulated from each theory. These hypotheses are then tested for each design on their tenability and on the question of whether their formulation needs to be adjusted or extended. This results at the end of the research in six partially reformulated and partially new hypotheses from the five hypotheses formulated at the beginning, which serve as the conclusion to this book.

The three theories (focusing on production, landscape and design process) are not independent of each other but merge into one another. An important evolution across all three is that from multidisciplinarity in the resource use efficiency theory to interdisciplinarity in the landscape theory to transdisciplinarity in the design process.

5.2 Theoretical production ecology and the De Wit curve[175]

Agricultural productivity, measured in dry matter yield per surface area, is constantly increasing thanks to the continuous progress in agricultural knowledge and economic stimuli. The use of fertilisers, water, pesticides, etc., per unit of output diminishes as yields per hectare rise. If growing conditions are improved by means of various measures such as soil improvement, irrigation, seedbed preparation, manure spreading, etc., then a crop responds better to the variable inputs. As such the right agricultural measures lead to a process in which the efficiency of the use of raw materials increases. A detailed analysis of the efficiency of raw material use was described by De Wit (1992)[176].

At the heart of the De Wit theory is a sigmoid curve (Figure 11) which represents the relationship between production costs and yield and is called the 'minimum cost curve'. Two points on this curve are interesting in an economic sense: the first is the intersection between the curve and the gross return line. On the left of this line the costs are higher than the returns and agricultural production makes no sense economically. The intersection marks the difference between marginal and profitable agriculture. The second point is the point at which the tangent to the curve, in relation to the x-axis is greater than 45°. This point represents the most optimal economic production, because beyond this point the marginal added returns fall off. A third point on the curve is interesting from an environmental standpoint: the point where a line from the intersection of the curve with the Y-axis (P_0) touches the minimum cost curve is the environmental optimum, where the highest production is achieved with the lowest external costs.

[175] This description of the De Wit theory has been taken from Rabbinge R., H.C. van Latesteijn and P.J.A.M. Smeets (1996). Planning consequences of long term land-use scenario's in the European union. In: Jongman R.H.G. (ed.) *Ecological and landscape consequences of land use change in Eeurope.* European Centre for Nature Conservation, Tilburg, the Netherlands.
[176] De Wit C.T. (1992). Resource use efficiency in agriculture. *Agricultural Systems* 40: 125-151.

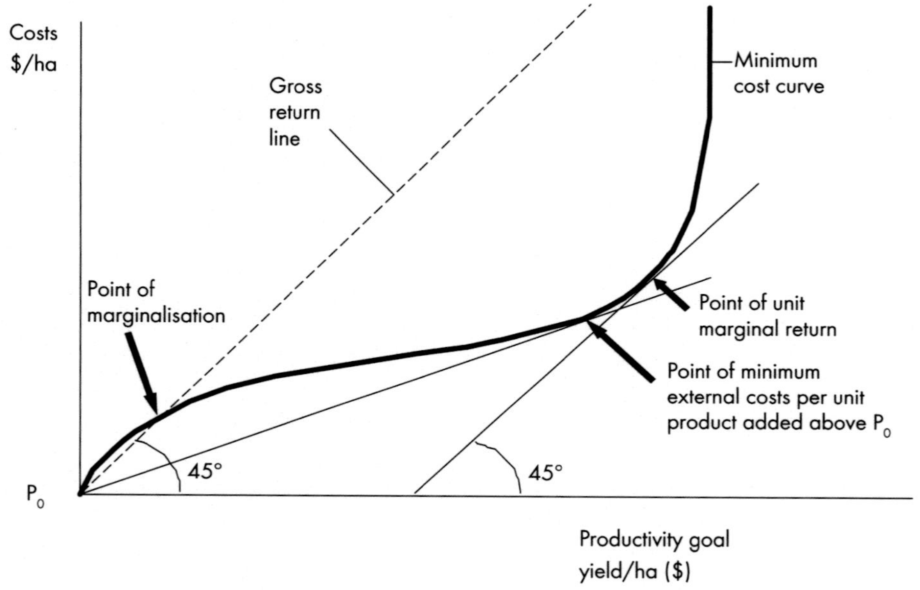

Figure 11. Link between yield and costs according to De Wit (1992).

For a more realistic representation, the fixed costs (for machines, buildings, interest, etc.) are added (Figure 12). The higher these costs, the further the point of marginal agriculture shifts to the right on the curve.

The De Wit curve leads to two generic conclusions: (1) Profitable agriculture is possible in the whole area between the gross return line and the curve. But most money is earned near the point of optimum economic production. The environment is least polluted near the point of optimum environmental production. (2) Both the environmental and economic optima are situated at a level of high inputs per ha and are therefore achieved with highly productive agriculture.

The De Wit theory was originally formulated for crop parcels and was validated for the important arable crops by De Koning *et al.*[177]. It has also recently been confirmed for a combined system of arable and ruminants by Glendining *et al.*[178].

[177] De Koning G.H.J., H. Van Keulen, R. Rabbinge and H. Janssen (1995). Determination of input and output coefficients of cropping systems in the european community. *Agricultural Systems* 48: 485-502.

[178] Glendining M.J., A.G. Dailey, A.G. Williams, F.K. Van Evert, K.W.T. Goulding and A.P. Whitmore (2009). Is it possible to increase the sustainability of arable and ruminant agriculture by reducing inputs? *Agricultural Systems* 99: 117-125.

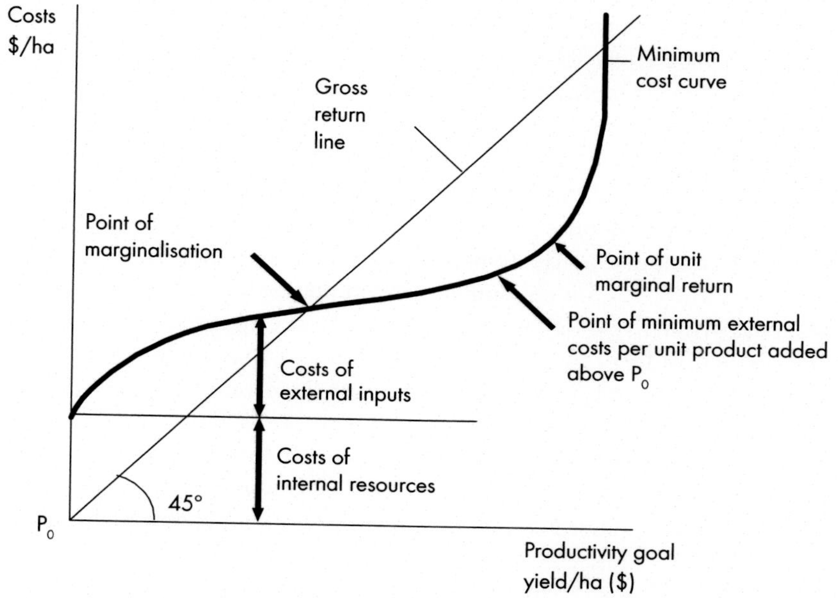

Figure 12. Link between yield and fixed and variable costs according to De Wit (1992).

The resource use efficiency theory represents an important relationship between the profit and planet aspects of sustainable development. It is not the quest to optimise single factors that is central, but rather the search for the minimum of each production factor whereby all others are used to their maximum. A plea for an integral approach, in other words:

> 'Whether external means of production are used at all depends of course on their price, but as soon as the farmer can afford them, they should be used in such a way that the production possibilities of all other available resources are fully exploited. It thus appears that with further optimizing of the growing conditions an increasing number of inputs gradually lose their variable character and the number of fixed operations on the farm increase. This makes more and more inputs not a variable cost element but a complementary cost element of the decision to farm a piece of land. Therefore strategic research that is to serve both agriculture and its environment should not be so much directed towards the search for marginal returns of variable resources, as towards the search for the minimum of each production resource that is needed to allow maximum utilization of all other resources.'[179]

[179] De Wit C.T. (1992). Resource use efficiency in agriculture. *Agricultural Systems* 40: 125-151.

Dutch glass horticulture can be regarded as the most far-reaching integral application of these principles in plant crops. It is the theory that forces natural sciences and agricultural sciences, including the economy, to work together in a multidisciplinary fashion.

De Wit formulated his theory for crops and thereby postulated that it was not only relevant for the mutual fine-tuning of the variable inputs, but that the system as a whole would work better if more use were made of better land. He also indicated that his theory could be applied to livestock farming, something that was later developed by Van de Ven *et al.*[180] for animal and combined plant-animal production systems. The biggest cost within livestock farming (both from an economic and environmental perspective) is that of feed, most of which is produced via arable farming. As with intensive arable farming and glass horticulture, both the economic and environmental outcomes of intensively operating livestock farms show that intensification and improvement of environmental performance complement each other. While the average size of the agricultural enterprise is steadily growing and the capital intensity of the systems is increasing, the damage to the environment is diminishing[181].

Van Ittersum and Rabbinge[182] further developed the resource use efficiency to the level of land use. Based on this further development, the WRR made it clear in the report Ground for choices[183] that by using the resource use efficiency theory to look for the best land use and to practise agriculture primarily in those places most suitable as regards physical conditions, productivity could again be dramatically increased. The title of the report referred to the viewpoint of the WRR, that important political decisions were needed about the use of land for farming in the 12 member states of the European Union at that time. In that report the resource use efficiency theory became the basis for political scenarios.

With that, according to the current insights, the theory domain covers not only crop parcels but crop and agro-systems and land use at various levels of scale, from local to continental.

---

[180] Van de Ven G.W.J., N. De Ridder, H. Van Keulen and M.K. Van Ittersum (2003). Concepts in production ecology for analysis and design of animal and plant-animal production systems. *Agricultural Systems* 76: 507-525..

[181] Brouwer F.M., C.J.A.M. de Bont, H. Leneman and H.A.B. van der Meulen (2004). *Duurzame landbouw in beeld*. Landbouweconomisch Instituut, The Hague, the Netherlands.

[182] Van Ittersum M.K. and R. Rabbinge (1997). Concepts in production ecology for analysis and quantification of agricultural input-output combinations. *Field Crops Research* 52: 197-208.

[183] Wetenschappelijke Raad voor het Regeringsbeleid (1992). *Ground for choices; four perspectives for the rural areas in the european community.* Wetenschappelijke raad voor het Regeringsbeleid, The Hague, the Netherlands.

See also Rabbinge R. and H.C. Van Latesteijn (1992). Long-term options for land use in the European community. *Agricultural Systems* 40: 195-210, and De Koning G.H.J., H. Van Keulen, R. Rabbinge and H. Janssen (1995). Determination of input and output coefficients of cropping systems in the european community. *Agricultural Systems* 48: 485-502.

So these insights of De Wit, Rabbinge and other production ecologists can be seen as the theoretical foundation of the mega-trend described in Section 4.1, which states that productivity increases in agriculture have been occurring for centuries and will continue, providing that the ensuing climate and environmental problems can be solved.

But increased agricultural productivity with simultaneously better environmental performance has so far developed in such a way that the various sectors have moved further and further apart, and parts of the production process have been subsumed in various separated and increasingly specialised companies. Arable, dairy and intensive livestock farming, open field vegetable cultivation, vegetable cultivation under glass, horticulture or mushroom production: these are, particularly in primary production, separate worlds, which are found only exceptionally in combinations in one company and are generally separated in space from one other. Only the combination of dairy farming and intensive livestock farming is seen with any frequency. Even within vegetable cultivation under glass, for example, seed production, the production of mother plants and the actual cultivation of the vegetable crop are undertaken in separate businesses. As in chicken meat production, within which the production of grandparental animals, parental animals, eggs, broods, one-day chickens and meat chickens in the primary line, the production of feed and feed components in the delivery and slaughter and further processing are all carried out in separate specialised companies.

The aim of agropark design is to remove this separation, where justified, via spatial clustering. Firstly to remove the need for all the transportation between these components, and to prevent the veterinary and phytosanitary risks which often go hand in hand with this transportation. Animal transportation is also a major source of animal suffering[184]. This suffering will be limited by reduced transportation. Secondly, to enable the processing of waste and by-products on a large industrial scale, and

---

[184] Leenstra F.R., E.K. Visser, M.A.W. Ruis, K.H. de Greef, A.P. Bos, I.D.E. van Dixhoorn and H. Hopster (2007). *Ongerief bij rundvee, varkens, pluimvee, nertsen en paarden. Inventarisatie en prioritering en mogelijke oplossingsrichtingen.* Report 71, Animal Sciences Group, Wageningen UR, Wageningen, the Netherlands. 'Animal welfare is not the exclusive domain of scientists and zoologists. Animal science is based on observations of the animal. However, however much we can scientifically establish about the animal, science cannot profess to speak exclusively for the animal. The social debate on animal welfare is also about ourselves, our values and convictions about what a good life is. In short, about – often controversial – values which (animal) scientists have no special authority on.
We therefore make a distinction in this report between 'animal welfare problems' and 'suffering'. All suffering is an animal welfare problem, but not all animal welfare problems mean suffering for the animal per se (...). 'suffering' concerns those forms of damage to the physical and psychological health of animals, whose nature and existence we can scientifically establish and substantiate' Strictly speaking animal suffering issues belong under *Matterscape* and the other animal welfare problems under *Powerscape*, insofar as they are represented in law, and in *Mindscape* as far as subjective interpretations are concerned.

to enable the re-use the products of this processing as raw materials in the same cluster. Thirdly, in processing the main products there will be much more emphasis on composite convenience products, whereby these different chains will ultimately come together.

Not all production will take place in agroparks. Agroparks also industrially process agricultural products that are produced elsewhere. Agroparks are connected to land-dependent production via collection centres. These can be in the direct surroundings of the metropolis with other independent industrial agricultural companies. But raw materials can also be supplied from the supply and transport regions that Taylor talks about (cited in Section 3.3).

The most important challenge in the coming years will be to reduce the amount of fossil fuel used, the 'universal natural labourer', according to Sloterdijk, cited in Section 2.1 of this work. In terms of production ecological theory, there are two arguments for this: the most important is the increasing scarcity of this raw material, which not only translates into direct fossil energy costs in primary production itself, but also means that all other raw materials and production means become more expensive, because part of the price is always determined by fossil fuel. But the environmental effects of the greenhouse gas CO_2 must also be reduced. However, since it is precisely in the industrial setting of vegetable production that some of this gas can be used as a raw material, this theoretical objective leads to two design requirements:

- Maximum reduction in the direct, and where possible, indirect use of fossil fuel.
- Where possible, the processing of CO_2 that is emitted from the use of fossil fuel in the park itself or as a result of neighbouring activities.

The integrated application of the principles of industrial ecology under discussion here also allows for the possibility of reducing all kinds of other environmental damage, because with a more integrated industrial and large-scale approach, environmental techniques can be used that are often unprofitable on a smaller scale: for example, air washers in stalls, manure processing via co-fermentation and water purification.

This is expected to result in cost reductions as well as a reduction in environmental pollution, while agroproduction itself continues to intensify, in terms of capital input, land use and labour. Expedition Agroparks reports on this cross-sector quest from production ecology to synergy from the following hypothesis in relation to production ecological theory, which states that the development of agroparks is a continuation of the basic idea behind resource use efficiency. The hypothesis which is being tested here, and which will be developed via research by design, states that:

An agropark realises lower costs, greater added value and lower environmental pollution per unit of output and space.

5.3 Landscape ecological theory and the three dimensions of landscape

From the theory of landscape ecology and landscape design it is logical to indicate Castells' space of places and Sloterdijk's 'non-reducible spherical expanses' with the synonym landscape:

> 'Land-scape associates people and space. The root "land" means both place and the people living there. In "scape" we can recognise both "to shape" and "-ship", meaning association, as in partnership or friendship. This linguistic background refers both to the "purposefully shaped" and to "the dynamic connection between place and those who dwell there".'[185]

The landscape is the starting point of the landscape dialogue. It provides the bond and the identity that enable people to work together. Working in a landscape (area, region) prevents people from thinking sectorally and in easy abstractions. A landscape enforces integration[186].

Truth, justice and trustfulness are criteria with which we can judge the world around us, including the landscape. Truth is about how the world works. Justice is concerned with the way in which people agree to act in this world. Trustfulness is about emotions and meanings that people can attach to the world.

The true landscape is the landscape as object, the landscape which we can measure and which we can put on a map and refer to in landscape types.

The just landscape is situated at the social level, the level on which there are agreements on valuation but also the level on which residents and users of the landscape organise themselves. They form a new type of organisation, which has an affinity with other scapes, like in the waterschappen (waterscapes) in the Netherlands, in which stakeholders jointly make arrangements on water management.

The trustful landscape is the landscape that we experience, the landscape of the anecdote that is becoming increasingly important in our current-day society, partly

---

[185] De Jonge J. (2009). *Landscape architecture between politics and science. An integrative perspective on landscape planning and design in the network society.* Wageningen University and Research Centre, Wageningen, the Netherlands: 42.

[186] A similar plea for an integrated approach to space is given by Asbeek Brusse W., H. van Dalen and B. Wissink (2002). *Stad en land in een nieuwe geografie. Maatschappelijke veranderingen en ruimtelijke dynamiek.* Wetenschappelijke Raad voor het Regeringsbeleid, The Hague, the Netherlands: 137. 'From these two thought steps it follows that an analytical distinction should be made between three mutually connected dimensions of space, or three spaces: social space, physical space and symbolic space.'

because this is the area where the communicative self-steering that Cornelis talks about plays out.

By looking at the landscape in its three-dimensional totality, quality can emerge without the need to restrict innovation.

5.3.1 Matterscape, powerscape and mindscape

In a further development of this theory of the three-dimensional landscape, Jacobs postulates[187] that:

> 'We could call the theory a landscape ontology. Ontology is the study of the nature of being. The three manifestations are three different ways in which the landscape exists, namely as material in the physical reality, as norms in the social reality and as thoughts and feelings in the inner reality. (...) More commonly used terms are the landscape as object, the landscape as organisation and the landscape as story. Although these terms are less exact philosophically than the true, just and trustful landscape, they cover the essence perfectly. (...)

	The true landscape	**The just landscape**	**The trustful landscape**
Domain	physical reality	social reality	inner
Position	objectivising	normative	expressive
Manifestation	object	organisation	story
To be described as	facts	standards	values
Science	natural sciences	social sciences	'experience sciences'

> This table (...) could lead to the misconception that social scientists and "experience scientists" are not concerned with truth, but with justice and trustfulness respectively. This is not the case: like all scientists they too are trying to find the truth. But not the truth about landscapes.
> If a sociologist examines the way in which a group of people interact with the landscape, this would provide data about the group of people, not the landscape. A sociologist is attempting to discover the truth about the group of people. Within this group of people the just landscape appears as a system of norms. Therefore, the scientist who registers the norms of an organisation with regard to the landscape, makes judgements about the just landscape as applicable within this group of people.

[187] Jacobs M. (2006). *The production of mindscapes. A comprehensive theory of landscape experience.* Wageningen University and Research Centre, Wageningen, the Netherlands: 9-12.

> Like a sociologist, an "experience scientist" (for example, an environmental psychologist) doesn't measure the landscape, but the experiences of people. The environmental psychologist studies the truth about the inner, the level of consciousness, of people. The trustful landscape emerges in people's experiences. Therefore, the environmental psychologist studying the experiences of a person in relation to the landscape, makes a judgement about the trustful landscape as applicable to that person. In short, social scientists and "experience scientists" don't measure physical landscapes, but groups of people or individuals, and thus are not concerned with the true landscape.
> Every landscape or element of landscape can be described from the perspective of the three manifestations, whereby none of these descriptions can be converted to another.'

Landscape ontology forces scientists from the domains of natural sciences, social sciences and humanities to set up far-reaching collaborations, and new concepts such as matterscape, powerscape and mindscape, which are only significant when interconnected, make this collaboration interdisciplinary.

Dirkx *et al.* also define landscape in these three dimensions[188]. The true landscape (matterscape) is the physical reality, the domain of natural sciences, the hardware. The just landscape (powerscape) is the social reality, the domain of the social sciences and economy, the orgware. The trustful landscape (mindscape) is the inner reality of each of the people involved and the communication thereof, the software. The boundaries of the landscape in which work is done are necessarily blurred. Sometimes a physical water boundary or a legal boundary helps, but this can easily be crossed by an engaged actor on the other side who wants to get involved. There is more clarity and unambiguity in the centres of the landscape, which don't have to be at the same place in the three dimensions, than at the edges.

This landscape in its three dimensions is constantly moving and changing. Therefore, the landscape dialogue is like working on a moving train, surfing a wave. It is important to make a distinction between the inner world of influence, which is more or less under control, and the outer world of engagement, with all events that are relevant

[188] Dirkx G.H.P., M. Jacobs, J.M. De Jonge, J.F. Jonkhof, J.A. Klijn, A. Schotman, P.J.A.M. Smeets, J.T.C.M. Sprangers, M. Van den Top, H. Wolfert and E. Vermeer (2001). Kubieke landschappen kennen geen grenzen. In: *Jaarboek Alterra 2000*, Alterra, Wageningen, the Netherlands.
See also Jacobs M. (2002). *Landschap3*. Expertisecentrum Landschapsbeleving, Alterra, Wageningen, the Netherlands, 68 pp., and Jacobs M. (2004). Metropolitan matterscape, powerscape and mindscape. In: Tress G., B. Tress, W.B. Harms, P.J.A.M. Smeets and A. Van der Valk (eds.) *Planning metropolitan landscapes. Concepts, demands, approaches.* Delta series, Wageningen University, Wageningen, the Netherlands, pp. 26-39.

but over which we have no control[189]. The landscape dialogue is actively involved with the world of influence and simultaneously develops a common insight into the world of engagement.

5.3.2 Networks and landscapes

Networks are the carriers of the landscape. Both in the physical reality (food webs, water systems, roads) and on the level of people and their organisations, and as such also in the way in which knowledge and stories are organised. A visualisation of the networks soon reveals the way in which the landscape works.

Networks prevent too strong a focus on a specific spatial scale because they relativise spatial boundaries. In a network a small-scale connection between individuals can occur but each individual member or the group as a whole can also be a hub in a worldwide network.

I will use the tripartition into the true, just and trustful landscape, which has developed in landscape theory to facilitate an interpretation of the integral character, to evaluate the agropark designs in relation to their real-life environment. As stated at the start of this chapter, popular synonyms for the true landscape are hardware or matterscape, for the just landscape, orgware or powerscape and for the trustful landscape, software or mindscape.

The way in which agroparks are incorporated in the global space of flows was outlined at the beginning of this chapter. But while they may be the dominant one, space of flows are not the only spatial logic in the network society. Spaces of flows are materially linked to spaces of places but according to their own logic and often in complete dissonance with the place where they actually flourish. Since our agroparks are designs that start from this knowledge and the social judgements thereof, it is not enough to regard them as completely footloose. After all, they are always designed in a place that is also a space of place and, according to the basic principles of landscape dialogue, the standard dissonance that often characterises space of flows should be inadmissible. Just on account of the many relationships that agroproduction, however footloose,

[189] See also Covey S.R. (2000). *De zeven eigenschappen van effectief leiderschap.* Uitgeverij Contact, Amsterdam, the Netherlands.

has with its surroundings, an agropark should also be defined as a space of places, as an 'inhabited expanse'[190], as a landscape.

From this point of view, every design for an agropark is a landscape design. Just as heathland, unfertilised meadowland, irrigated meadows and ashes are regarded as the building blocks of the former agricultural landscape, so agroparks are modern agricultural man-made landscapes composed of greenhouses, intensive livestock farms, storage, logistic elements, etc. They are unmistakeably metropolitan landscapes.

The link between the agropark as part of the global agro-network and its concrete site must be fleshed-out via various different aspects. First and foremost, there has to be some building and in this respect the physical quality of the site matters: What are the soil properties? How does the hydrology of the area function? At the same time one characteristic of many spaces of flows is that their economic weight is often so big that major adaptations have to be made to the physical surroundings. For example, airports for which islands have been constructed, harbours in the sea, etc.

Secondly, the site has to offer logistical advantages. There has to be a good infrastructure in the vicinity, with different provisions, with a strong preference for water and rail transport because of the many heavy loads.

Thirdly, the agropark must be a place that has 'expanded within itself', where people can work, live and enjoy themselves, and which is a neighbour to many inhabitants in its metropolitan environment. In some of the foreign examples in particular, the recreational and educational value of all that an agropark has to offer has been worked out and they could become home to many thousands of people. In the Dutch examples agroparks are seen as industrial sites, that need licences, potentially pollute their surroundings and are opposed by local residents. In the Netherlands there are arguments about (supposed) environmental pollution but also arguments related to animal welfare. These aspects belong to the domains of powerscape and mindscape.

---

[190] Sloterdijk P. (2006). *Het kristalpaleis. Een filosofie van de globalisering.* Uitgeverij Boom/SUN, Amsterdam, the Netherlands. As examples of inhabited expanses, Sloterdijk refers to 'biological reproduction (...), the upbringing and education of children, the handing down of culture and the acceptance of this offer by the recipient generations' as 'the most convincing paradigms for the non-compressible, that unfolds in obstinate asymmetric processes. Learning to live means learning to be in places; places are by definition not reducible spherical expanses, that are surrounded by a circle of omitted things that remain at arm's length' (See Section 1.2). Where agroproduction is concerned, it is logical to look for Sloterdijk's 'inhabited expanses' primarily in diverse forms of pluri-activity agriculture, organic farming in all its variants as well as allotments and other forms of small-scale farming, because these forms of agriculture can often easily be slotted into a residential area. But that would be too easy. The challenge in the metropolitan environment is to arrange and design business parks, industrial terrains and thus agroparks too so that they cause a minimum of inconvenience. Better still: they should be able to function as a source of inspiration, as recreational objects.

This necessitates design criteria, that far exceed the demands made by production ecology, and that were described above. The theory of production ecology is situated mainly in the domain of the true landscape. However, it primarily delivers pronouncements on natural science processes. Where the consequences of this are translated into economic terms, into reduced costs, it also relates to the world of the just landscape: powerscape.

But the economic aspect, too, extends beyond business economics, which is a matter for discussion in resource use efficiency theory. Agroparks also make a significant contribution to shifting the balance between the haves and have-nots, the inhabitants of the 'Crystal Palace' and the excluded, and thereby help to combat conflicts resulting from this second mega-trend described in Section 2.5. Agroparks make a contribution here in three ways:

- By producing enough good food, the contrast between the haves and have-nots is reduced beyond the level of hunger or no hunger.
- By reducing the use of fossil fuel in food production, this aspect also becomes a less dominant driving force behind this trend and food becomes cheaper.
- By creating employment in the metropolitan areas, the 'Crystal Palace' can accommodate many more of those on the outside looking for access.

The position in the field of tension of spaces of flows and spaces of places thereby determines the quality of the agropark design. We can translate this into the following working hypothesis:

> **An agropark can only come into being on the basis of an integral design of matterscape, powerscape and mindscape both on the global scale of Intelligent Agrologistic Networks and on the local scale of a landscape.**

5.4 The design process during design and implementation

While the two previous theories addressed the agropark itself, which can be evaluated as agroproduction facility and landscape, during design, implementation and also during the operational existence, the third theory addresses the design process. Some cases involve systematic monitoring and evaluation of the process, therefore we can begin with theory development in relation to this work process on the basis of the acquired experiences, via the method of research by design.

5.4.1 Designs for agroparks as system innovation

A more generic development of a social innovation process is evident across the seven examples. In order to describe this properly, I use the distinction made by Groen *et al.*[191] between inventions, innovations and transitions.

- Inventions are new technical applications, new knowledge, new hardware or new procedures, usually developed at knowledge institutions like universities, research institutes or company research laboratories.
- An innovation is the generation of value based on one or more inventions. Innovations as value creation are not limited to the economic system, in which new market value is created on the back of an invention. An innovation can also take place in the political system, whereby the new value created in the currency (power, power relations) of that system must be discovered. There are three types of innovations:
  - Market-driven or rapid innovations use the improvement capacity within a business, without using external knowledge. ('Doing things better').
  - Innovation leaps use knowledge from another area or combine knowledge from different areas in completely new applications ('Doing better things').
  - Knowledge-driven innovations are situated somewhere in-between, whereby inventions are put on the market with immediate success. In order to increase the number of knowledge-driven innovations, knowledge institutions are increasingly formulating part of their research agenda on the basis of practical demands, coming from market-driven entrepreneurs and thereby shifting the emphasis from curiosity-driven to market-driven research.
- A system innovation consists of one or more innovations whereby the interest groups in question form fundamentally different relationships. System innovations focus on innovations of a system itself. System innovations are implemented innovations at the level of the hardware, (technological innovations), orgware (organisational, institutional-administrative innovations), but also the software (visions, perceptions and creations). System innovations transcend the level of a discipline, sector and organisation.
- A synergetic combination of innovations and system innovations can lead to a transition that is defined as a structural social and technological change over a long period[192]. Transitions are the result of changes on various levels of scale: the macro, meso and micro levels. Shifts on the macro level mean changes in policy, world views, culture and paradigms. Changes on the meso level mean changes in the existing regimes: systems of dominant practices, rules and interests that are shared by groups of actors. At the micro level niches can evolve in which

[191] Groen T., J.W. Vasbinder and E. Van de Linde (2006). *Innoveren. Begrippen, praktijk, perspectieven.* Spectrum, Utrecht, Sloterdijk P. (2006). *Het kristalpaleis. Een filosofie van de globalisering.* Uitgeverij Boom/SUN, Amsterdam, the Netherlands.

[192] Schot J. (2005). Transities: veranderen met het verleden en de toekomst. *De Eerste Verdieping* 1 (1).

derivations of the existing regime can emerge, such as new technologies or new forms of administration. A transition requires the changes on the different levels to connect up to each other in a positive way. Changing agriculture so that it fits in with the perspective of sustainable development is a transition.[193]

5.4.2 Designs for agroparks as contribution to sustainable development

I regard the work on agroparks, that is central to this publication, as work on inventions, innovations and system innovations which fit into a larger context of work on the transition to sustainable development, in which organisations like TransForum are involved.

Sustainable development has been described by many authors as the attempt to find a better balance between the value domains of People, Planet and Prosperity (triple P)[194]. This requires the ability to adapt and be flexible where changes in markets, consumer demand and consumer and citizen experience are concerned. With regard to agriculture these changes relate to food and other agroproducts, but also to other social economic domains, as is plain to see at this moment with the possible introduction of biofuels and all the possibilities and problems that this entails. Agriculture and regional development of the outer area are part of a big complex adaptive system and it is therefore vital to see sustainable development as a dynamic system characteristic, for which there are many potential implementations from many different perspectives[195]. The collection of examples provides a picture of agroparks that can be defined in these terms. That leads to the following hypothesis:

An agropark is a knowledge-driven system innovation and makes a significant contribution to sustainable development.

---

[193] Rotmans J. (2003). *Transitiemanagement: Sleutel voor een duurzame samenleving.* Koninklijke van Gorcum, Assen, the Netherlands, 243 pp.

[194] Veldkamp A., A.C. Van Altvorst, R. Eweg, E. Jacobsen, A. Van Kleef, H. Van Latesteijn, S. Mager, H. Mommaas, P.J.A.M. Smeets, L. Spaans and H. Van Trijp (2008). Triggering transitions towards sustainable development of Dutch agriculture: Transforum's approach. *Agronomy for Sustainable Development* 29: 87-96.

[195] Van Ittersum M.K. and R. Rabbinge (1997). Concepts in production ecology for analysis and quantification of agricultural input-output combinations. *Field Crops Research* 52: 197-208; Wetenschappelijke Raad voor het Regeringsbeleid (1994). *Duurzame risico's: een blijvend gegeven.* Report 44, Wetenschappelijke Raad voor het Regeringsbeleid, The Hague, the Netherlands.

5.4.3 Landscape dialogue, design dialogue, co-design

In recent years regional or landscape dialogue[196] appears to have successfully generated coherent developments, actions and local projects.

For the landscape dialogue it is essential to organise the KENGi parties in the social network right from the beginning. Knowledge workers, enterprises, NGOs, governments, (and citizens) and can only come up with innovations if they act together. The process has to start with people from all these groups, otherwise they will seriously slow down the process with vetoing power, unwillingness or a 'not invented here' attitude[197]. The initial phase of a landscape dialogue consists of informal meetings in the social networks in order to visualise and define these networks in the world of influence and engagement and of the networks themselves.

The KENGi partners have different currencies that define success in their own world[198]: Knowledge institutions, subject to the constraints of science, are evaluated on peer-reviewed publications in scientific journals. Enterprises are subject to the constraints of the market and will have to make a profit in order to be able to continue their activities in the longer term. In NGOs there is political influence and membership figures. Finally, in government there are the constraints of policy, the need for majority votes, against the background of power and ideology. Therefore, system innovations generate their value proposition not only in the market, where entrepreneurs function, but also in the real-life world of science, of government policy and of the social field of influence, in which environmental groups, animal protection societies or agricultural lobbies operate, and often in a combination of all of these. A system innovation is a shared value creation.

---

[196] Smeets P.J.A.M. and M.J.M. Van Mansfeld (2002). The landscape dialogue: interactive planning as a way to sustainable land use in metropolitan areas. Cases from Northwestern Europe. In: *The International Engineering Consultancy Forum on Sustainable development of Shanghai.* Shanghai Investment Consulting Corporation, Shanghai, P.R. China; Van Mansfeld M., M. Pleijte, J. De Jonge and H. Smit, 2003a. De regiodialoog als methode voor vernieuwende gebiedsontwikkeling. De casus Noord-Limburg. *Bestuurskunde* 12: 262-273; Van Mansfeld M., A. Wintjes, J. De Jonge, M. Pleijte and P.J.A.M. Smeets (2003). *Regiodialoog: Naar een systeeminnovatie in de praktijk.* Report 808, Alterra, Wageningen, the Netherlands.

[197] The abbreviation KENGi was first used in Innovatienetwerk Groene Ruimte en Agrocluster (2000). *Initiëren van systeeminnovaties.* Report 00.3.002, Innovatienetwerk Groene Ruimte en Agrocluster, The Hague, the Netherlands.

[198] Groen T., J.W. Vasbinder and E. Van de Linde (2006). *Innoveren. Begrippen, praktijk, perspectieven.* Spectrum, Utrecht, the Netherlands.

5.4.4 Process design of the landscape dialogue

The content of a landscape dialogue can vary. In many examples from recent years the problem of city landscapes in a metropolitan environment has dominated. But the specific combination of problems is different in every area. The constant factor in the landscape dialogue is not the content but the process.

The first explicit step after the informal initial phase, in which the social network of KENGi parties takes shape, involves accurately defining the problem. In this phase the different participants start from their own individual problem in the area and work towards a joint problem definition. To get to that point, a resolute relativisation of their own problem is important, not on a permanent basis, but to give other parties room to articulate their problem in the joint problem defining process. Finally, the urgency of the common problem is a key driving force in determining the success of the landscape dialogue as a whole.

The second step in the landscape dialogue consists of joint fact finding. Communality is again the key element in this process. It is a matter not only of demonstrating what a certain participant knows but primarily of finding out what the others know and discovering what the participants collectively don't know.

The focus of the landscape dialogue lies in the third phase, the design phase. It is here that the collaborative network formulates future projects in a creative process. These can be projects that individual actors may already have been occupied with before the process. The criterion is not that the projects are new but that the participants demonstrate energy and engagement. They are virtually always enhanced in the network. Completely new projects also frequently emerge in new collaborations.

Landscape visions have no major focus in the landscape dialogue. Visions often appear too noncommittal or simply lead to endless intellectual nitpicking. Projects require concrete steps and because they have been implemented by the united KENGi parties in an area, an integral feasibility and workability test is often carried out in passing.

After the design phase, in which creativity predominates, the landscape dialogue moves into the phases of support and implementation. Testing and fine-tuning take place at this stage and the projects are actually put into operation via business plans and financing. In these phases it is essential that other people from the participating parties now take part. Creative designers are often bad at estimating support for their design and those who can bring support to a project are not the ones who make decisions on funding and implementation. In the support phase projects fall by the wayside or their implementation is postponed. Sometimes they have to be radically redesigned in order to get to the implementation stage. Distinguishing between and not separating the three phases of design, support and decision-making is essential in the landscape dialogue[199].

5.4.5 Quality of the design process

There are also various criteria by which the quality of the landscape dialogue can be measured during the journey. A few intrinsic criteria are generic, for example different characteristics of sustainable development. If a design leads to a reduction in fossil fuel consumption or to more efficient use of scarce minerals like phosphate, this is almost always regarded as a success in terms of quality. The status can be constantly examined on the question of whether the aspects of planet, people and profit have been tackled sufficiently ambitiously and are in harmony with one another. Other criteria can be taken from the spatial development policy. In this respect, however, governments that operate on a higher level of scale (European Union, national governments) provide parameters within which a region has to develop. These parameters are partly generic, for example the competition policy of the European Union, and partly region-specific, for example a specification in water supply for a region in the river area. But because the landscape dialogue is a process and not a predetermined intrinsic development direction, most criteria for success are process characteristics.

The involvement of the different actors is the most important measure of success. This must be accomplished before the landscape dialogue starts – hence the often long informal preparation period – and should be regularly tested during the journey and explicitly declared. The need to have and keep on board all KENGi parties was referred to earlier. The involvement is not only reflected in terms of time and money. Experiences to date frequently show examples of organisations that only participated to keep an eye on their own interests. A good measure of involvement is the extent of dissemination of the thought process developed by means of the landscape dialogue throughout the organisations where the participants come from. Communication between participants and the people in their own organisations therefore requires a lot of attention throughout the entire process.

The landscape dialogue is a transdisciplinary process[200], in which different actors share individual and scientific knowledge with each other and reproduce it. This knowledge is not mono- but multi- and sometimes inter-disciplinary. Transdisciplinarity is a strict prerequisite. If it is absent, the dialogue descends into a monologue of scientists. It is stimulated by operating not only with scientific facts and figures, but by laying the

---

[199] See also Rotmans J. (2003). *Transitiemanagement: Sleutel voor een duurzame samenleving.* Koninklijke van Gorcum, Assen, the Netherlands: 75.

[200] Tress G., B. Tress and M. Bloemmen (eds.) (2003). *From tacit to explicit knowledge in integrative and participatory research.* Delta series 3. Alterra, Wageningen, the Netherlands, 147 pp.

emphasis on subjective forms of knowledge transfer like stories, images, in short by seeking a balance between trustfulness, justice and truth in the process.

After a time, a Community of Practice[201] emerges from the landscape dialogue, a network of participants who have built up a social learning environment with each other and learn to learn better with each other. Such a learning community is then in a position to react autonomously and dynamically to changes that occur in the world of engagement. In process terms and in terms of spatial development policy an important goal has been reached with this.

In her dissertation Jannemarie de Jonge[202] has given the work processes – like the regional dialogue, referred to by her as co-design and design dialogue – a scientific basis, by describing social discussions (simultaneously major planning assignments in recent decades) like the reconstruction of the sandy areas and the development of the river areas, and by analysing work processes as co-design.

One of the important conclusions of this is that no prescription for a successful working model can be derived from this scientific description. That is consistent with Rotmans' observation[203] on the limited possibilities of transition management.

This does not mean that nothing useful can be said about a certain method. This has already happened, in theory, above and will be confirmed in Chapter 6. But there are so many uncertainties, so many factors during the work process, that don't come from the world of influence but still have such a huge effect that every design is largely a matter of improvisation. The concept of *Kairos* is an essential part of this. This is the ability to act strategically and at the right time in a complex dialogue with many partners. Spotting and then acting on a strategic opportunity in a complex design process is a typical example of knowledge that can only be acquired through experience:

> 'An expert (...) has skills or tacit knowledge that cannot easily be verbalised. It is a kind of intuitive understanding that comes primarily from practical

---

[201] See Gordijn F. (2004). *Communities of Practice als Managementinstrument: Over de meerwaarde en het faciliteren van CoP's.* MSc Thesis Communication & Innovation, Wageningen University, Wageningen, the Netherlands.
Kersten P. and R. Kranendonk (2002). *CoP op Alterra; 'Use the world around as a learning resource and be a learning resource for the world'.* Report 546. Alterra, Wageningen, the Netherlands.
Wenger E. and W. Snyder (2000). Communities of practice: the organizational frontier. *Harvard Business Review* January-February 2000: 139-145.

[202] De Jonge J. (2009). *Landscape architecture between politics and science. An integrative perspective on landscape planning and design in the network society.* Wageningen University and Research Centre, Wageningen, the Netherlands, 233 pp.

[203] Rotmans J. (2003). *Transitiemanagement: Sleutel voor een duurzame samenleving.* Koninklijke van Gorcum, Assen, the Netherlands, 243 pp.

experience; it is embodied knowledge (...) applied in the context of the particular.'[204]

There is no preset recipe[205] but essential characteristics or, if necessary, parameters can be derived from it for the work process:

- ambition that goes beyond compromise;
- the involvement of experts, including 'professional amateurs', with a range of expertise, skills and practical wisdom, performing a 'multilingual' conversation using imaginative, graphical language, verbal, narrative language and the language of facts and figures, all with their own rationality;
- the creation of new insights through a design approach: the iterative process of creative imagination and reflective judgment', making design moves that integrate a wide range of expertise and interests and represent various levels of scale and detail;
- dialogue takes place in a 'free space', implying a state of mind and atmosphere without obligations in terms of interests, cognitive frameworks and time, as a condition for learning and creativity;
- participants have an open mind, allowing them to seize opportunities outside the 'dialogue space' as key players who can connect conceptual ideas to implementation power.

De Jonge's dissertation was finished in 2009. It is obviously not possible to use the insights from it as parameters ahead of the design process of agropark projects as a leitmotif. But they were present as tacit knowledge in the whole work process and because they are now synthesised as explicit knowledge on the basis of work processes such as regional dialogues, these parameters can later be used for a meaningful evaluation of those processes.

5.4.6 Transdisciplinarity in the design process

In the work on system innovations, the distinction between all the participants from the various stakeholders blurs when it comes to supplying knowledge. This enters the process as scientific knowledge from scientific institutions but also frequently as individual knowledge (tacit knowledge) from all the other participants.

This mingling of interdisciplinary scientific knowledge with the tacit knowledge of experts from other domains is the essence of transdisciplinarity. According to

---

[204] De Jonge J. (2009). *Landscape architecture between politics and science. An integrative perspective on landscape planning and design in the network society.* Wageningen University and Research Centre, Wageningen, the Netherlands: 149.

[205] *Ibid.*: 164. One of the statements in De Jonge's thesis is that: 'A prescriptive model for reflective practice is a contradiction in terms'.

Nonaka and Takeuchi[206], two Japanese researchers who investigated the innovation strategies of Japanese businesses in the 1980s, the continuous cycle of individual and explicit knowledge (Figure 13) is even the essence of knowledge development as such. This cycle has in their view been denied for too long in the Western world, with the consequence that the importance of individual knowledge has been underestimated and the importance of scientific knowledge overestimated.

The design and implementation of such a system innovation therefore goes way beyond the traditional scientific process that was appropriate in the traditional innovation model. That model was founded purely on the perspective of knowledge-driven innovation, whereby time and again the role of knowledge institutions was to put new knowledge on the market as inventions via research stations and information, and wait for entrepreneurs to turn these inventions into innovations as value creations.

However, the knowledge-driven system innovation is an activity in which all KENGi parties have to participate. Then the role of the knowledge institute changes, it is not only simply the supplier of scientific knowledge, of which the value is increased by innovations (for example, in the case of new energy technology in glass horticulture or on relationships between animal welfare and productivity). Other parties will bring their knowledge to that process and the knowledge institutes can assess the quality of that knowledge. All participants can make a contribution to the process in the process design, which, as described above, is not necessarily fixed but has to evolve

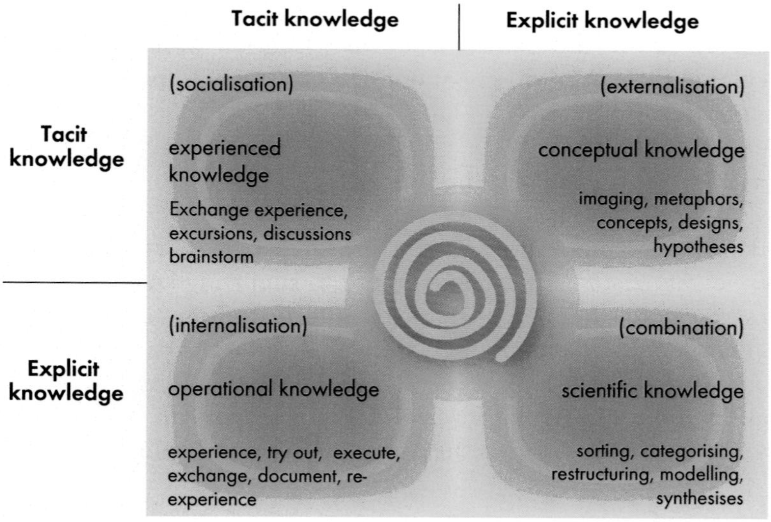

Figure 13. The knowledge cycle according to Nonaka and Takeuchi (1995).

[206] Nonaka I. and H. Takeuchi (1995). *The knowledge-creating company: How Japanese companies create the dynamics of innovation.* Oxford University Press, New York, NY, USA.

along the way. In this way the innovation process is rationalised and becomes a system innovation, whereby the role of knowledge institutions, expand from knowledge suppliers to innovative entrepreneurs, particularly in the design and implementation process. In so doing the boundary between inter- and transdisciplinarity is also crossed because the explicit interdisciplinary knowledge laid down in scientific publications continuously interacts with the tacit knowledge, the individual knowledge of the other KENGi partners[207].

Dealing with the different currencies of the KENGi parties requires much attention to the process. Consequently, many authors[208] suggest placing a fourth P for Process at the centre of the triple P triangle. Without careful attention to the process, a system innovation will not materialise. Organisations like TransForum and KnowHouse define themselves as facilitators who focus in system innovations on the bringing together and mutual conversion of the currencies of the various KENGi partners[209] (Figure 14). But as will become clear in the later discussion of the projects, and as is also the case in the practical projects like *Regiodialoog Noord Limburg*[210], the knowledge institution Wageningen-UR also repeatedly plays this role.

This can all be verbalised in the following hypothesis, which relates to the long line of development of the work process across the seven examples, that has materialised in the last few years:

---

207 Tress B., G. Tress and G. Fry (2003). Potential and limitations of interdisciplinary and transdisciplinary landscape studies. In: Tress B., G. Tress and G. Fry (eds.) *Interdisciplinary and transdisciplinary landscape studies: potentials and limitations.* Alterra, Wageningen, the Netherlands, pp. 182-192.

208 Rotmans J. (2003). *Transitiemanagement: Sleutel voor een duurzame samenleving.* Koninklijke van Gorcum, Assen, the Netherlands, 243 pp; Groot A.M.E. and P.J.A.M. Smeets (2006). Transitie en transitiemanagement. In: O. Oenema, J.W.H. van der Kolk and A.M.E. Groot (eds.) *Landbouw en milieu in transitie.* Wettelijke Onderzoekstaken Natuur & Milieu, Wageningen, the Netherlands. De Jonge J. (2009). *Landscape architecture between politics and science. An integrative perspective on landscape planning and design in the network society*, Wageningen University and Research Centre, Wageningen, the Netherlands, 233 pp., allocates the role of broker to the landscape architect as director of the co-design process.

209 Smeets P.J.A.M. (2009). Transforum en de innovatie van de kennis-infrastructuur in de Nederlandse landbouw. In: Smulders H., M. Gijzen and F. Boekema (eds.) *Agribusiness clusters: Bouwstenen van de regionale biobased economy.* Shaker Publishing, Maastricht, the Netherlands, pp. 83-94.

210 Groot A.M.E. and P.J.A.M. Smeets (2006). Transitie en transitiemanagement. In: O. Oenema, J.W.H. van der Kolk and A.M.E. Groot (eds.) *Landbouw en milieu in transitie.* Wettelijke Onderzoekstaken Natuur & Milieu, Wageningen the Netherlands.

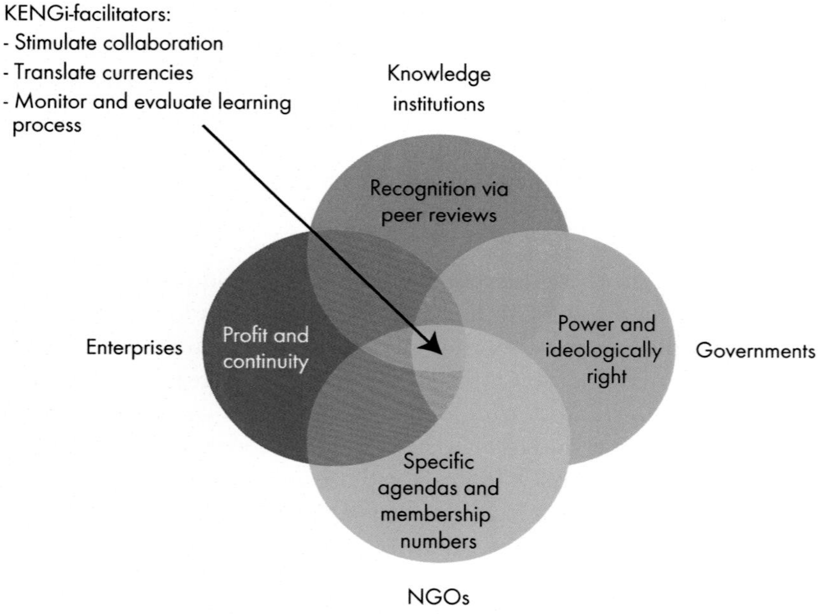

Figure 14. The role of the KENGi facilitator in system innovations.

The design and implementation of system innovations like agroparks presupposes the participation of knowledge institutions, enterprises, NGOs and government (KENGi). It is a transdisciplinary process in which the explicit knowledge of knowledge institutions and the tacit knowledge of the other partners are constantly evolving. KENGi brokers are the facilitators of this transdisciplinary process.

5.4.7 The social context of the design process

When looking at the work process in the seven projects individually, it must first be noted that by definition this occurs in a metropolitan environment, to use Sloterdijk's terminology (see Section 1.4) in a 'post-historic dense situation'. In this, 'long before the first spade enters the ground, everything that wants to move forward, that challenges the boundaries, that wants to build, is reflected in protests, objections, counter-proposals, swan songs; measures are overtaken by counter-measures.'[211]

As soon as it goes beyond simple thought processes on a piece of paper, every step of the design and implementation process can expect to encounter a lot of resistance.

[211] Sloterdijk P. (2006). *Het kristalpaleis. Een filosofie van de globalisering.* Uitgeverij Boom/SUN, Amsterdam, the Netherlands: 207.

In the social polemic that has arisen in the Netherlands around the development of agroparks, production ecology arguments are drowned out by arguments from the world of justice (powerscape) and trustfulness (mindscape). This is further reinforced because the arguments from production ecology theory that support the development of agroparks (in short: cheaper and better for the environment and more efficient in use of space), are precisely the arguments in the space of flows and where, at least in the Netherlands, there is no powerful government that can stimulate, or wants to enforce, this kind of development.

The transdisciplinary nature of the design process means that it doesn't take place in a remote workshop, far from actors in the community. The involvement in the process of actors who are not scientists means that communication also has to be taken into account. That happens in the spaces of flows, whereby the Greenport logo has been taken from national spatial planning and developed into a global brand for the Intelligent Agrologistics Network in which agroparks are key.

But particularly in its powerscape and mindscape, every agropark will have to respond to the demands made from the specific spaces of place by the local population in the metropolises.

As an example, at the level of nation states, the Dutch government makes very explicit and meticulous demands in relation to animal welfare. In India because of the dominant Hindu culture there is a taboo on the slaughter of cows and as a consequence of the Muslim culture pig-rearing hardly occurs. But in the dealings with local citizen groups too, as will be seen in the examples like Amsterdam Westpoort and *Nieuw Gemengd Bedrijf* [New Mixed Farm] in Horst, communication is crucial even in the design process. Because the citizens in the metropolitan 'Crystal Palace', detached as they are from traditional concepts like nation state or religion (see mega-trend 3 in Section 4.1) 'are constructing their own meaning'. Some do so from a position of well thought-out and substantiated communicative self-steering. Agroparks, incorporated in the metropolitan inhabited expanses are in direct contact with these citizens in spaces of place. Part of the design must therefore be the dialogue with this local environment. Because the local environment should deliver the 'practicable test', arising from the collective intelligence of individuals with the wisdom of communicative self-steering, which can be found in the metropolitan density of the 'Crystal Palace'. The reality, as will be seen, is less poetic.

It is interesting to compare how this discussion proceeds in other metropolises, and I will do this using the several examples of agropark development in this publication. I will conduct this comparison using a provocative hypothesis based on the balance between matterscape, powerscape and mindscape in the Netherlands, in which, for the sake of argument, I will assume that the capacity for self-steering among the citizens in question is fully developed and is expressed as follows:

> In all decision-making on the implementation of the integral agroparks design, with matterscape, powerscape and mindscape aspects, arguments from the world of justice and trustfulness take precedence over arguments from the world of truth.

5.4.8 Test framework for research by design to agropark projects

Thus there emerges a test framework which I will use in the methodical reflection on the seven examples of agropark projects. It shows that the driving forces behind the development of agroparks and the basic characteristics thereof are the same, because they arise from the production-ecological and economic demands that are made on modern agriculture from the perspective of resource use efficiency and also because this agriculture is integrated in the spaces of flows of the global network society.

But the implementation of every agropark is necessarily different and adapted to the space of place where it is put into practice. That concerns both adaptations to physical conditions like water, climate, substrate and infrastructure but also to the social and cultural environment, in which the park has to take shape. This is all summarised in the working hypotheses which I have previously formulated.

> An agropark realises lower costs, greater added value and lower environmental pollution per unit of output and space.

> An agropark can only come into being on the basis of an integral design of matterscape, powerscape and mindscape both on the global scale of Intelligent Agrologistic Networks and on the local scale of a landscape. An agropark is a knowledge-driven system innovation and makes a significant contribution to sustainable development.

> The design and implementation of system innovations like agroparks presupposes the participation of knowledge institutions, enterprises, NGOs and governmental organisations (KENGi). It is a transdisciplinary process in which the explicit knowledge of knowledge institutions and the tacit knowledge of the other partners are constantly evolving. KENGi brokers are the facilitators of this transdisciplinary process.

> In all decision-making on the implementation of the integral agroparks design, with matterscape, powerscape and mindscape aspects, arguments from the world of justice and trustfulness take precedence over arguments from the world of truth.

6. Research by design on agroparks

In an essay written on the occasion of his departure as Minister of Agriculture in 2006, Cees Veerman predicts a great future for agroparks.

> 'In order to effectively satisfy the nature and scale of the need [for food and raw materials], a combination of activities and types of process is necessary. Combinations of vegetable-based and animal production in the form of "new mixed farms", primarily focused on energy efficiency, will emerge on a greater scale. The high level of knowledge and technology in Europe will facilitate the development of totally new combinations of businesses and processes in this area.'[212]

6.1 Definition of agroparks

I use the term agropark for the 'New Mixed Farm' referred to by the minister. An agropark is a cluster of agro- and non-agro functions on or around a location[213]. This concept combines five important benefits. Firstly, the use of waste flows (waste and by-products, minerals, CO_2, energy and water) from the different interconnected chains. Secondly, efficient use of space by means of physical clustering in one location. Thirdly, efficient logistics by limiting transportation, but also important gains in the area of animal welfare and reduction in veterinary and phytosanitary risks that coincide with transportation. Fourthly, the clustering of entrepreneurs from different sectors which stimulates all kinds of inventions and innovations. Finally, the concept provides for the supply of agroproducts from transparent chains that facilitate quality management.

The concept of agroparks as defined here also appears in the literature under the synonym agroproduction parks. The term agribusiness parks is occasionally used, although this also implies office areas with many agribusiness-related services.

---

[212] Veerman, C. (2006). *Landbouw verbindend voor Europa. Van vrijheid in gebondenheid naar vrijheid in verbondenheid.* Ministerie van Landbouw, Natuur en Voedselkwaliteit, The Hague, the Netherlands.

[213] The *Innovatienetwerk groene Ruimte en Agrocluster* gave in De Wilt, J. G., H. J. van Oosten & L. Sterrenberg (2000). *Agroproductieparken perspectieven en dilemma's.* Innovatienetwerk Groen Ruimte en Agrocluster, The Hague, the Netherlands. as definition: a purposeful clustering of agro- and non-agro-production functions on an industrial terrain or in a specific region.

The first design activities on agroparks took place in parallel in 1999 and 2000 in the *Meervoudig Ruimtegebruik* project in the province of Zeeland[214] and in the Regional Dialogue North Limburg[215]. In the Zeeland project a proposal was drawn up for a cluster of combined glass horticulture and intensive livestock breeding that could be implemented for the port areas of Vlissingen and Terneuzen. But in an initial response to these designs the Zeeland provincial council rejected both the idea of glass horticulture and intensive livestock farming. In North Limburg a similar conceptual sketch was drafted for an agropark, which would later form the basis of the New Mixed Farm project, to be discussed in more detail later in this chapter.

In 2000, at the request of the Green Space and Agrocluster Innovation Network (Innonet), the discussion memorandum 'Agroproductieparken: perspectieven en dilemma's'[216] was drawn up, in which a team of several different designers came up with various concepts for agroparks:

> 'In designing the agroproduction parks, there has been an attempt to come up with various combinations of sectors and functions, positioned in regions with a clearly different character. This has resulted in the following designs:
> 1. Deltapark: non-land-dependent production with industrial processing in an urban environment (Rotterdam docklands).
> 2. Agro-specialty park: land-dependent production with industrial processing in Eem docklands.
> 3. Greenpark: land-dependent vegetable and animal production with industrial processing in a rural environment (Noordoostpolder).
> 4. Multipark: non-land-dependent and land-dependent production, interwoven with other functions in the rural area, such as agrotourism (Gelder Valley).
>
> In the first two designs, the aspect of technological innovation is dominant, while in the last two designs the regional perspective dominates.'[217]

The question is whether the spatial clustering at the regional level aspired to in the last two examples of this report should still be characterised as an agropark. However, not characterising these examples as agroparks means in fact that the presence of clustered

---

[214] Bakema, A. H., R. G. W. Dood, G. J. Manschot, C. W. M. v. d. Pol, J. M. T. Stam, I. d. Fries & M. P. Wijermans (1999). *Eindrapport; pilotproject meervoudig ruimtegebruik Zuid-West Nederland.* Ministerie van Verkeer en Waterstaat, Directoriaal-Generaal Rijkswaterstaat, The Hague, The Netherlands/RIVM, Bilthoven, the Netherlands; Overlegorgaan voor Vastgoedinformatie; DLO, Wageningen, the Netherlands.

[215] Van Mansfeld, M., A. Wintjes, J. De Jonge, M. Pleijte & P. J. A. M. Smeets (2003b). *Regiodialoog: Naar een systeeminnovatie in de praktijk.* Report 808. Alterra, Innonet, WISI, Wageningen, the Netherlands.

[216] De Wilt J.G., H.J. van Oosten and L. Sterrenberg (2000). *Agroproductieparken perspectieven en dilemma's.* Innovatienetwerk Groen Ruimte en Agrocluster, The Hague, the Netherlands.

[217] *Ibid.:* 11.

primary production becomes decisive in the definition and this constraint would be too strong. It is precisely this levelling of the boundaries between components in agrochains which have become separate – because of extensive specialisation – that is one of the attractive options under discussion with the emergence of agroparks.

The following conclusions were drawn in a summary of this report:

> 'The concept of agroproduction parks, a purposeful clustering of agro and non-agro production functions on an industrial terrain or in a specific region, potentially offers interesting possibilities to close cycles, reduce transportation and use scarce space efficiently (...).
> The system innovation of agroproduction parks allows for very diverse implementations, ranging from a concentration of intensive, non-land-dependent (agro)production functions on an industrial terrain, to a blend of land-dependent and other functions in a specific region. (...)
> The stakeholders take account of the ecological and economic potential, but point to the risks in relation to the image, the spatial suitability, the administrative feasibility and the consequences for entrepreneurial freedom in specific forms of agroproduction parks. (...)
> Given the potential and the possible social resistance, a two-pronged strategy is needed: firstly the stimulation of the debate on normative trade-offs and choices in relation to agroproduction parks, and secondly the stimulation of the development of agroproduction parks in interactive design processes.'[218]

Many scientists, government employees, entrepreneurs, and employees from various social organisations, have been working on these design processes over the last few years. This process is the Expedition Agroparks, which is the central theme of this publication. In this chapter seven agropark designs are systematically discussed and evaluated, and working hypotheses are developed on the basis of this evaluation. For each example, the discussion begins, under the heading 'design', with the history, actual status and future perspective. There then follows a detailed description of the project, divided into the components of matterscape, powerscape and mindscape.

Without exception, the designs can be characterised as research by design. In line with this definition, each design contains evaluative research on how the design in question was meant to work in the opinion of the designers. I used this research in the designs as much as possible for the evaluation of the hypotheses on matterscape, powerscape and mindscape.

In the matterscape description the physical characteristics of the design come under discussion. That is a description of the way in which the park operates and

[218] *Ibid.*: 2-3.

the environmental, veterinary and phytosanitary performances thereof, which are then highlighted by the research by design itself. The 'comfort' aspect is also part of matterscape, but because the other animal welfare aspects belong partly to powerscape or mindscape, animal welfare is always discussed under the heading mindscape.

In the description powerscape, the assigners of the project are first discussed. These include the parties that initiate the design process and in that sense this overview belongs in the section on the work process. But when the designs enter the implementation phase, these assigners are the initial organisers, the basic powerscape of the plan. Powerscape then refers to the organisational model of the park and on the basis of that, also to the economic performances of the design, obviously to the earning possibilities, but also employment opportunities and improved working conditions- and sometimes to the business planning, feasibility studies, cost-benefit analysis and risk analysis. In powerscape I describe the way in which the authorities deal with the plan and the power issues, insofar as they emerge during the designs.

In the description of mindscape I look at the question of how subjective aspects are addressed in the design in question, for instance the anthropocentric animal welfare[219], and problems such as landscape suitability and architecture of buildings are also discussed. The mindscape theme also looks at aspects of human resource management and communication. Since Wageningen-UR has been involved as a knowledge institution in all the designs discussed, it is possible to discuss a long line of knowledge management across all the designs. This includes a discussion on the way in which a body of knowledge has been built up from project to project on the basis of the learning process that has been explicitly monitored and assessed in a number of cases. It also includes the question of how the transdisciplinarity took shape, whereby each subsequent project was able to build on the experiences of the last. That culminates in a knowledge infrastructure that, strictly speaking, should be described as part of the powerscape.

Last but not least, mindscape is about people's individual choices during the design, which later appeared to be of strategic importance. That is the underlying motives of stakeholders as well as the essential moments at which strategic choices were made- which were decisive for the progress of the process and which according to De Jonge[220] determine the *kairos* of the work process, and thereby pave the way for the *chronos*, and which by definition cannot be planned in advance.

---

[219] Animal comfort will be assigned separately.

[220] De Jonge, J (2009). *Landscape Architecture between Politics and Science. An integrative perspective on landscape planning and design in the network society*. Thesis Land Use Planning & Landscape Architecture, Wageningen University, Wageningen, the Netherlands, 233 pp.

After the descriptions of matterscape, powerscape and mindscape, there is a discussion of the work process that gave rise to the design. According to the co-design theory formulated by De Jonge and tested here, there is no fixed template for such a process. Moreover, because there was no uniform process design in the various design processes and because process evaluations didn't take place systematically, it is not possible to distil such research results according to a fixed template from the research by design projects. Some of the projects conducted a systematic evaluation of the work and learning process and we can use this as a reference in this test. In other cases the test in this publication is the first to take place systematically and because this has to be done on the basis of explicit factual descriptions, the representation of the process is necessarily very detailed in some cases and at first hand journalistic in character.

Recurrent themes in the evaluation of the work process are; the position of the design in the range of invention-innovation-system innovation; the KENGi partners that together shaped the design; the transdisciplinarity between these parties in the design and the question of whether these people could maintain this collaboration in subsequent designs- in other words, whether a Community of Practice emerged. Where possible there is an analysis of the characteristics of the learning process in the project and the roles of the KENGi parties and KENGi facilitators. In addition, when evaluating the work process, the question was addressed as to the extent to which the process adhered to the characteristics of the Landscape Dialogue and Co-design, which as a set of parameters (not a process model) is a prerequisite for successful design processes in the context of agriculture in the network society.

Every example discussion ends with conclusions in which the hypotheses formulated in the previous chapter are tested and the design is evaluated on its contribution to sustainable development.

The first hypothesis, related to the De Wit theory on Resource Use Efficiency will be tested by relating the performances in the area of matterscape and the economic performances as described under powerscape to each other.

I use the description of matterscape, powerscape and mindscape together in order to be able to test the second and third hypotheses, again as much as possible on the basis of the results of the research by design that took place in the designs themselves.

For the fourth hypothesis the process design takes centre stage and the testing of the fifth hypothesis occurs on the basis of the description of the work process but also on the basis of the way in which matterscape, powerscape and mindscape are implemented.

The designs being dealt with are grouped on the basis of the start date of the project in question. These are:

- (Section 6.2) Deltapark[221], a theoretical design for an agropark in Rotterdam docklands, part of the Innonet study published in 2000;
- (Section 6.3) Agrocentrum Westpoort[222], a design for an agropark in the Amsterdam docklands, which was discussed in detail between 2002 and 2006 with many stakeholders involved, but which was ultimately deemed to be unfeasible in this area;
- (Section 6.4) New Mixed Farm[223], one of the results of the Regional Dialogue North Limburg, which has been worked on since 2004 and which will be implemented in 2010.
- (Section 6.5) WAZ-Holland Park[224], in the city of Changzhou in China. The master plan was produced in 2004 and parts of that have since been implemented.
- (Section 6.6) Biopark Terneuzen[225], started as a project in 2005. The implementation started in July 2007.
- (Section 6.7) Greenport Shanghai[226], a design created in 2006 and 2007, the first parts of which has to be delivered for World Expo 2010.
- (Section 6.8) IFFCO-Greenport Nellore, part of a bigger Intelligent Agrologistics Network intended for the Chennai metropolis. The first stone of IFFCO-Greenport Nellore was laid in March 2008.
- The chapter ends with two concluding sections (6.9 and 6.10).

---

[221] Broeze J., A.E. Simons, P.J.A.M. Smeets, J.K.M. te Boekhorst, J.H.M. Metz, P.W.G. Groot Koerkamp, T. van Oosten-Snoek and N. Dielemans (2000). Deltapark: Een haven-gebonden agroproductiepark. In: De Wilt J.G., H.J. van Oosten and L. Sterrenberg (2000). *Agroproductieparken perspectieven en dilemma's.* Innovatienetwerk Groen Ruimte en Agrocluster, The Hague, the Netherlands.

[222] Breure A.S.H., P.J.A.M. Smeets and J. Broeze (2005). *Agrocentrum Westpoort: utopie of innovatie? Reflecties en leerpunten rond een systeeminnovatief project.* Alterra, Wageningen, the Netherlands.

[223] Anonymous (2005). Nieuw gemengd bedrijf Horst: 'Neem alle belanghebbenden mee in het proces'. *Syscope: kwartaalblad van systeeminnovatieprogramma's* 3: 6-7.

[224] Smeets P.J.A.M., M. Van Mansfeld, R. Olde Loohuis, M. Van Steekelenburg, P. Krant, F. Langers, J. Broeze, W. De Graaff, R. Van Haeff, P. Hamminga, B. Harms, E. Moens, R. Van de Waart, L. Wassink and J. De Wilt (2004b). *Masterplan WAZ-Holland Park. Design for an eco-agricultural sightseeing park in Wujin polder, Changzhou, China*, Alterra, Wageningen, the Netherlands.

[225] Boekema F., M. Gijzen, F. Timmer and J. Dagevos (2008). Biopark Terneuzen. Een innovatief en duurzaam cluster. *Geografie* 17: 17-20.

[226] Smeets P.J.A.M., M.J.M. Van Mansfeld, C. Zhang, R. Olde Loohuis, J. Broeze, S. Buijs, E. Moens, H. Van Latesteijn, M. Van Steekelenburg, L. Stumpel, W. Bruinsma, T. Van Megen, S. Mager, P. Christiaens and H. Heijer (2007). *Master plan Greenport Shanghai agropark.* Report 1391, Alterra, Wageningen UR, Wageningen, the Netherlands.

6.2 Deltapark

6.2.1 The design

Three Wageningen-UR institutes were asked by the Green Space and Agrocluster Innovation Network to design an agropark for a dockland area. Deltapark was drawn by BBOI landscape architects and bundled together with three other designs in the report '*Agroproductieparken, perspectieven en dilemma's*'[227] (Agroproduction parks, perspectives and dilemmas), with the aim of stimulating a discussion on such system innovations.

> 'Deltapark provides for a regional clustering of different sectors (...) in which agricultural production (...) is integrated with the chemical industry in a sea port in the vicinity of a population concentration. In this way transport flows for (half)manufactures are minimised, waste flows are used in an economically responsible way, and emissions are more effectively controlled.'[228]

The Deltapark design (Figures 15, 16 and 17) was the first agropark design and did not get further than an invention on the drawing board. The objective was to start a debate on agroparks and this objective was accomplished. Deltapark was a source of inspiration for Agrocentrum Westpoort and Biopark Terneuzen, the latter of which was actually implemented.

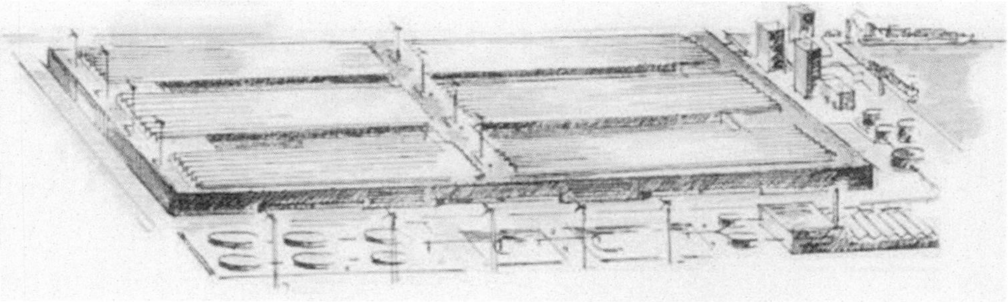

Figure 15. Bird's eye view impression of Deltapark in the Rotterdam docklands.
De Wilt et al. 2000.

[227] De Wilt, J.G., H.J. van Oosten and L. Sterrenberg (2000). *Agroproductieparken perspectieven en dilemma's.* Innovatienetwerk Groen Ruimte en Agrocluster, The Hague, the Netherlands.
[228] *Ibid.*: 47.

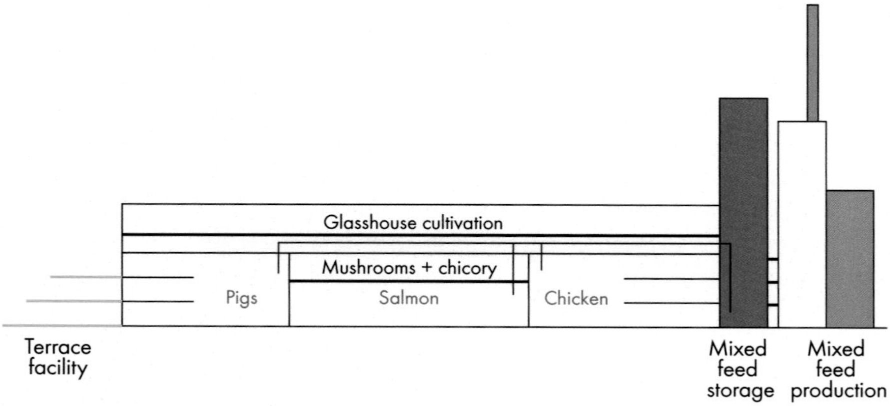

Figure 16. Cross section of the Deltapark building with the different functions in layers.
De Wilt et al. 2000.

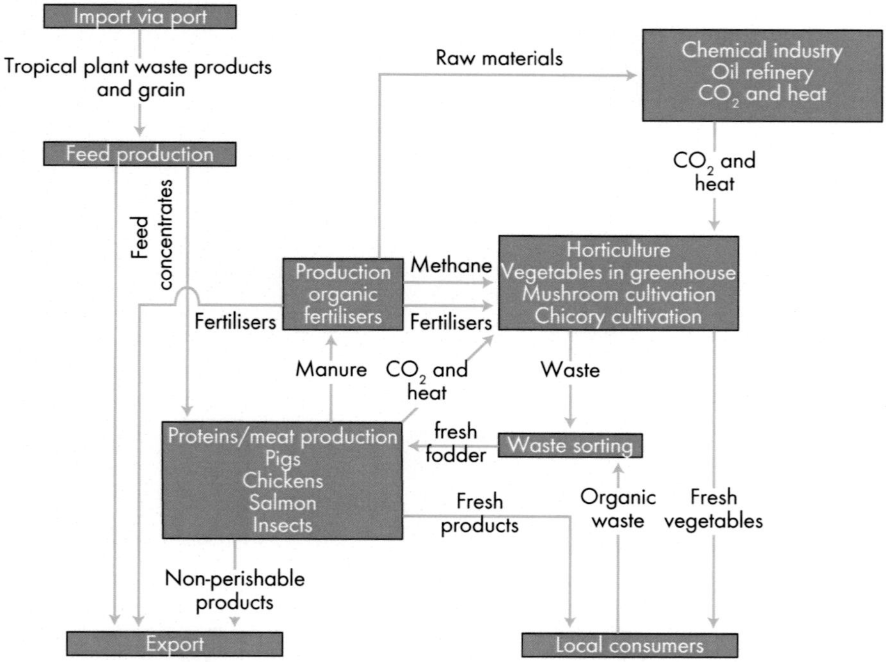

Figure 17. Global process flow diagram in Deltapark.
De Wilt et al. 2000.

6.2.2 Matterscape

> 'The stacked cultivating and breeding systems (fish, pigs, chickens and horticulture) are situated in a rectangular building with open interior spaces (patios) for sufficient light, air, and external space. External spaces (terraces) have been placed along the patios so that the animals can go outside. The fish breeding areas are situated on the lower floors, partially underground, and above them are a few floors for pigs and chickens. The top floors are for horticulture, and are constructed according to the need for light-no light for chicory and mushrooms – while the roof layers are filled with vegetable and fruit crops.'[229]

The various functions are stacked:

- Production and processing of animal proteins: pigs: 300,000 places; poultry: 250,000 laying hens and 1 million meat chickens; 0.5 ha salmon breeding and insect breeding; slaughtering and meat processing.
- Horticulture and other forms of closed cultivation: 25 ha in several layers.
- Waste sorting: production of organic fertilisers, including anaerobic fermentation (the most important by-product of which is methane; this enables the park to be largely self-supporting in terms of energy).
- Production of concentrates.

The size of the total building was huge: 1 km × 400 m × 20 m high. The design agency proposed positioning Deltapark in an area in the western dockland area of Rotterdam on the site where the ECT container terminal is currently situated. There was no discussion with the Local Port Authority about this during the design process. Only when the report was presented did it transpire that the local regulations of the Rotterdam municipality prevented the establishment of agricultural activities in the dockland area.

Industrial ecology is incorporated in this design by the closure of chains via the linking of concentrates production, slaughter, and meat processing. Links with existing industry in the port consist of CO_2 and waste heat processing.

The designers made global estimates about the likely environmental gain compared to similar quantities of animal and vegetable production in conventional conditions. The lower greenhouse gas emissions amount to 1,700 tonnes CO_2 and 50 ton methane per year. Manure is processed by the separation of thin and thick fractions. The thin fraction, in water with minerals, is used in horticulture, while the thick fraction is fermented. With the use of the biogas produced from this fermentation and heat

[229] *Ibid.*: 16.

savings, 32,500 GJ/year can be saved on heating the greenhouses and stalls compared to conventional production. Since the CO_2 used in the greenhouses does not need to be produced by burning natural gas, an additional saving of 600,000 GJ/year is also possible. Additional energy savings come from minimising transport between chain components and by exporting and importing voluminous cargo by ship whenever possible. Expectations regarding disease suppression have not been calculated in the design.

6.2.3 Powerscape

The assigner for Deltapark was Green Space and Agrocluster Innovation Network. Stakeholders in the design were various Wageningen-UR institutes as assigner, and the BBOI architect agency who produced the drawings. The Minister of Agriculture received the report. Non-governmental organisations were not involved in the design beforehand but did take part in the discussion. Entrepreneurs were also not involved in the design.

No economic return calculations or cost-benefit analyses were carried out. NIB-Capital carried out a feasibility analysis for Deltapark[230]. It was thereby assumed as an organisation model that the park would be run by a development company leasing space to primary producers. The results of the feasibility study were cautiously optimistic: The calculated savings were big enough to compensate for a higher lease price. The higher lease price arose from higher building costs (as a result of stacked floors) and from higher capital costs. In addition to this positive economic feasibility, the study indicated major social benefits. A risk analysis (Table 5) showed that the biggest risks occur in the preparation phase.

6.2.4 Mindscape

The design included important ambitions for animal welfare: pigs would have 1.5-2 m^2 per animal and terrace space to walk around in. The space for laying hens was designed to provide 0.1 m^2 per animal and the 1 million meat chickens would be kept on 2 ha.

Neither human resource management nor the knowledge value chain were included in the design. Innonet carried out a lot of work on communication about the design. But there was no well developed communication strategy, other than the willingness to have as much debate as possible about the design. Discussions were conducted on the design with all kinds of stakeholders (conservation and environmental organisations, representatives from primary producers, regional authorities, national government, consumer organisations, animal welfare organisations). The reactions were obviously

[230] Van Gendt S., G. De Groot and C. Boendermaker NIBConsult B.V. (2003). *Globaal Businessplan van een Agro-center*. InnovatieNetwerk Groene Ruimte en Agrocluster, The Hague, the Netherlands.

Table 5. Overview and assessment of risks during the preparation, implementation and operation of Deltapark.

Risk assessment	Risk	Effect (time)	Effect (money)
Preparation phase			
1 No social acceptance	High	High	Low
2 No proven commercial success on similar scale	High	High	Low
3 Risk that no willing entrepreneurs will be found	High	High	Low
4 Risk that agro-entrepreneurs will not want to lease	High	High	Low
5 Required procedures for activity allocation model and (building and environmental) permits	High	High	Low
6 Pre-financing does not materialise	High	High	High
Design phase			
7 Inadequate integration of agro-functions	High	Low	Medium
8 Construction more expensive than forecast	High	Medium	High
Utilisation phase			
9 Changes in rules and regulations (safety, animal health, animal welfare, environment) which affect investments and utilisation	Medium	Medium	High
10 Leasee pulls out later and no appropriate substitute found in time	Medium	High	High

Van Gendt *et al.*, 2003.

pending the global design, but in many cases were positive about the concept. Conservation, environmental and consumer organisations pointed out the expected image problems. Only the animal welfare organisations declared that 'health and well-being of animals (..) benefits most from extensive farming in a natural environment'[231] and rejected the design.

Innovation Network decided to adopt a low-profile approach to the discussion upon publication of the report. But Minister Brinkhorst of Agriculture responded by greeting the report extremely enthusiastically and promised to get into direct contact with the Rotterdam Port Authority to look at how the design could be implemented there. This attracted much press attention. The design was stigmatised by the press as a 'pig flat'. In this publicity climate, many social organisations decided to drop the original nuance in the first reactions, which were still included in the report, and to sharpen the debate further. As a result Deltapark became the subject of an adverse publicity campaign by the collective animal welfare organisations. Abroad,

[231] De Wilt, J.G., H.J. van Oosten and L. Sterrenberg (2000). *Agroproductieparken perspectieven en dilemma's.* Innovatienetwerk Groen Ruimte en Agrocluster, The Hague, the Netherlands: 17.

Deltapark was received primarily as a spectacular invention, a logical development in a long tradition of Dutch agriculture[232]. Innovation network instructed Daalder and Koopman[233] to investigate the reactions in the press to the Deltapark. They list the following arguments in order of frequency:

- animal comfort – aversion to image of stacked animals 'in a flat';
- pig flat is caricature of intensive livestock farming gone mad;
- fierce, emotionally charged reactions, mostly directed at animal comfort;
- farmers want to keep independent family farms in trusted rural environment;
- administrative complexity, obstruction and lack of support;
- risk of catastrophic viral outbreak with concentration in a flat;
- agroproduction parks: great idea, but not 'in a flat', preference for organic/ecological agriculture.

In this analysis the most common arguments from supporters were:

- improvement on conventional intensive livestock farming;
- better for the environment: benefits of closed ecosystem;
- animal transportation no longer necessary;
- minimises risk of spreading diseases (food safety);
- space gain;
- the concept is good, but it shouldn't be stacked, focus on space gain;
- scale or technology, better directed at animal welfare.

The report on the communication about Deltapark draws the following conclusion:

> '...the emphasis in the communication about Agroparks should not be placed univocally on information about the high-tech nature and the advantages of scale. In order to invite faster acceptance among the general public, aspects such as environmental improvements, animal friendliness and food safety should always be at the forefront of the communication. According to the authors, the central positioning should be as follows: Agroparks can offer an adequate solution in a broader perspective, whereby considerable improvements are obtained in relation to current intensive livestock farming. This should consequently dispel the myth of the romantic idyll of farming life in the countryside. Two types of communication were identified: firstly a general support wave campaign about agroparks, in which government and knowledge institutions set a positive tone and emphasise the urgency of clustering. Secondly, individual initiators must be aware of the importance of adequate communication.'[234]

---

[232] De Wilt J.G. and T. Dobbelaar (2005). *Agroparken. Het concept, de ontvangst, de praktijk.* InnovatieNetwerk Groene Ruimte en Agrocluster, Utrecht, the Netherlands: 22.

[233] Daalder A. and J. Koopman (2004). *Verguld en verguisd. Agroparken in de media.* InnovatieNetwerk Groene Ruimte en Agrocluster, The Hague, the Netherlands.

[234] *Ibid.:* i.

6.2.5 Work process

Deltapark is an invention, a multidisciplinary collection of new technologies from different scientific Wageningen-UR institutes, integrated in one design, that has never been developed into a value proposition. The KENGi facilitator Innovation network played a decisive role in the initiation of this invention, by acting as paying assigner for the various design agencies (Figure 18).

But seen from the relationship between the KENGi partners and most of all between the directly involved stakeholders, the media, politicians and the public, this invention immediately sharpened the debate. There appeared to be explicit opponents and proponents of the concept among each of the KENGi partners and it is impossible to say whether most KENGi partners were opposed or supportive. The communication that followed the completion of the design was not managed.

Without a well-calculated communication plan, publicising a revolutionary invention with the aim of starting a social debate will lead inevitably to the plan being shot down. That is the most important learning experience of the media discussion and the public debate which resulted from the design. Sloterdijk predicts this social reaction in his description of the inhibiting context in the 'Crystal Palace'.

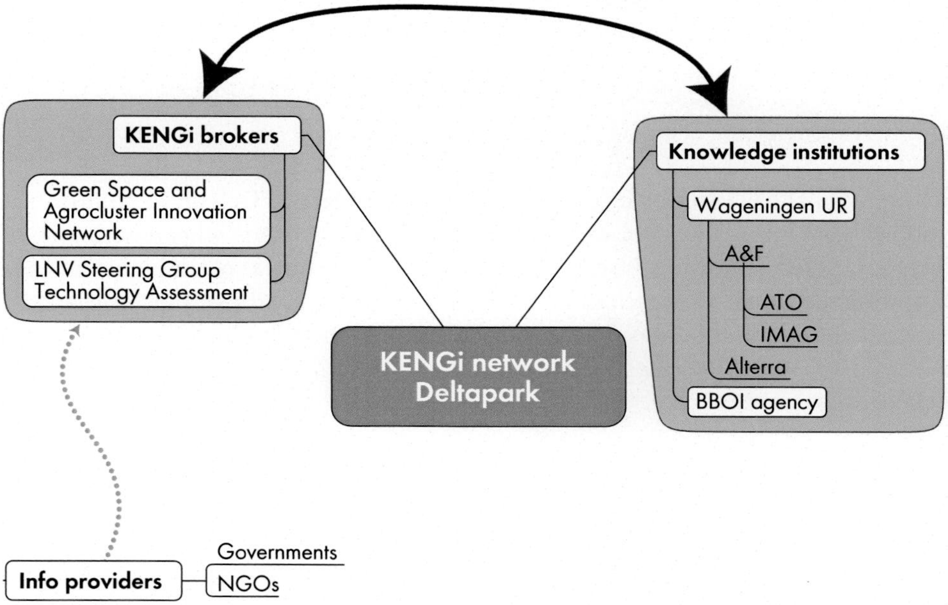

Figure 18. The Network around Deltapark. In the design and social debate only the clients and implementing knowledge institutions were members of the Community of Practice. Governments, enterprises and NGOs acted as information providers via interviews with the clients. Enterprises were not involved in the design process.

Transdisciplinary knowledge development about social communication was the main outcome of the design. The fierce reactions to the report came as a surprise to the authors[235] and put them on the defensive in the media in successive years. But the learning experience was that social debate cannot be avoided and that professional help is necessary here. The failures (according to Deltapark supporters) in the communication process did, however, lead to a good analysis of this communication later. As a result, communication around later agropark designs was more adequate and was usually included strategically from the start as an important component of the design.

6.2.6 Community of practice

There was little point in a comparison with the steps in the design process of the landscape dialogue in the case of Deltapark. After all, the design was not intended to be implemented but to provoke a discussion and was therefore still a classic invention of scientists. Deltapark occurred mainly in the phases of Design (3) and Support (4). The research by design appeared to be limited to the researchers and designers who created the design. There was no question of forming a local or regional group of stakeholders and therefore no joint fact-finding took place between these stakeholders either.

6.2.7 Testing of the hypotheses

An agropark realises lower costs, greater added value and lower environmental pollution per unit of output and space.

The first global calculations of the environmental effects of Deltapark confirm this hypothesis: integration of the various sector boundaries leads to higher environmental efficiency. At the same time the calculations made in the context of the global business plan show that the higher economic costs of the building and extra facilities, which are reflected in a higher lease price, can be compensated for by lower costs in the production process. The resource use efficiency of this further integration is therefore positive on balance.

[235] De Wilt J.G. and T. Dobbelaar (2005). *Agroparken. Het concept, de ontvangst, de praktijk.* InnovatieNetwerk Groene Ruimte en Agrocluster, Utrecht, the Netherlands: 22.

An agropark can only come into being on the basis of an integral design of matterscape, powerscape and mindscape both on the global scale of Intelligent Agrologistics Networks and on the local scale of a landscape.

Deltapark was almost completely designed in the Space of flows of the Rotterdam Port, whereby it was assumed that the physical local environment could be completely adapted to the design. Virtually no attention was paid to Powerscape and Mindscape. A serious social debate arose, to the great surprise of the designers, and resulted primarily in major reticence on the part of politicians about the concept of agroparks. In a negative sense this completely confirms this hypothesis: without attention to powerscape and mindscape, the matterscape of a groundbreaking system innovation is not only ultimately without obligation, it also invokes negative reactions. The fact that this is even being discussed in an ultimate space of flows location like the Rotterdam Port, is an indication of the social weight that is attributed to the Powerscape and Mindscape of these designs.

An agropark is a knowledge-driven system innovation and makes a significant contribution to sustainable development.

As a result of the negative publicity, Deltapark has not been developed further than a knowledge-driven invention. But the learning experiences, acquired from the Power- and Mindscape aspects, delivered important practical experiences, which could be used in future projects.

An evaluation of Deltapark on the triple-p aspects of Sustainable Development was only performed on a limited basis. Different processes are included in the design which hold the promise of favourable environmental effects and far-reaching energy savings. In the first discussion round, outlined in the report, these promises are sometimes approached in an anticipatory fashion.

- As far as the people aspect is concerned, only extensive ambitions in the area of animal welfare are explicitly declared.
- The global business plan produced on the basis of the Deltapark design is positive about the profit aspect. The expected cost savings justify the higher investment costs required for the implementation of Deltapark.

The design and implementation of system innovations like agroparks presuppose the participation of knowledge institutions, enterprises, NGOs and governmental organisations (or KENGi). It is a transdisciplinary process in which the explicit knowledge of knowledge institutions and the tacit knowledge of the other partners are developed in a process of continuous iteration. KENGi brokers are the facilitators of this transdisciplinary process.

There was no fully-fledged collaboration between KENGi parties in the design of Deltapark. Authorities, NGOs and enterprises were consulted as information providers and reviewers of the design but were never asked to make any commitment. But the discussion, which arose after publication of the design, did start a transdisciplinary learning process which, while not registered by Innonet as KENGi broker, was exploited as a learning process and registered in reports.

In all the decision-making concerning the realisation of the integral agroparks design, involving matterscape, powerscape and mindscape aspects, subjective decisions of individuals are ultimately decisive. The world of trustfulness (mindscape) is therefore dominant in the last instance.

The events around the presentation of the Deltapark design confirm this hypothesis. While the original design concentrates on matterscape and a global business plan is only added in the second instance, the first reactions, which are part of the report, are mainly about image (the desirability of intensive livestock farming, animal welfare, etc.). An avalanche of, often negative, publications ultimately leads to the stigma of 'pig flat'. Decisive individual decisions were taken:

- by Innovation Network, to adopt a low profile approach to the discussion about the report;
- by Minister Brinkhorst, who in his naïve enthusiasm wanted to turn a design intended only to provoke discussion, into reality in Rotterdam port;
- by different social organisations (*Natuur en Milieu, Stichting Varkens in Nood*) which decided in this for them favourable publicity climate to drop the original nuance – in the first reactions, still included in the report – and to sharpen the discussion.

The key learning point to be distilled from this example as a consequence of this hypothesis is the importance of a good communication strategy as part of the mindscape design, on which the possibility to influence the crucial individual decision-makers on their subjective opinions stands or falls. The stigma 'pig flat', with its highly negative connotations, which arose in the poorly prepared communication about publicity on this design, is still playing a prominent role today in the Dutch debate and stops people connecting with this concept.

A new hypothesis, which can also be formulated as a conclusion from this example, emerges from the analysis of the communication about Deltapark. This shows that the opposition of social groups is focused primarily on intensive livestock farming and is motivated by animal welfare issues. The new hypothesis, which is added to the original list from Section 4.1 in the following examples, states that:

> Social opposition to agroparks is directed primarily at intensive livestock farming.

If this hypothesis is true, a design without intensive livestock farming should evoke much less social opposition.

6.2.8 Conclusion

The Deltapark design definitely achieved its objective. It started a social debate about the system innovations that were foreseen by the assigner in the future of agricultural development. It also instigated an interactive design and development process that is being continued in the Agrocentrum Westpoort and New Mixed Farm projects.

During the design process itself, the work on Deltapark was a scientific exercise for the researchers involved that did not go beyond generating an invention. It was research by design within the limited circle of researchers and designers and was therefore in that sense not transdisciplinary. Only the storm that broke in the media can be seen with hindsight as a forewarning of what it means to turn these inventions into actual innovations. The limited attention to the communication strategy while designing the plan provoked the fierce reaction of a number of NGOs (who focused on environment and animal welfare) after the presentation. The plan therefore obtained a stigma (pig flat) which is still causing problems for agropark designers to this day.

6.3 Agrocentrum Westpoort

6.3.1 The design

Between 2002 and 2006, the Amsterdam Port Authority, Green Space and Agrocluster Innovation Network and Wageningen-UR worked on a design for an agropark with 300,000 pig places, fish cultivation and vegetable production in modular buildings each comprising six floors and combined with an abattoir and co-digestion installation (Figure 19). The design was initially intended for the Amsterdam port area. The design was discussed in the period 2002-2006 with potential stakeholders from all KENGi parties. However, none of these parties came forward as a pioneer to take up an assignment. The concept then lost the support of the Amsterdam Port Authority. In 2006 there was another attempt to implement the design in the Zaan region of the North Sea Canal area, but the Zaan municipal council rejected the plan on formal

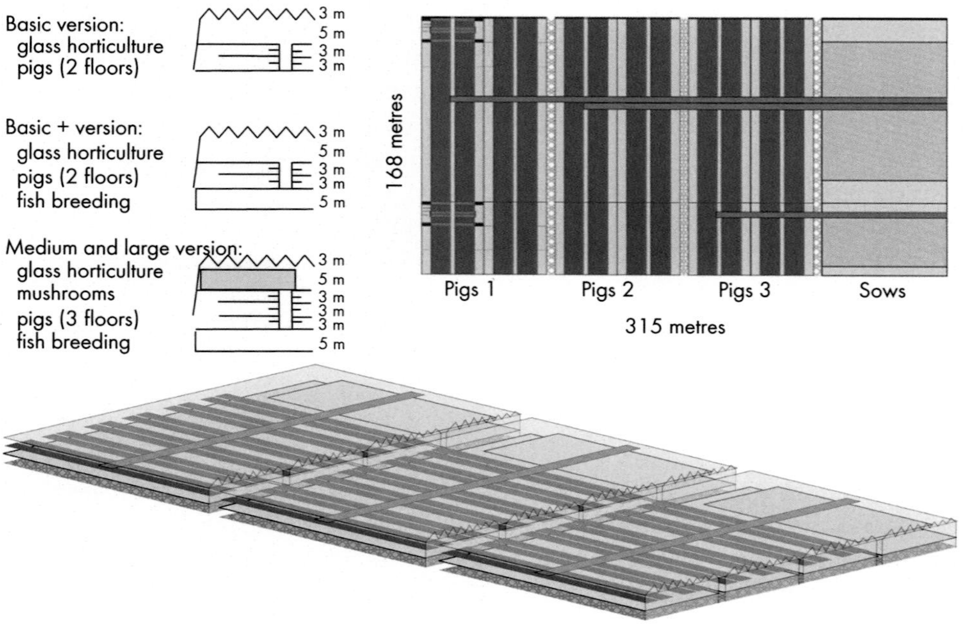

Figure 19. Design for an agropark in Amsterdam docklands.
Broeze et al. 2005.

grounds because it did not comply sufficiently with employment norms per ha to which new port developments are subject.

Agrocentrum Westpoort has not been implemented. Not yet at least[236]. It has however become the basis for an Agrologistics Community of Practice which has generated successful new designs. The design has amassed large amounts of fundamental knowledge which can be successfully used at a later date.

6.3.2 Matterscape

Port locations, including the Amsterdam docklands, are regarded by the designers as ideal locations for the development of an agropark because of

[236] During a symposium on Mega-stables on 18 June 2008, Jan de Wilt, from *Green Space and Agrocluster Innovation Network,* announced to the project leader of Agrocentrum Westpoort that Innonet was collaborating with a group of pig farmers on a plan for an agrocentre which would bring together pig breeding, a biofuel plant, an abattoir and suppliers. Groningen and West-Brabant were being investigated as prospective sites (De Graaff M. (2008). Agrocentrum varkens in zicht, Agrarisch Dagblad).

'...the presence of relevant agro-related businesses for intensive agricultural production (e.g. Cargill, Amfert), the presence of enterprises that offer the possibility to close cycles (Amfert, AfvalEnergieBedrijf, Nuon), the available port function for effective import and export of bulk flows, the large number of consumers and the labour potential in the area, the favourable location of the region for glass horticulture [and] the isolated location in relation to other livestock businesses, which reduces the risk of spreading animal diseases.'[237]

Agrocentrum Westpoort was designed in different versions, which were to be implemented consecutively over time, on the basis of a modular construction. Each module is 315 m × 168 m wide (5.3 ha) and offers space for 33,000 porkers and 4,500 sows on 11.5 ha gross floor area, 5 ha fish breeding and 5.3 ha glass horticulture.

The smallest or basic version comprises three such modules. With manure processing and slaughtering included, this version comprises a built surface area of 23 ha.

Agrocentrum Westpoort should be able to expand at the intended location to 5 times the basic version, i.e. into 15 modules. If necessary, an extra floor can be added for mushroom production. At maximum size, the agropark consists of 70,000 sow places, 500,000 pig places, 83 ha fish breeding and 80 ha available surface area on roofs and, potentially, 80 ha mushroom production.

The key internal relationships between the various production components are shown in Figure 20.

In manure processing, thin and thick fractions are followed separately by co-digestion of the thick fraction. The digestate is processed into fertiliser. In the pig abattoir and meat processing section, 300,000 to 1.5 million slaughters are planned to take place per year. Intensive livestock farming and glass horticulture are interlinked via co-digestion. Co-digestion provides heat for fish breeding. Table 6 shows the environmental advantages of Agrocentrum Westpoort compared to conventional production[238].

The principles of industrial ecology can also be exploited within the port between different enterprises operating in the port, as shown in Figure 21 and 22. The design has attempted, particularly in relation to pig breeding, to reduce the veterinary risks in comparison to the conventional situation. This starts with the isolated location of

[237] Anonymous (2004). *Agrocentrum Westpoort. De haalbaarheid verkend.* Gemeentelijk Havenbedrijf Amsterdam, Amsterdam, the Netherlands: 2.
[238] Anonymous (2004). *Agrocentrum Westpoort. De haalbaarheid verkend.* Gemeentelijk Havenbedrijf Amsterdam, Amsterdam, the Netherlands: 29.

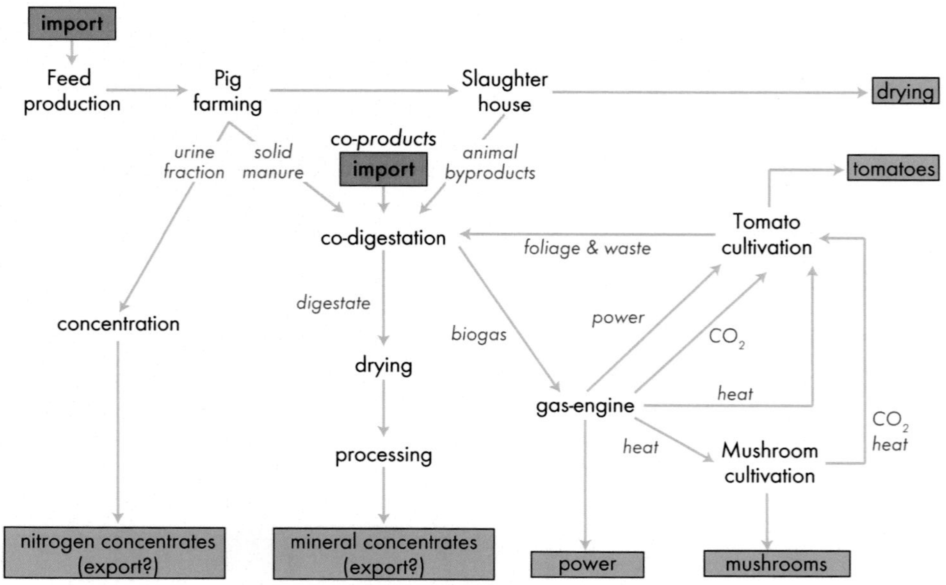

Figure 20. Diagram of industrial ecological relationships between the components of Agrocentrum Westpoort.
Broeze et al. 2005.

Table 6. Overview of the various environmental savings in Agrocentrum Westpoort compared to conventional production.

Effect	**Basic**	**Medium**	**Large**
Use of space	23 ha	39 ha	83 ha
Number of pigs			
porkers	100,000	200,000	500,000
sows	14,000	28,000	70,000
Electricity production from biogas per year			
mln KW h	60	120	300
size of generator MW	7	14	35
Biogas production for energy company per year			
mln m^3	25	120	125
TJ	625	1,250	3,125
Cuts in CO_2 emissions by substituting natural gas use (kton)	40	80	200
Amount of recoverable phosphate per year from the gasification of manure residue (kiloton)	0.5	1.0	2.5
Transport savings on lorries (km/year)	100,000	200,000	500,000

Broeze et al. (2005).

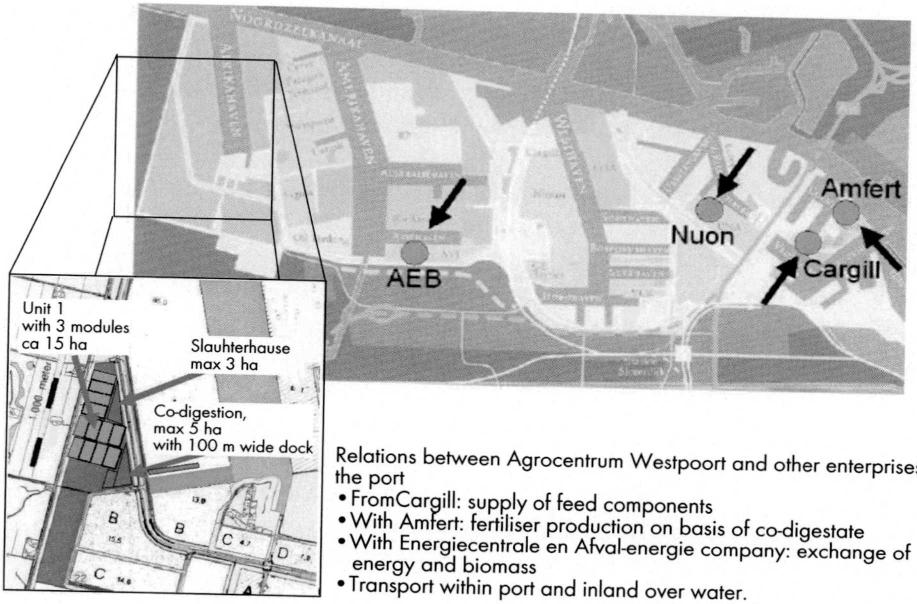

Figure 21. Relationships between Agrocentrum Westpoort and other businesses in the Amsterdam docklands.

Anonymous (2004).

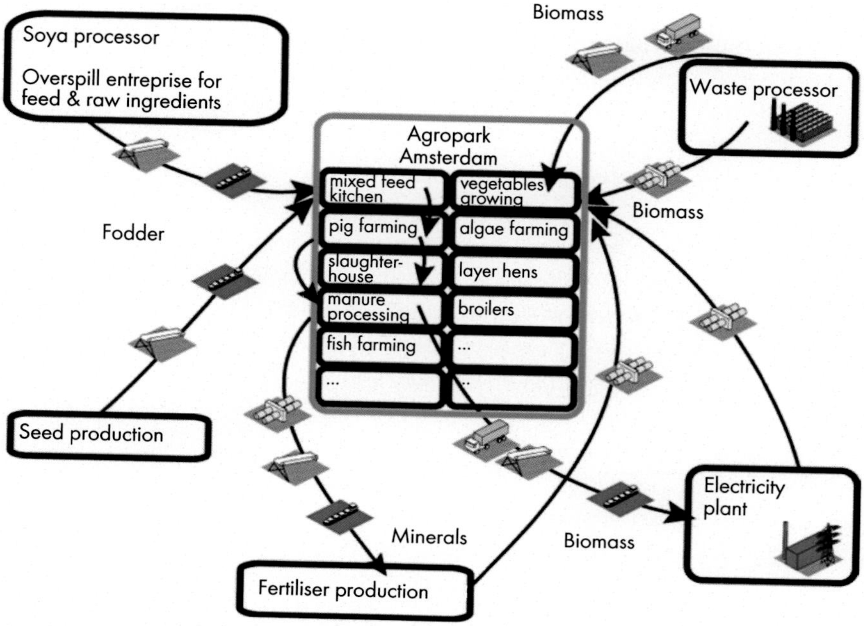

Figure 22. Substance flows between Agrocentrum Westpoort and other businesses in the Amsterdam docklands.

Anonymous (2004).

this business in relation to other piggeries, and in relation to the waste energy company in the port and in the closed chain, whereby it is no longer necessary to transport pigs during the various breeding stages. But within each module too, the different departments are separated from each other by hygiene sluices.

6.3.3 Powerscape

The assigners for the design were Amsterdam Port Authority, Green Space and Agrocluster Innovation Network and Wageningen-UR. Table 7 gives an overview of the stakeholders that invested a substantial amount of time in the project, some of whom also made a financial contribution.

The calculations made by NIB-Capital in the global business plan for Deltapark were finalised by employers at Ballast Nedam for Amsterdam Westpoort[239]. They carried out a minimal cost-benefit analysis, starting with the costs and benefits of the various industry branches involved in reference situations and only including those savings agreed upon by all the experts with whom they had spoken for the purposes of making their calculations. They also only looked at the basic version. The expected investment for this version is comparable with the reference situation because the higher investments in buildings and installations are compensated for by much lower investments in land purchase and infrastructure. For a calculated investment of € 60 mln for the basic version, the calculated savings on an annual basis amount to € 4 mln. That represents an increase in the return on investment of 6.6% per year. It should be noted here that these savings concern, primarily, lower energy costs that are calculated on an oil price of € 45 per barrel.

Table 7. Stakeholders in Agrocentrum Westpoort.

Knowledge institutions	Green Space and Agrocluster Innovation Network[1], Wageningen UR[1], HAS-Den Bosch, 10-Wizards
Government	Amsterdam Port Authority[1], AfvalEnergieBedrijf[1], Province of North Holland, Ministry of Agriculture, Agrologistics platform[1]
NGOs	Environmental Federation North Holland, WLTO
Enterprises	Nuon bv[1], Cargill[1], Amfert bv[1], Dumeco, Dura Vermeer Vastgoed[1], ABN-AMRO, Prinsenland bv[1], Govera

[1] Stakeholders who also made a financial contribution.
Anonymous (2004). Table taken from various pages.

[239] Bijpost S. and R. Overdevest (2005). *Beoordeling haalbaarheidskansen agrocentrum vanuit economisch perspectief.* Ballast Nedam. Bouw Speciale Projecten, Nieuwegein, the Netherlands.

In addition to the annual savings, a one-off saving would also be made in the piggery as a result of a 50% discount on procurement of the necessary pig rights, which is applicable when manure is processed and exported. Table 8 gives a detailed overview of the calculated savings.

The form of association used in the Ballast Nedam calculations was the same as that postulated by NIB-Capital in its global business plan, drawn up for the Deltapark case (Section 6.2). It was the so-called 'shopping centre formula', whereby a project developer builds an agropark with the backing of banks and institutional investors and leases the production space to entrepreneurs, who themselves make the investments in their plant material or livestock. For this, additional arrangements are required on truck systems and back-up facilities. There are other formulae, in addition to this, such as the cooperative property of the producers, a total company that administers all functions or a limited partnership with silent partners.[240]

But despite the favourable financial forecasts, Ballast Nedam and the project developer, with whom this company implements many of its projects, decided not to act as a pioneer in implementing the project. The key reasons for this were again the supposed image problem and the uncertainty of finding sufficient entrepreneurs to take part in Agrocentrum Westpoort.

The project was discontinued after the design phase because without a clear pioneer with sufficient investment power, none of the port authorities involved (first in Amsterdam, then in Zaanstad) was prepared to make space for Agrocentrum Westpoort.

Table 8. Cost-benefit analysis, conducted by Ballast Nedam.

	Economic benefits of agrocentre	
A.	Lower fixed and variable energy costs	€ 1,735,000/year
B.	Extra profit fromco-dogestion by own use of energy	€ 747,000/year
	CO_2 savings for glass horticulture	€ 250,000/year
C.	Lower transport costs due to chain integration	€ 1,325,000/year
	Total benefits	Ca. € 4,000,000/year
D.	Reduced cost of pig rights	€ 14,200,000 one-off

Anonymous (2004). Table taken from various pages.

[240] *Ibid.*: 17.

Various partners in this project have continued to work together as a result of the work on Agrocentrum Westpoort. They have done so as an Agrologistics Community of Practice (CoP), which under the auspices of the Agrologistics Platform has had a number of meetings, in which various aspects of agropark design were discussed and further developed with the help of external experts[241]. A consortium arose from this CoP, that started work on international projects, and received backing for the New Mixed Farm project.

6.3.4 Mindscape

Explicit attention is devoted to animal welfare in Agrocentrum Westpoort. Again the important improvement in relation to animal comfort is due to the complete exclusion of animal transportation. At the time of the design process, considerably more space was given to each animal in the piggery than laid down by the European Union norms (1.1 – max 1.9 m^2 instead of 0.65 m^2)[242]. No attention was paid to working conditions within the business in this phase of the design.

Agrocentrum Westpoort has been an important impulse for further knowledge development, particularly in the area of quantitative implementation of internal industrial ecology[243], communication, and process management of design and implementation.

During the design process, the communication strategy was frequently discussed within the project group. The proactive communication, that Daalder and Koopman proposed in their analysis of Deltapark, was further developed for Agrocentrum Westpoort with the Think Big campaign (Figure 23), which was nevertheless not brought to public attention because the project was discontinued after the design process.

---

[241] Kranendonk R., F. Gordijn, P. Kersten and P.J.A.M. Smeets (2003). *Cop agrologistiek; verslag van werkatelier (6-7 november, Venraij)*, Alterra/WING, Wageningen; Kranendonk R., P. Kersten, P. Smeets and F. Gordijn (2004). *Cop agrologistiek; verslag van werkatelier (7-8 april, Zaandam)*, WING/Alterra, Wageningen, the Netherlands; Kranendonk R., P. Kersten and P. Smeets (2005). *Cop agrologistiek. Verslag van cop bijeenkomst (14 december, kasteel Groeneveld Baarn)*, Alterra, Wageningen, the Netherlands; Kranendonk R.P., P.H. Kersten, P. Smeets and F. Gordijn (2005). *Cop agrologistiek: Verslag van masterclass (12 januari 2005, Den Bosch)*, Alterra/WING, Wageningen, the Netherlands; Kranendonk R., P. Kersten and P.J.A.M. Smeets (2006). *Cop agrologistiek verslag van de copbijeenkomst (14 juni 2006, living tomorrow Amsterdam)*, Alterra/WING, Wageningen, the Netherlands.

[242] Broeze J., I.A.J.M. Eijk, K.H. de Greef, P.W.G. Groot Koerkamp, J.A. Stegeman and J.G. de Wilt (2003). *Animal care. Diergezondheid en dierwelzijn in ruimtelijke clusters*, Report 03.2.028, InnovatieNetwerk Groene Ruimte en Agrocluster, The Hague, the Netherlands.

[243] Broeze J., M.G.N. Van Steekelenburg and P.J.A.M. Smeets (2005). *Agrocentrum amsterdam. Ontwerpen voor agroparken in havengebieden*, Report 05.2.106, InnovatieNetwerk Groene Ruimte en Agrocluster, Utrecht, the Netherlands.

Figure 23. One of the Think Big posters, part of the proposed communication strategy stating: Agriculture in the city: it can be done!!
Daalder and Koopman (2004).

During the various meetings of designers with the other stakeholders, the objections to the design, witnessed by the representatives of some organisations among their own grassroots also became clear. Breure *et al.*[244] mention the following:

'...

- Farmers, who appear on the surface of it to be the most important party in the discussion on agroparks, are turning out to be opponents, particularly via their interest organisations. They have every reason to be. Agroparks will put farmers out of work, in particular as independents, but also because they represent an innovation leap in work productivity. (...)
- An agropark is an innovation leap in agrologistics. In many cases road transport will be switched to water transport, but much of the transport will disappear altogether because of spatial clustering. The logistics sector in the Netherlands is also looking at the developments with mixed feelings.
- An agropark takes a radically different approach to animal welfare. Most solutions for problems in the area of animal welfare use the "back to nature" principle and try and maximise animal productivity while striving to create as natural a living environment as possible. In an agropark the welfare problem is resolved separately from this fundamental approach. Much animal comfort is related to transportation or too little space or light. So an agropark makes an effort to minimise transportation and maximise the amount of space and light. But it will be years before animal

244 Breure A.S.H., P.J.A.M. Smeets and J. Broeze (2005). *Agrocentrum Westpoort: Utopie of innovatie? Reflecties en leerpunten rond een systeeminnovatief project*, Report 1394, Alterra, Wageningen, the Netherlands.

welfare organisations will be ready to discuss this approach. Until then, an agropark is regarded as a disposal pig flat.

- The meat processing industry has been confronted in recent years with the consolidation of retail chains; in the struggle between the major players in the chain, both are trying to take control of the administrative functions so as to establish a new position of power. Since primary production is more intensely up-scaled in an agropark than in the existing agricultural sector, the agropark appears to be threaten this administrative position of the major meat processors.'

Under pressure because of the threatened disinvestment in the Container Terminal of Amsterdam port, the Board of Directors of the Amsterdam Port Authority too could ill afford any adverse publicity at the time the project was unfolding, and thus never really supported the project.

6.3.5 Work process[245]

Employers at Ballast Nedam conducted many discussions with experts to substantiate their feasibility study. In their discussions with pig farmers it became obvious that the latter, the most important owner of the problem for whom Agrocentrum Westpoort was trying to find a solution, had not been at the discussion table during the entire design process. Breure *et al.* have this to say on the subject:

> 'A key problem-holder, the pig farm, is located a long way from the dockland area. The problems of this sector are scarcely understood in Amsterdam, let alone acknowledged, and the local and regional governments will not set themselves up as a generic problem-holder. The government that could do this (the national government, more particularly the Ministry of Agriculture) is still so busy considering the issue of whether agroparks should be allowed in the Netherlands, that their initial enthusiasm has now been replaced by extreme reticence. Representatives of the sector and also other chain parties such as abattoirs and the Rabobank have been present at many meetings in various compositions but more as observers than problem-holders. Only the representative from the WLTO, [the regional farmers organisation] (which incidentally has very few pig breeders among its members) entered half way through the process and thereafter continued to participate with great commitment.
>
> From the beginning Agrocentrum Westpoort was a solution for a problem not explicitly formulated by relevant stakeholders in the sense that the problem-holders from the intensive pig-breeding sector were not involved in the design. Not only is Amsterdam not the place where they are looking

[245] *Ibid.*

for solutions to their problems, the scale of this solution far exceeds the capacity of most pig breeders in the Netherlands (...). From the perspective of the project developer this creates a dilemma: either he has to join forces with a pig breeder who can single-handedly make a one-off investment in keeping 300,000 pigs, but that would have to be a partner who is involved in the project right from the start. Or the park is implemented with a group of pig breeders, immediately creating a problem of mutual competition, which is difficult to overcome in the design phase.'[246]

In their analysis of the work process around Agrocentrum Westpoort, Breure *et al.* came to the following conclusions about the strategy used to commission stakeholders:

'An (...) important obstacle arose from the strategy we adopted, partly in response to the myth of the negative image (...). We rarely looked seriously at the issue of commitment. The strategy focused on keeping potential stakeholders around the table by gradually building up rational arguments. Despite this, various stakeholders left during the process. But more importantly for the outcome of the design process, this strategy never enabled us to find a party willing to carry the load in the implementation phase. That effectively brought the project to a standstill, once the Local Port Authority had ceased to be the pioneer in the design process. As of this moment, the future value of the design remains too uncertain to provide enough motivation for the formation of a consortium to start on the implementation as an investor. Nor is there a dominant party in the port to take on this role alone. The design, which integrated different aspects of sustainable development, bears too little relation to the mission of one single party. (...) The "not no strategy" meant that the design was made on the basis of compromises between the parties present. No work was done on more appealing spectacular concepts, which are often just what is needed to bring doubters round in the early phase of many uncertainties. Innovation leaps therefore had little chance beforehand, because a minimum programme was being worked to. In hindsight, we can state that this was neither a groundbreaking intrinsic design nor a focused design process. However, given the situation in which the project was created, there was little choice. Obviously it is important to find a good balance between the appeal of the design for those involved on the one hand, and the feeling of "uniqueness" of the design for those not involved in the decision-making process, on the other. But as the saying goes, no guts, no glory.'[247]

The KENGi network around Agrocentrum Westpoort is shown in Figure 24.

[246] *Ibid.*: 44.
[247] *Ibid.*: 37.

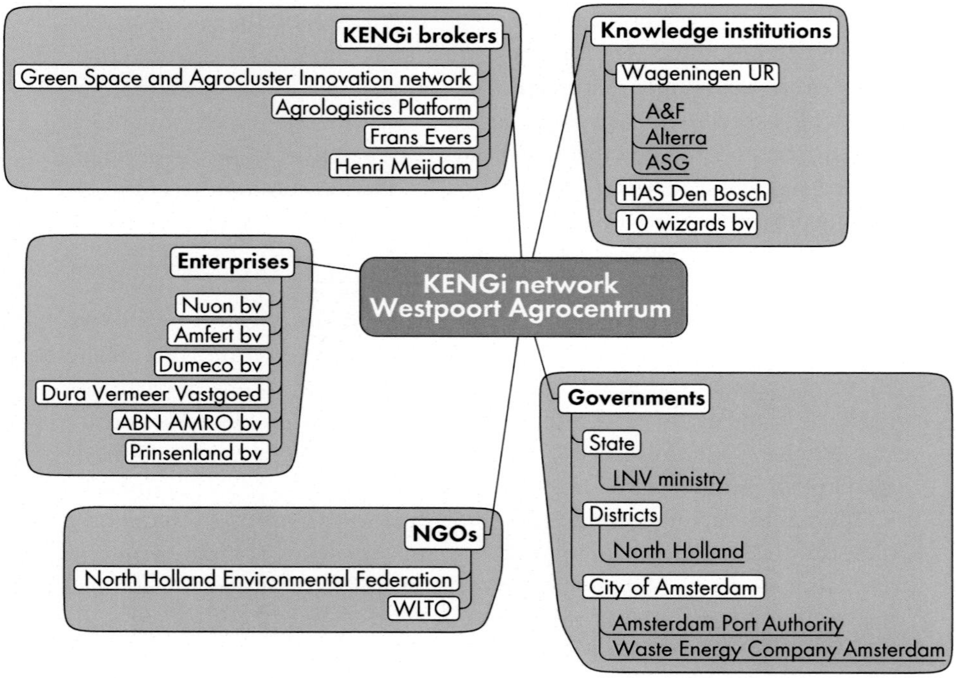

Figure 24. The KENGi network around Agrocentrum Westpoort.

'Many of the participants met for the first time during the project, resulting in new connections between the agricultural and port sector. Therefore, institutional learning took place implicitly. But the extent of this differed among the participants, partly because of the frequency of participation and exchange that occurred outside the formal meetings. A few of them were in regular contact, and this intensified further after March 2004 with the joint participation in the Agrologistics Community of Practice (...). [This] CoP acted as a channel for a social learning process, bringing together people from different backgrounds, disciplines and functions. What they had in common was their involvement in pilot projects on agrologistics and to a certain extent they wrestled with the same problems. These shared issues were central during the CoP meetings (for example, marketing, knowledge management, the corridor concept, monitoring of results, etc.).'[248]

[248] *Ibid.*: 41. See Van Gendt S., G. De Groot and C. Boendermaker NIBConsult B.V. (2003). *Globaal Businessplan van een Agro-center*. InnovatieNetwerk Groene Ruimte en Agrocluster, The Hague, the Netherlands: 39, for the reports from the Agrologistics Community of Practice.

In the network around the Agrocentrum Westpoort the problem of the relationship with the grassroots supporters of the different participants came up several times and it came to a head – yet again – on the image problem of intensive livestock farming. Breure *et al.* talk about the myth of the negative image[249]. The experience from the design process shows that the participants themselves were not troubled by the image problem, but they believed that THE Dutch consumer would not want to buy meat from the bio-industry and THE Amsterdammer would not want a 'pig flat' and for that reason there would be no support for an agropark in Amsterdam. Breure *et al.* have this to say on the subject:

> 'From the minute they took part, we saw a process of showing interest and seeking information among these people. The myth was dispelled and often over time people became more enthusiastic about the idea. But whenever they had to take the plan back to their own grassroots, a little of that scepticism reappeared. The problem arose time and again in organisations that changed participants. Thus it happened that nearly every meeting in which all stakeholders sat together had to start by discussing this myth again, dispelling it or at least neutralising it before moving on to the actual agenda. But the main consequence was that the design was treated with increasing enthusiasm by the group of insiders who were intensively occupied with it, while the scepticism within their support ranks remained stuck at the level of the myth. This is why there was no public support.'

This evaluation shows that some key criteria in De Jonge's co-design process, described in Chapter 5, were not met in this design process. Compromises must be sought; there was no free space. There was a large group of professional amateurs at the table but too many of them were tied to the opinions of their company or organisation.

Transdisciplinary knowledge development was a point of discussion throughout the entire design process because the consecutive design steps were constantly examined in the Community of Practice in which scientists and participants from each of the parties involved were present with their own tacit knowledge. Scientists, government workers and a few knowledge brokers in particular, who were involved in this or that agropark project, came together in the Agrologistics Community of Practice, which was set up as the first case with Agrocentrum Westpoort.

---

[249] *Ibid.*: 36.

6.3.6 Testing of the hypotheses

An agropark realises lower costs, greater added value and lower environmental pollution per unit of output and space.

As in the Deltapark example, this hypothesis is also confirmed in this example, not only from the calculations made by the designers themselves, but mainly as a result of the feasibility tests that Ballast Nedam carried out on the design, and which promised major annual cost savings if the design were ever to be implemented in practice.

An agropark can only come into being on the basis of an integral design of matterscape, powerscape and mindscape at both the global scale of Intelligent Agrologistics Networks and the local scale of a landscape.

Like Deltapark, Agrocentrum Westpoort was designed mainly to function in the global space of flows of the Amsterdam docklands. But as with Deltapark too, the discussion and the decision to abandon implementation centred primarily around arguments that belong to powerscape and mindscape. There was admittedly a lot of discussion on precisely these aspects, but there was no design. However, the joint fact finding process was continually having to be restarted, because of the constant flow of new people from certain organisations joining the process. There was no clear distinction between the design phase and support and decision-making. These shortcomings in the process meant that the design phase was actually a continuous search for compromise and support to maintain the continuity, and therefore the feasibility, of the project. As a result, the Agrocentrum Westpoort design failed to deliver the appealing image of innovation leap that is so essential for communication in the support and decision-making phase. It was precisely this lack of local support that caused both the Amsterdam Port Authority and the investors behind Ballast Nedam to withdraw their support for the project at crucial moments. The Zaanstad local council used the formal argument of the employment perspective.

An agropark is a knowledge-driven system innovation and makes a significant contribution to sustainable development.

Agrocentrum Westpoort was developed by the employees at Ballast Nedam into a value proposition which promised to be lucrative, in terms of return on investment and one-off benefits, resulting from the legislation concerning pig rights, but for which there appeared to be no investor. In that sense, this design too remained on the level of an invention.

The possibilities of Agrocentrum Westpoort as regards the triple p aspects of sustainable development are summarised by Breure *et al.* as follows:

> 'The cost-benefit balance drawn up for Agrocentrum Westpoort shows that there is economic space for giving aspects such as environment and animal welfare in the design a major role. Conversely, it is the attention to these aspects that may bring with it extra financial benefits. Size of scale is crucial here – in contrast to current assumptions – in order to combine a profitable operation with both favourable environmental effects and good conditions for the animals.'[250]

The design and implementation of system innovations like agroparks necessitates the participation of knowledge institutions, enterprises, NGOs and government (KENGi). It is a transdisciplinary process in which the explicit knowledge of knowledge institutions and the tacit knowledge of the other partners are developed in a process of continuous iteration. KENGi brokers act as the facilitators of this transdisciplinary process.

While some enterprises in the Amsterdam docklands were actively involved in the design process for Agrocentrum Westpoort, the key problem-holders, for which Agrocentrum Westpoort was generating a solution, namely the Dutch pig breeders, were not at the table. The question is, what would have happened if a project developer (for example, a construction company) with enough clout had decided to start the project, with or without other key parties in the port, and only after completion had looked for pig breeders to participate in the agropark? The question is less hypothetical than it seems, as the Biopark Terneuzen example will show. There is also a matter here of the scale on which many entrepreneurs in agroproduction are currently practising (that of the medium and small business) versus the level of scale on which the benefits of an agropark really become noticeable.

The hypothesis must also be modified in relation to another important KENGi party, the NGOs and in particular the environmental and animal welfare organisations. There is no point in discussing agropark designs with organisations which reject any further evolution in highly productive agriculture towards further scale increases and more intensification.

250 *Ibid.*: 51.

In all the decision-making concerning the realisation of the integral agroparks design, involving matterscape, powerscape and mindscape aspects, subjective decisions of individuals are ultimately decisive. The world of trustfulness (mindscape) is therefore dominant in the last instance.

As with the Deltapark design, it seems from the design for Agrocentrum Westpoort that agroparks are perfectly suitable from a physical perspective in the industrial and logistical setting of a dockland area and that many synergetic advantages can be obtained from this. In other words, the design could be realised in the area of matterscape. The financial-economic feasibility test (powerscape) also points to good possibilities. But as with the Deltapark example, the reason for not implementing the design is to be found in the area of mindscape: firstly, the subjectively experienced image problem in various versions, and secondly the myth on that same image, were the main reasons why decision-makers in the participating parties, both project developers, pig breeders and the most directly involved entrepreneurs, chose not to continue. The conflicts of interest that result from the innovation leap of the agropark and which threaten farmers, road hauliers and the meat industry, play out in the background. In the Deltapark example the internal rift among animal and environmental activists has already been mentioned. Some of them support the improvements in the area of the environment and animal welfare which could be accomplished by agroparks, while others radically reject any further development of the 'bioindustry'.

The position of the Amsterdam Port Authority was a specific example of this. They struggled with a temporary image problem because of the threatened disinvestment in the container terminal. With that in mind they could ill afford a second controversial decision on an anchor investment.

Social opposition to agroparks is directed primarily at intensive livestock farming.

This hypothesis, which was formulated in the Deltapark example, is not only confirmed in the Agrocentrum Westpoort example, it became mythically loaded. In the discussions during the design the hypothesis was highlighted by many stakeholders as a myth, although they didn't mention experiencing this resistance themselves. 'I can see the benefits but the citizens and the consumers don't want it.'

6.3.7 Conclusion

In the end, the co-design process on Agrocentrum Westpoort was only a learning process. It produced a lot of new knowledge about matter-, power- and mindscape aspects of agroparks, whereby the cost-benefit analysis by Ballast Nedam in particular showed spectacular cost savings for the agropark in comparison with conventional agriculture. The environmental advantages were also convincingly portrayed. But the most important lesson was in the design process itself. Only during this process did it become clear that active communication campaigns were necessary to take the public debate on intensive livestock farming off the defensive and to do something about the social stigma attached to intensive livestock farming. This stigma and the lack of well-formulated counter-arguments seriously hindered the design process, because it was constantly fed by a broadbased forum of KENGi parties.

6.4. New Mixed Farm

6.4.1 The design

Efforts to establish an agropark have been ongoing in North Limburg since 2001. This is now taking specific shape in the form of the New Mixed Farm (Figure 25). It is an initiative of three entrepreneurs in the Agricultural Development Area (LOG) in the Horst aan de Maas municipality in the North Limburg region and aims to be an example of modern agricultural management on all fronts. In combination with a biofuel plant, this operation next to the A73 motorway should soon provide space for 35,000 pigs and 1.2 million chickens. The sustainability scan, carried out by independent researchers for the Horst aan de Maas municipality, shows that this form of large-scale operation is sufficiently sustainable and innovative. New Mixed Farm is expected to open in 2010.

Figure 25. Bird's eye view (top left) and some artists' impressions of New Mixed Farm made by TRZIN, Amsterdam. Top right the chicken business and below the pig business.
KnowHouse (2006).

The idea can be traced back to the Regional Dialogue North Limburg[251], where it was developed as one of the most promising projects. Crucial for the further development of this project was the decision by Wageningen-UR, as compensation for the closure of some regional test stations in Horst, to invest in KnowHouse bv, a company that as a public/private liaison organisation[252], will act as a knowledge broker between the local agricultural businesses and the knowledge suppliers (removed to a distance because of the reorganisation). KnowHouse seized the idea for the agropark as one of its spearheads and has stood by the development ever since. In 2003 New Mixed Farm became an A-project of Agrologistics Platform. During a conference held by this Platform in October 2004, Minister Veerman assigned the project 'separate status':

> 'In order to ensure that existing legislation on sustainable initiatives in the Greenports or the agricultural development areas doesn't inhibit the process, Minister Veerman wants to experiment with a "separate status" for certain areas, whereby certain rules can be parenthesised, so to speak. (...) If this is successful, other projects can benefit from this in the future. This experiment may form the basis of a new, integral approach, whereby the rules reinforce instead of obstruct each other.'[253]

In 2004 it became one of the first Innovation Practical Projects of TransForum. Under the initiative of Agrologistics Platform and TransForum, a Steering Group was set up around the project team, with the entrepreneurs, a few scientists and the coordinator KnowHouse. Also included were representatives from Agrologistics Platform, TransForum, the LNV Ministry, Horst Municipality and a former director of the Limburg Environmental Federation.

The original plan[254], in which, as well as chicken and pig breeding, a mushroom business, and the Californië glass horticulture area were to work together and manage their waste and by-products via a joint materials factory, failed because of a lack of

---

[251] Van Mansfeld M., A. Wintjes, J. De Jonge, M. Pleijte and P.J.A.M. Smeets (2003). *Regiodialoog: Naar een systeeminnovatie in de praktijk*, Report 808, Alterra, Innonet, WISI, Wageningen, the Netherlands. See also Termeer C.J.A.M. (2008). Barriers for new modes of horizontal governance. A sensemaking perspective. In: Proceedings Twelfth Annual Conference of the international Research Society for Public Management Queensland University of Technology, Brisbane, Australia: 7.

[252] Leeuwis C.C., R.R. Smits, J.J. Grin, L.L.W.A. Klerkx, B.B.C. Mierlo and A.A. Kuipers (2006). *Equivocations on the post privatization dynamics in agricultural innovation systems. The design of an innovation-enhancing environment.* Transforum working papers. TransForum, Zoetermeer, the Netherlands.

[253] Platform Agrologistiek (2004). Ministers Veerman en Peijs zetten greenports op de kaart. In: Agrologistiek in uitvoering 2004, 5 oktober 2004, Amsterdam. Ministerie van LNV, The Hague, the Netherlands, pp. 4.

[254] For the original layout, a design was drawn up by Van Weel P. (2003). *Ontwerpen van geïntegreerde concepten voor agrarische productie in het kader van een agro-eco park in Horst aan de Maas*, Report PPO 588, Praktijkonderzoek Plant & Omgeving B.V. Sector Glastuinbouw, Wageningen, the Netherlands.

support from the new Californië glass horticulture area on the south side of the A73 motorway. Horticulturalists were afraid of the smell and dust and had difficulties with the poor image of intensive livestock farming. The project developer of Californië was worried about delays. The STIDUG rules to which he was subject, meant that he had to hurry. A slimmed down version was then worked on. In 2006 the participating mushroom cultivator went bankrupt and dropped out. The remaining entrepreneurs decided to continue work on a robust, open structure, enabling the parties, both on the supply and on the delivery side to link up with the core design, which from that moment was made up of a chicken business, a pig business and a manure processing installation, and for which a location north of the A73 in the LOG-Witveldweg was being sought.

Right from the presentation of the joint vision of the development by the LOG-Witveldweg[255], there were protests by local inhabitants. A few political parties and national NGOs joined these protests against the development of intensive livestock farming. The Project Group and the Steering Group explicitly chose to communicate openly with these protesters, consistently emphasising sustainable development, innovation, the open structure and the socio-economic development of the area.

In February 2008 the Witveldweg Area Vision was accepted by the Horst aan de Maas Municipal Council. This included a space for New Mixed Farm on condition that the sustainable nature of the business could be demonstrated. To that end an additional sustainability test[256] was conducted, which reached a positive conclusion and was adopted by the municipal council.

The Steering and Project Groups profile New Mixed Farm as the beginning of the cross-sector collaboration between businesses in agriculture. They accept that extreme care is required of them in its implementation and in dealing with job opportunities. Implementation is therefore proceeding slowly. The actual opening is not expected before 2010.

After that, not only will there be internal growth but also expansion possibilities in relation to other enterprises. The biofuel plant can also be expanded with gasification and/or composting, so that mushroom businesses can join. The link with glass horticulture can be made via the supply of electricity, heat and CO_2, and with non-agro functions via the processing of green waste. But the entrepreneurs are also finding that they are busy with new management techniques in the area of communication, environment and energy and want to catch the value of their knowledge at this level.

[255] Gemeente Horst aan de Maas (2007). *Informatiedocument gebiedsvisie Witveldweg.* Gemeente Horst aan de Maas, Horst aan de Maas, the Netherlands.
[256] Kool A., I. Eijck and H. Blonk (2008). *Nieuw gemengd bedrijf. Duurzaam en innovatief?* Blonk Milieu Advies, SPF Gezonde Varkens, Gouda, the Netherlands.

6.4.2 Matterscape

New Mixed Farm will initially be built at two locations in LOG-Witveldweg. The intended position at two locations in the LOG-Witveldweg is shown in Figure 26. The poultry business at one location, of 3.2 ha, will consist of six adjacent stalls with 70,000 mother animals, a hatchery, 1.2 million meat chickens and meat processing. The manure processing will be situated next to the chicken business on a parcel size of 2.5 ha. The piggery will be located on 1.9 ha with 2-storey stalls of 2,500 sows, 10,000 piglets and 20,000 meat pigs.

The industrial ecology takes shape among these operations, as shown in the diagram in Figure 27, whereby the pig and chicken businesses are connected via manure processing and composting. Various scenarios have been worked out for the entrepreneurs, and these will evolve in the course of time. In the basic scenario, a combination of co-digestion and then composting is used. Co-digestion allows for the pig manure to be mixed with organic waste (available in abundance from the direct surroundings of Horst). This co-digestion process releases a substantial amount of biogas which, with the aid of co-generation, can be converted into electricity, thereby releasing heat and CO_2. The CO_2 can potentially be used to 'fertilise' the adjacent

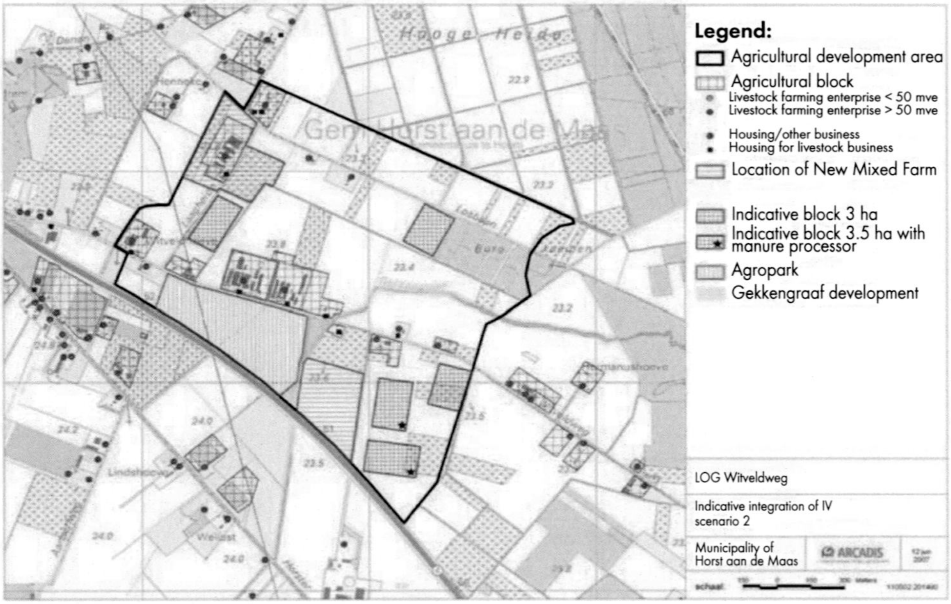

Figure 26. The LOG-Witveldweg within which are the locations for New Mixed Farm. *Gemeente Horst aan de Maas (2007: 17). Three scenarios are created in the document. In the scenario shown here, there is extra space for 'innovative agri-related businesses', which could set up in the area indicated as 'agropark'.*

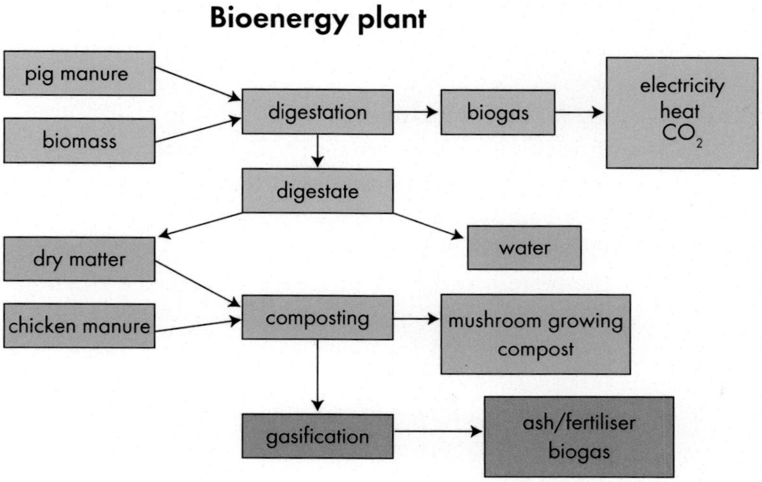

Figure 27. Process diagram of the basic scenario of industrial ecology in New Mixed Farm. Trudy van Megen, KnowHouse bv, project leader of New Mixed Farm.

greenhouse area (horticulturalists normally use natural gas to produce CO_2) and the heat can also be used in the greenhouses. Heat may also become available for other businesses and/or residents in the area. This biofuel plant will ferment between 60,000 and 120,000 tonnes of manure a year. The exact amount depends on the desire of surrounding livestock farmers to participate. Any remaining manure will go to a compost installation. This produces high-value and clean (mushroom) compost which can also be sold outside the agricultural sector and easily exported. The water is evaporated from the manure partly with the aid of some of the heat from co-digestion. This water can then be purified for use in, for example, the greenhouses. This part of the biofuel plant has a maximum capacity of 50,000 tonnes.

The aim is to proceed to the next step as quickly as possible and complete the digestion and composting with burning or gasification. More sustainable energy is released via this extra step, leaving only ash, which is very small in volume and even easier to sell outside the farming sector. However, the techniques in question are currently insufficiently reliable.

The following scenarios for the development of the biofuel plant have been worked out:

'...

- standard: co-digestion of available pig manure with co-products; separation of the digestate into thick and thin fractions; the thin fraction is purified along with the slaughter waste water and then released into the sewers;

the thick fraction is dried along with the poultry manure and burned or gasified;
the ash is exported.

- scenario 1a: big with technical drying:
in addition to the internally produced manure, pig manure from the surroundings is imported and processed;
the quantity of co-products increases proportionately.
- scenario 1b: big with biological drying:
the technical drying process is replaced by biological drying.
- scenario 2: no co-digestion (no purchase of co-substrates either), otherwise "standard".
- scenario 4: composting of the thick fraction into mushroom compost + poultry manure is processed into mushroom compost.
- scenario 5: export thick fraction:
the thick fraction of the digestate is dried + exported;
poultry manure is dried + burned/gasified.
- scenario 6: potting compost version:
after drying, the thick fraction of the digestate is used as peat substitute in potting compost;
poultry manure is dried + burned/gasified.'[257]

The different flows under discussion in the biofuel plant are quantified in Table 9 according to Kool (2008)[258].

> 'The establishment of the poultry business is accompanied by the decontamination of four existing locations (...) in North Brabant (...) situated in an extensification area around nature reserves; one site is situated in a pluri-activity area. Concentration of animals and the complete sealing off of the poultry business means some reduction in intensive livestock farming in extensification areas and some concentration in agricultural development areas.'[259]

Kool *et al.* write about the external environmental impacts of New Mixed Farm (NMF) in the Sustainability Test:

[257] Broeze J., S. Schlatmann, M. Timmerman, A. Veeken, L. Bisschops, D. Kragić, J. van Doorn and A. Boersma (2006). *Uitwerking ontwerp bioenergiecentrale ngb bij het integraal project transforum agro & groen: nieuw gemengd bedrijf*, Agrotechnology & Food Sciences Group, Wageningen, the Netherlands: 4.
[258] Kool A., I. Eijck and H. Blonk (2008). *Nieuw gemengd bedrijf. Duurzaam en innovatief?* Blonk Milieu Advies, SPF Gezonde Varkens, Gouda, the Netherlands: 11.
[259] Gemeente Horst aan de Maas (2007). *Informatiedocument gebiedsvisie Witveldweg.* Gemeente Horst aan de Maas, Horst aan de Maas, the Netherlands: 10.

Table 9. Flows in the biofuel plant according to Kool (2008).

Process	Input	Quantity (1000 ton)	Output	Quantity (1000 ton)
Digestion	pig manure NMF	38	digestate	106.5
	pig manure external	24		
	co-products	62		
Separation	digestate	106.5	thick fraction	27
			thin fraction	80
Ultrafiltration	thin fraction	80	aqueous fraction UF	72
			concentrate UF	8
Composting[1]	thick fraction	27	compost	20
	poultry manure	12		
	straw	6		
	concentrate UF	8		
Drying[1]	thick fraction	27	dry manure product	19.5
	poultry manureNMF	12		
	concentrate UF	8		
DAF/UASB	abattoir waste water	88	aqueous fraction SAW	88
Struviet Precipitation/SBR	aqueous fraction UF	72	drainable water	160
	aqueous fraction SAW	88		

[1] Composting or drying will be chosen in NMF. Both options are analysed in later calculations.

'The ammonia emissions from the NMF pig site are less than half the ammonia emissions from a conventional pig business. (..) The completion of the NMF will result in reduced ammonia emissions elsewhere in the Netherlands. (...) With the use of air washers, the NMF is making an important step in the direction of the policy objective of a 75-85% reduction by 2030. (...) The NMF processes all of its own manure (in addition to imported manure and co-products) and exports the residual organic fertilisers. This means a reduction in the animal manure surplus on the Dutch manure market. (...) The production of "sustainable" energy (energy from renewable sources) by means of manure processing is a positive and genuine difference compared to the reference. This difference is biggest if New Mixed Farm opts for composting (saving of more than 80%) instead of manure drying (saving of 59-66%). Composting process produces the most heat for supplying to third parties. An important distinction is that 90% of the energy generated in the NMF can be attributed to the co-products supplied. (...) The reduction in transport in the poultry section of NMF is a positive and genuine difference but makes less of an impact on the total energy savings (...) With the

> production of a considerable amount of "sustainable energy" (82% and 61% of the total energy consumption with composting and drying respectively), the NMF exceeds the Government's objective of supplementing 20% of total energy consumption with sustainably generated energy by 2030.(...)
> Within the area of greenhouse gas effects, the NMF makes a positive and genuine difference compared to the reference. This is due, on the one hand, to the generation of sustainable energy (and with it a reduction in the use of fossil fuels) and on the other, to the reduction in methane emissions created by storing pig manure. The NMF produces 30-40% fewer greenhouse gas emissions than the reference. (....)
> The expansion and rehousing of the pigs at the current pig location from 7,000 meat pigs to a total of 33,000 breeding and meat pigs in the NMF creates a comparable odour emission on site (at the NMF 2% lower than current situation). This is achieved by installing combi-airwashers which reduce the odour emissions by 75%. The establishment of the poultry business and the manure processing constitutes a drastic on-the-spot increase in odour emissions. The environmental impact of this will be judged by the legal norms in the Environmental Impact Assessment. In contrast, the odour emissions at the current poultry location will disappear. (...) Compared on the national scale, the low-emission accommodation of pigs and poultry in New Mixed Farm results in a drastic reduction in odour emissions. That means almost two-thirds less odour emission produced per unit of pigs (on the basis of the current and existing animal rights).'[260]

It is expected that the concentration of animal production in a closed operation combined with several links in the chain on site (feed preparation, hatchery, poultry slaughter, manure digestion, manure burning, composting) will help reduce veterinary risks[261]. Kool *et al.* have this to say on the matter:

> 'Thanks to the closed nature of the businesses (integration of several chain links) the business scores well (particularly the poultry section) on reducing the spread of and introducing pathogens. In the pig section there are a few aspects concerning introduction and spread which do not differ from the conventional sector. The NMF could make improvements in this area in order to obtain an optimal health score (...) The scale of the operation offers advantages because personnel can be separated and tasks (which risk spreading disease) can be automated. (...) The coupling of the poultry business with an abattoir offers an advantage in that the poultry can be

[260] Kool A., I. Eijck and H. Blonk (2008). *Nieuw gemengd bedrijf. Duurzaam en innovatief?* Blonk Milieu Advies, SPF Gezonde Varkens, Gouda, the Netherlands: 61-62.
[261] Gemeente Horst aan de Maas (2007). *Informatiedocument gebiedsvisie Witveldweg.* Gemeente Horst aan de Maas, Horst aan de Maas, the Netherlands: 25.

> cleared out by this abattoir in the event of an outbreak of a contagious disease.'[262]

Kool *et al.* also included aspects of public health in their sustainability test:

> 'The risk of airborne transmission of MRSA from the NMF to the outside world is very slight. (...) Reducing the risk of MRSA in the piggery is best accomplished by reducing the use of antibiotics. The pig section of the NMF begins with a higher health status, and good management will ensure that the use of antibiotics is reduced to a minimum. (...) As far as particulate matter is concerned, the picture is positive across all livestock farming in the Netherlands (and outside) because animals are housed in low-emission stalls and transportation is diminishing. However, at the LOG-Witveldweg site this (lower) emission is concentrated, which means an increase in particulate matter emissions locally.'[263]

6.4.3 Powerscape

Assigners for the design are the companies Heideveld, Kuijpers Kip and Christiaens bv. KnowHouse bv is the project leader of design and implementation and together with TransForum acts as a KENGi broker (Figure 28).

Broeze *et al.* have provided cost-benefit analyses for all scenarios created for the development of the biofuel plant. The key conclusions arising from this are:

> '...
>
> - The standard scenario is the most attractive in financial terms; the yield would be further increased with up-scaling (scenario 1a).
> - Subsidy (MEP or a more or less comparable subsidy) is required. (...)
> - In all scenarios the manure is kept off the Dutch manure market; doubling of the animal procurement rights is therefore possible (on basis of approval from LNV Minister Veerman). The profit to be made in this respect is considerable.'[264]

In their feasibility test, Kool *et al.* also looked at working conditions and employment and came to the following conclusions:

---

262 Kool A., I. Eijck and H. Blonk (2008). *Nieuw gemengd bedrijf. Duurzaam en innovatief?*, Blonk Milieu Advies, SPF Gezonde Varkens, Gouda, the Netherlands: 60.

263 *Ibid.*: 60-61.

264 Broeze J., S. Schlatmann, M. Timmerman, A. Veeken, L. Bisschops, D. Kragić, J. van Doorn and A. Boersma (2006). *Uitwerking ontwerp bioenergiecentrale ngb bij het integraal project transforum agro & groen: nieuw gemengd bedrijf.* Agrotechnology & Food Sciences Group, Wageningen, the Netherlands: 5.

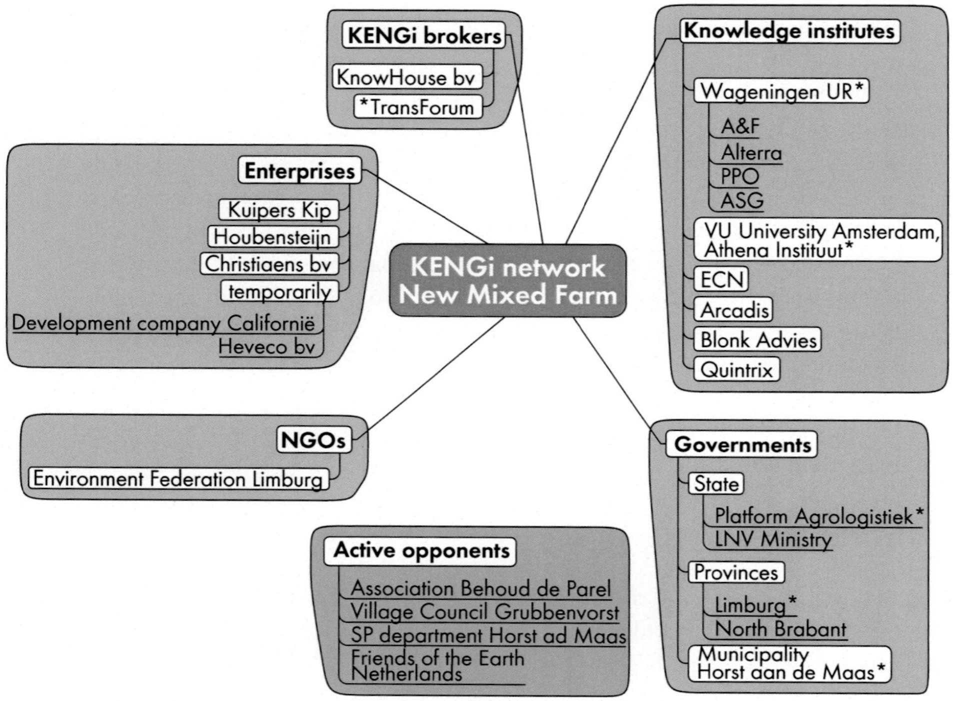

*Figure 28. Stakeholders, with substantial time investments in the project, some of whom have made a financial contribution to the design process (indicated with *).*

'New Mixed Farm has been set up in such a way as to avoid some of the serious and unpleasant activities that take place in conventional intensive livestock farming. For instance, manually catching meat chickens, hanging the meat chickens while alive on the slaughter hook, and cleaning out the poultry stalls. (...) Furthermore, the sheer size of the business will allow personnel to get involved in more specialised work. This can be a positive (more specialisation in a specific aspect of livestock farming) and negative (less alternation) factor. (...) The problems personnel encounter with particulate matter in the stalls will not differ greatly from those in a conventional stall.
(...) New Mixed Farm represents an expansion of economic activity in the Horst aan de Maas municipality, with the increase in the number of animals and by adding value to the end products: manure processing and the slaughter of poultry. The added value of the end products creates the most additional employment, i.e. 42-44 jobs.'[265]

---

[265] Kool A., I. Eijck and H. Blonk (2008). *Nieuw gemengd bedrijf. Duurzaam en innovatief?* Blonk Milieu Advies, SPF Gezonde Varkens, Gouda, the Netherlands: 63.

In February 2008 the Horst aan de Maas Municipality Council endorsed the area vision for LOG-Witveldweg and approved the content of the supplementary sustainability test. On the basis of this opinion, the municipal government decided on 8 July 2008 to join forces in the establishment of New Mixed Farm. Thus began the procedures for the establishment of the NMF. That process contains all kinds of conditions which the business still have to satisfy:

- The permit applications must be consistent and in compliance with the concept that was the subject of the research.
- The use/funding of the NMF concept will be tested by independent financial experts with regard to reality levels.
- The traffic situation and attention for the safety of cyclists.
- An external expert has been asked to test the technology used in the NMF on reality levels and practical feasibility and applicability.
- The design of the building and its suitability must be high quality and should fit in with the surroundings.
- The Environmental Impact Statement will have to show whether the business stays within the legal parameters concerning particulate matter and such like.[266]

The 'separate status' once assigned to it by Minister Veerman has in fact not been elaborated, with the exception of the LNV Ministry giving exemption for manure rights. But for the most part, the separate status is in reality precisely the opposite of what Minister Veerman intended. In view of the innovative character of New Mixed Farm, because of the large scale and as a result of the political discussions that accompanied the decision-making, all kinds of extra tests and occasional rules with which the project must comply, have been thought up. This means diverging, sometimes outrightly conflicting, demands from the various government departments. For example, at the national level the LNV Ministry requires proof that the technology to be used is actually innovative, but the VROM Ministry threatens to withhold approval for the stall system of the poultry breeder because it has not yet been used in practice. The local authority requires an extra test on the innovative and sustainable character, but at the same time it wants assurance beforehand on the financial feasibility and reality levels. The province declares a desire to work via three separate permit processes for the three parts of New Mixed Farm, but the municipality wants to settle the whole procedure via a combined application.

6.4.4 Mindscape

Attention is explicitly devoted to animal welfare in New Mixed Farm. Again the important improvements in relation to animal comfort apply by reducing or completely excluding animal transportation in the pig and poultry operations respectively. On

[266] Gemeente Horst aan de Maas (2008). Persbericht bij het verschijnen van de duurzaamheidstoets nieuw gemengd bedrijf. Gemeente Horst aan de Maas, Horst aan de Maas, the Netherlands.

top of that, manual collection of chickens is no longer necessary in the meat chicken business. As a consequence of legal directives, the pig business has announced that it will no longer castrate piglets, but instead send the pigs to slaughter at a lighter weight. This prevents the occurrence of boar taint and improves the quality of the meat. The piglets' teeth will no longer be filed either. Kool *et al.* (2008)[267] assessed the animal welfare in New Mixed Farm:

> '...positive compared to the conventional sector because of the serious reduction in animal transportation in the poultry chain, higher chance of survival among chicks in the stalls and animal-friendly slaughter of meat chickens (...); positive on the layout of the pig sties and poultry stalls (...). [but] no difference concerning the use of a conventional fast-growing chicken variety.'

With the participation of TransForum, an alternative model of knowledge generation was experimented with in the process of knowledge development. The scientists thereby defined their problem on the basis of practical questions from the entrepreneurs involved. The support from TransForum was also conditional on the monitoring and evaluation of the learning trajectory during the project, via a parallel project. In one of the outcomes of this evaluation process, the evolution of the collaboration between entrepreneurs and scientists was described:

> 'The initiative of the project was by three entrepreneurs who wanted to develop a new farm system (...). During the start-up of the project much attention was paid to creating an intensive bond between these entrepreneurs. The project leader explicitly utilized the setup of a business plan to pose questions that needed intensive deliberation. (...) Only when the entrepreneurs had formulated their first requirements for the design, did the project leader contract scientists that could develop the farm design. Because knowledge on farm designs is fragmented over several scientific disciplines, an interdisciplinary research group had to be formulated. (...) Although many meetings were organized within the scientist and entrepreneurs group, fewer meetings were organized between the scientists and entrepreneurs.(...) In the first meeting between scientists and entrepreneurs, the entrepreneurs posed a well thought through question to the scientists. This research question appeared to fit rather well within the scientific domain. The scientific project leader responded: "the strength of the question the entrepreneurs posed is that it has a central idea, but enough openness to execute explorative research". Therefore no adjustment was needed to create agreement about the research question. However, in order to assess what topics need further exploration, the scientists needed in-depth understanding of the requirements for the desired design as identified by

[267] Kool A., I. Eijck and H. Blonk (2008). *Nieuw gemengd bedrijf. Duurzaam en innovatief?* Blonk Milieu Advies, SPF Gezonde Varkens, Gouda, the Netherlands: 60.

> the entrepreneurs. These requirements did not solely concern technical desires, but also organizational and ethical choices, such as sustainability, innovativeness and open design, which influence the design options that are applicable. (...) As a result, the initial doubt of the entrepreneurs towards the scientists transformed into high expectations. From the description above, we see that rather than one joint heterogeneous learning trajectory of scientists and entrepreneurs, two homogeneous learning trajectories were formed. Both the entrepreneurs and the scientists formed their own learning community, following its own learning trajectory. It seems that these homogeneous trajectories facilitated the development of the project. The open and reflective attitude that emerged within these homogeneous trajectories helped to overcome expected challenges when participants from different institutional background collaborate. As such, a shared goal could be developed.'[268]

Subjective decisions that have played a crucial role in the development of the project are:

- The decision by different authorities at the national, regional and local level to support the project, actively with people power and subsidies but also verbally with, for example, the 'separate status' that LNV Minister Veerman declared in 2004. The 'separate status' did not lead to a legal exemption position but did oil the wheels in the decision-making process by organisations and individuals to support the project. It also resulted in the exclusive position that businesses processing manure have in terms of obtaining pig rights.
- Mushroom producer Heveco, who had to withdraw after bankruptcy.
- Glass horticulturalists and development company Californië, which pulled out because of expectations concerning particulate matter and odour, but also because of the image of intensive livestock farming.
- The Kuipers brothers, who played a really active role in the Agrologistics Platform and later in the public debate on New Mixed Farm, together with the other entrepreneurs involved.
- The entrepreneurs in their decision to pay extra attention to the design of the different operations buildings and to that end had a landscape design drawn up for New Mixed Farm.
- Director, board and employees of KnowHouse bv, in their decision to profile the project as the spearhead of innovation in North Limburg agrologistics.
- Institutes and employees from Wageningen-UR, in their decision to invest their own money to support the knowledge development of New Mixed Farm.

[268] Hoes A., B. Regeer and J. Bunders (2008). Transformers in knowledge production. Building science-practice collaborations. *Action Learning: Research and Practice* 5: 207-220.

The Project Group and the Steering Group have invested a lot in communication. Since 2006 a professional communications advisor has been involved in all project group meetings and public debates. In the public debate on New Mixed Farm there is a continuous exchange of arguments in the world of justice and trustfulness. On the one hand, the Project and Steering group and the various authorities, who both at the national, municipality and local levels, have made the political decision to support the project. The Project Group and the Steering Group explicitly chose to communicate openly, consistently emphasising sustainable development, innovation, the open structure and the socio-economic development of the area. After making explicit their decision, the authorities strictly abided by the formal requirements of the various procedures. On the other hand the action groups used a mixture of arguments from the world of truth, justice and trustfulness:

As an example of the latter, a citation of the open letter written by two members of the Working Group *Behoud de Parel* to Ger Driessen, member of the provincial government:

> 'This is not only about the arrival of New Mixed Farm, but also the arrival of a soil processing station on the Maas, the expansion of the auction market, the development of Trade Port North, the development of the Californië horticultural area and the Floriade. Grubbenvorst will be surrounded by all kinds of industrial activities, all of which will be accompanied by a huge volume of traffic. All these activities therefore constitute a source of particulate matter. (...) Of course the people of Grubbenvorst are getting emotional about this. Emotional, because they are under the impression that a number of ambitious leaders at the local, regional and national level are setting developments in motion over their heads, and that if realised, these will have an enormous impact on the environment and will considerably affect the quality of life. Wouldn't you get emotional if you were forced to look on helplessly while fundamental values of openness, peace and healthy air which the countryside offers, were undermined?'

6.4.5 Work process

New Mixed Farm can be regarded as an innovation in the sense that on the basis of a combination of inventions – within the chicken and pig operations and the invention of manure processing – an integrated value proposition was defined by the entrepreneurs, who began with preliminary investments and made substantial headway in obtaining the necessary permits and government authorisation.

But the project still has to prove its 'right to exist' as a system innovation. For that, indeed, 'approval' is required in the various currencies of the KENGi parties involved, and this can only come from a realignment of their mutual relationships. The characteristics of these relationships that can already be observed, are:

- the intensive involvement of the KENGi brokers KnowHouse bv and TransForum;
- the long-term investment of knowledge institutions in this innovation process;
- the proactive attitude of the entrepreneurs towards the other KENGi partners, underpinned throughout the process by a well thought-out communication strategy;
- the status of the project in the government's innovation strategies at the national, regional and local level.

As far as three of the four KENGi groups are concerned, it can already be established that in terms of their own currency they regard the project as an innovation: in addition to the entrepreneurs, this includes the knowledge institutions. Looked at from a positive perspective, despite all the ensuing conflict, the support that the project has received – at least in a verbal sense – from politicians and public servants, may also be seen as an attempt to let this innovation flourish in terms of permit policy and legal rules.

Only a few NGOs (local mayor initiatives, backed by a few political parties and rural environmental and animal welfare groups) rejected the project. But where it really counts, i.e. in the political deliberations of the municipality council, they are in a minority.

The breeding ground for establishing a network around New Mixed Farm was the Regional Dialogue North Limburg.[269] An integrated agropark was developed as an invention by collaborating entrepreneurs and employees from knowledge institutions and government as one of the design projects in the Regional Dialogue. This invention, which was the breeding ground for the innovation project, had therefore already been made possible on the basis of a KENGi process. The Regional Dialogue had all the characteristics of a co-design according to De Jonge (Section 5.3).

---

[269] Dammers E., F. Verwest, B. Staffhorst and W. Verschoor (2004). *Ontwikkelingsplanologie. Lessen uit en voor de praktijk.* Ruimtelijk Planbureau, NAi Uitgevers, The Hague, the Netherlands; Van Mansfeld M., A. Wintjes, J. De Jonge, M. Pleijte and P.J.A.M. Smeets (2003b). *Regiodialoog: Naar een systeeminnovatie in de praktijk*, Report 808, Alterra, Innonet, WISI, Wageningen, the Netherlands. See also Termeer C. (2006). *Vitale verschillen. Over publiek leiderschap en maatschappelijke innovatie. Oratie, 7 december 2006.* Wageningen Universiteit en Researchcentrum, Wageningen, the Netherlands, 48 pp. In her description of the Regional Dialogue as a breeding ground for projects like NGB, Termeer ignores the role of the knowledge institutions in starting up this regional development process. It was in the context of the preparation on the North Limburg pilot, in the context of the reconstruction, that staff from Wageningen-UR went to talk with the directors at Rabobank Maashorst, and thereby started the Regional Dialogue. The first step in that process was many bilateral talks with other KENGi parties, which then jointly set up the Regional Dialogue Foundation.

The process from the beginning of the New Mixed Farm innovation process, can be characterised as the implementation phase of the project. But in itself and anew it retains the character of a co-design. By 2004 there were already several scientific publications substantiating the appeal of an agropark in North Limburg[270]. But only with the efforts of the innovation brokers KnowHouse and TransForum was the step made from invention to system innovation.

KnowHouse itself, although set up by Wageningen-UR as an experimental alternative to traditional, regional test stations, functions in the region as a programme agency, whose primary activity is to informally coordinate all the innovation experiments being undertaken[271]. In addition to provinces, cities and knowledge institutions, various businesses in the region have a share in the operation.

The KENGi parties (Wageningen-KnowHouse, TransForum Agrologistics Platform, city, province, state, and Environmental Federation Limburg (not directly but somewhat remotely via their former director)) intensively guided the work process in the project group via the New Mixed Farm Steering Group.

The whole work process in the research by design of Regional Dialogue and New Mixed Farm can therefore be summed up as a transdisciplinary process, in which, via consecutive iteration attempts, individual knowledge of the participants from the KENGi parties can time and again be anchored via scientific tests and internalised by all these parties. Then the process begins another cycle. From 2004 this work process as a whole was also monitored and evaluated as a so-called second-order learning process by the staff at the Athena Institute[272].

New Mixed Farm was frequently discussed in the Agrologistics Community of Practice that operated from 2003 onwards[273]. These meetings gave rise to a second project entitled 'California Streaming', which aimed to shape the communication on New

[270] For example, Van Weel P. (2003). *Ontwerpen van geïntegreerde concepten voor agrarische productie in het kader van een agro-eco park in horst aan de maas*, Report PPO 588, Praktijkonderzoek Plant & Omgeving B.V. Sector Glastuinbouw, Wageningen, the Netherlands; Van Eck W., R. Groot, K. Hulsteijn, P.J.A.M. Smeets and M.G.N. Van Steekelenburg (eds.) (2002). *Voorbeelden van agribusinessparken*, Report 594, Alterra, Wageningen, the Netherlands.

[271] The importance of a programme agency was highlighted in the Regional Dialogue as an organisation that should safeguard the synergy between the various projects that arose from the dialogue over the course of their development. See Van Mansfeld M., A. Wintjes, J. De Jonge, M. Pleijte and P.J.A.M. Smeets (2003). *Regiodialoog: Naar een systeeminnovatie in de praktijk*. Report 808, Alterra, Innonet, WISI, Wageningen, the Netherlands.

[272] Hoes A., B. Regeer and J. Bunders (2008). Transformers in knowledge production. Building science-practice collaborations. *Action Learning: Research and Practice* 5: 207-220..

[273] Kranendonk R., F. Gordijn, P. Kersten and P.J.A.M. Smeets (2003). *Cop agrologistiek; verslag van werkatelier (6-7 november, Venraij)*, Alterra/WING, Wageningen, the Netherlands.

Mixed Farm with the parties in the surroundings as an iterative design process[274]. This project was also supported by the Agrologistics Platform and TransForum. But the focus was on a higher spatial level of scale, when, with the publication of the most recent National Policy Document on Spatial Planning with its allocation of the Venlo region as one of the Greenports[275], the region set about looking for a way to flesh-out this identity. In the same period the Venlo region was given the task of organising Floriade 2012[276], increasing the urgency to implement Greenport Venlo. At the same time, partly inspired by the developments in the Venlo region, the focus in the Agrologistics Community of Practice also shifted – from the local level, at which work was being done on the agroparks – to the regional level, at which the discussions on Greenport development and agrologistics were being held[277]. The TransForum project then acquired the name 'Streamlining Greenport Venlo'. This meant that a structured participatory plan around the suitability of New Mixed Farm, which was the original aim of this second project, was not drawn up, and the residents from the centre of the village of Grubbenvorst were first involved in the plans via the formal lines of the reconstruction process around LOG-Witveldweg. That can be viewed as a crucial failure in the very carefully prepared design process around New Mixed Farm. But, on the other hand, the question remains whether it would have been possible to have a meaningful discussion with these principle opponents, who operate from a 'not-in-my-backyard' perspective.

In its opinion on mega-stalls, the Council for the Rural Area summed up the field of tension succinctly:

> 'The mega-stall discussion is not confined to pig breeding but also extends to the poultry sector, dairy farming and the increased scale of glass horticulture. This discussion may throw the peaceful co-existence of these two different claims on the countryside into sharp relief. Most of the Dutch population wants to use the rural areas for recreation, living, relaxation and rest. The quality of this Arcadian landscape is becoming an increasingly important part of the quality of the living environment. From the moment it becomes

[274] Kranendonk R., P. Kersten, P. Smeets and F. Gordijn (2004). *Cop agrologistiek; verslag van werkatelier (7-8 april, Zaandam)*, WING / Alterra, Wageningen, the Netherlands.
[275] Ministerie van Volkshuisvesting Ruimtelijke Ordening en Milieu (2004). *Nota ruimte, ruimte voor ontwikkeling*. SDU Uitgevers, The Hague, the Netherlands.
[276] The Floriade is also one of the ideas launched in the Regional Dialogue North Limburg. See Van Mansfeld M., A. Wintjes, J. De Jonge, M. Pleijte and P.J.A.M. Smeets (2003). *Regiodialoog: naar een systeeminnovatie in de praktijk*, Report 808, Alterra, Innonet, WISI, Wageningen, the Netherlands.
[277] Kranendonk R., P. Kersten and P. Smeets (2005). *Cop agrologistiek. Verslag van cop bijeenkomst (14 december, kasteel Groeneveld Baarn)*, Alterra, Wageningen, the Netherlands; Kranendonk R., P. Kersten and P.J.A.M. Smeets (2006). *Cop agrologistiek verslag van de copbijeenkomst (14 juni 2006, living tomorrow Amsterdam)*, Alterra/WING, Wageningen, the Netherlands; Kranendonk R.P., P.H. Kersten, P. Smeets and F. Gordijn (2005). *Cop agrologistiek: verslag van masterclass (12 januari 2005, Den Bosch)*, Alterra/WING, Wageningen, the Netherlands.

difficult to distinguish the landscape (locally) from an industrial terrain, because of the trend towards concentration and up-scaling, other ways of bringing business to the landscape in a wider region may dry up and the quality of the living environment will deteriorate for large groups in the community.'[278]

The process that took place around the permit procedures for NMF, can subsequently be typified as the sum of what Termeer[279] describes as two extremes in the way authorities can deal with variation.

'The first extreme concerns wanting to reduce variety by wanting to control and manage it. (...) Uncertainty and crisis reinforce the political pressure to come up with a picture and to fixate on it. A concept for Greenports then emerges, a picture of sustainable agriculture (...) The paradox of control is that it seems manageable. In practice, it is often a time bomb. Development is blocked when limits are imposed from outside, when there is a lot of variation. A lot of energy is required to preserve the stable situation. People are frenetically trying to maintain the status quo, while at the same time realising that this is not possible. An example of this is the rejection of customised legislation because that would bring down the entire policy house of cards, so carefully built up. (...) For example, unlike in the world of business, the emphasis on justification within the government has not resulted in more scope for, or more air and space for innovation, but in more rigidity (...).

The second extreme is the bringing together of differences with the express aim of reaching a consensus. That results in the caricature of talking for just long enough so that compromise is reached which everyone can live with but nobody is really happy with. As a result nobody appears to be capable of judging that the outcome is pretty much "negotiated nonsense" or simply nonsense (...) There is then the risk that new variety is frantically held at bay for fear of having to spoil the nicely forged compromise. A situation that is aptly described as "escalated harmony". (...) The biggest danger for democracy is not the eternal discord, but the choking consensus.'

Despite Minister Veerman's pledge to create a 'separate status' for the NMF in the legislation, all permit procedures with regard to the Nuisance Act, activity allocation models, etc., will be meticulously scrutinised. Since the project was partially financed by TransForum, the intrinsic criteria for making a contribution to sustainable

---

[278] Raad Landelijk Gebied (2008). *Het megabedrijf gewogen. Advies over het megabedrijf in de intensieve veehouderij*, Report 08/03, Raad Landelijk Gebied,, Amersfoort, the Netherlands.

[279] Termeer C. (2006). *Vitale verschillen. Over publiek leiderschap en maatschappelijke innovatie. Oratie, 7 december 2006.* Wageningen Universiteit en Researchcentrum, Wageningen, the Netherlands: 6.

development applied here, as well as the complex administrative rules, inspired by the BSIK criteria, which these innovation programmes have to meet.

When the decision was finally made on the Witveldweg location, previously recommended as an Agricultural Development Area in the context of the Reconstruction Act, the highly 'negotiated nonsense' nature of this legislation became apparent. While the LOGs were meant to provide space for the expansion aims of intensive livestock farming in the Netherlands, residents in the area, supported by rural action groups, led intensive campaigns in opposition to the new establishment. In the end, the area vision of Witveldweg was accepted in the local council with the smallest possible majority (11 to 10 votes).

6.4.6 Testing of the hypotheses

An agropark realises lower costs, greater added value and lower environmental pollution per unit of output and space.

Compared to the other agroparks discussed in this publication, the ambitions of New Mixed Farm in terms of production-ecology are modest. In fact, the efforts were confined to scale increase, chain integration and manure processing. Because the integration with vegetable production was abandoned, after the initial ambitions to include glass horticulture in the plan, and thereafter to collaborate extensively with horticulturalists in the Californië glass horticulture area. The reasons for this modesty lie primarily in the world of justice and trustfulness; physical synergy advantages, that could have realistically been obtained, were lost. But the decision was dictated by glass horticulturalists who had no wish to be associated with the image of intensive livestock farming; who were afraid of particulate matter and odour pollution; and who preferred to reap the benefits of a CHP installation independently.

That doesn't mean it will remain like this. Heat and CO_2 will be offered to glass horticulture businesses in the neighbouring Californië and there are expected to be buyers for it there too. A structural link with mushroom cultivators is logical via the composting of the digestate from manure processing. This is where the New Mixed Farm entrepreneurs see the biggest opportunities for short-term collaboration. In their view that also has something to do with the fact that, like intensive livestock farming businesses and unlike glass horticulturalists, mushroom cultivators and composters are holding a similar social debate with their environment on odour, environmental problems, etc.

But the hypothesis is confirmed by the New Mixed Farm example. All publications and approval procedures which assess the intended effects of this modest step in production-ecological terms, confirm the enormous possibilities for reduced costs, more added value and less environmental pollution.

An agropark can only come into being on the basis of an integral design of matterscape, powerscape and mindscape at both the global scale of Intelligent Agrologistics Networks and the local scale of a landscape.

This hypothesis is also confirmed, but the New Mixed Farm work process shows that the design process is relatively simple with regard to the innovation in the space of flows. The complexity arises when it has to become part of the space of places.

In the context of the Regional Dialogue North Limburg, the New Mixed Farm project is a textbook example of what is regarded in this publication as a well-implemented participatory, transdisciplinary landscape design, as a co-design for an agropark. This is so by definition, because the models have been developed in the same Regional Dialogue.

We do see that in practice, from regional to project design and to implementation, the successive process steps were repeatedly followed, so that the introductory phases of the network building, problem formulation and joint fact-finding, naturally received less focus in the second and subsequent iteration attempts. But the state of affairs during the New Mixed Farm project reveals how design phases, the search for support, and decision-making go in cycles; first among the entrepreneurs themselves, and then in discussion with the authorities and finally with the authorities and residents. In that sense, the design is not finished, and will never be finished, because the participating businesses themselves never stop developing and also because other businesses will join them.

We can conclude from the example of New Mixed Farm that the task of implementing a system innovation in the 'Crystal Palace' of at least the Dutch part of the Northwest European Delta Metropolis is very complex, and may well be impossible. A design for this purpose must be flexible, because during the journey it has to adapt to rules that are changed or come into existence, and therefore must be developed partly as powerscape, precisely on the basis of the same system innovation. The same applies to the anecdotes and other subjective expressions in the mindscape, which on the one hand are used as a motive, as a utopia, for the entrepreneurs and other system innovators, and on the other hand give opponents a reason to resist the innovation.

The position of the government appears to be crucial: Whatever happens, it has to arm itself against the 'inhibiting context' (see Section 2.2) in which it operates by definition as a government of the 'Crystal Palace'. In the description of New Mixed Farm there are three positions that the government can take. The first two are those described by

Termeer as extreme control or extreme consensus leading to escalated harmony. They lead to nothing. The third is that of 'separate status', as defined by Minister Veerman, in the sense of 'momentarily put the rules aside'. Only with this third position is it possible to put system innovation into practice.

An agropark is a knowledge-driven system innovation and makes a significant contribution to sustainable development.

This hypothesis is also confirmed in the sense that all *ex ante* sustainability tests to which the New Mixed Farm was subjected, point to major progress which the design will yield in many aspects of people, planet and profit. But it must also be noted that the level of system innovation has not yet been reached, because New Mixed Farm is still not operational.

The New Mixed Farm project has been tested on multiple occasions and by different organisations on its contribution to sustainable development. This contribution has not gone unchallenged in the tests. What remains is the contrast between the improvement that will occur at the local, regional and national level, and the increase in environmental pollution which will take place because of the concentration on the local scale.

The design and implementation of system innovations like agroparks necessitates the participation of knowledge institutions, enterprises, NGOs and government (KENGi). It is a transdisciplinary process in which the explicit knowledge of knowledge institutions and the tacit knowledge of the other partners are developed in a process of continuous iteration. KENGi brokers act as the facilitators of this transdisciplinary process.

The permanent designs in matterscape, powerscape and mindscape of New Mixed Farm rests on the continuous exchange of knowledge of the four KENGi parties. In the positive but also in the negative sense, because opponents also introduce new arguments that need to be dispelled or that will lead to the design being modified. As far as that is concerned, there is definitely a process of transdisciplinary knowledge development taking place. Interestingly, the evaluation of the learning process, carried out by the Athena Institute, clearly demonstrates that the scientists' agenda is often determined by problems in the real-life operations of the entrepreneurs. In other words, the knowledge in the knowledge-driven system innovation does not necessarily have to come from a scientific institute. KENGi brokers are closely supervising the development process of New Mixed Farm. TransForum is also conducting ongoing monitoring and evaluation of the learning process in parallel to the work process.

In all the decision-making concerning the realisation of the integral agroparks design, involving matterscape, powerscape and mindscape aspects, subjective decisions of individuals are ultimately decisive. The world of trustfulness (mindscape) is therefore dominant in the last instance.

Of the seven works examined in this publication, New Mixed Farm is the clearest example of an innovation process in which the decision-making is carried out virtually separately from the world of truth (matterscape). The physical arguments in the world of truth, which indicate the innovative nature of the project, are actually clear and undisputed right from the start of the project. They have been confirmed on numerous occasions in recent years in the research by design itself and thereafter via various tests for sustainable development: via the procedures to approve the project in the Agrologistics Platform, at TransForum and recently again, via the sustainability test, that the government of Horst aan de Maas had performed.

But it is in the world of justice (powerscape) and trustfulness (mindscape) that serious discussions are being held. Firstly, between the entrepreneurs themselves, focusing on their collaboration model, as well as on the damage done to the image, that the glass horticulturalists believed would be incurred right from the start, if they became too closely involved with intensive livestock farming. Then in the public domain on the basis of formal procedures which had to run their course in relation to reconstruction, activity allocation model, Environmental Impact Assessment and Nuisance Act, and informal, local extra procedures devised purely for this project.

The crucial arguments in this discussion, which will also continue to play a role in the future development of New Mixed Farm, are:

- On the local level: the question of whether LOG-Witveldweg (but essentially the whole region between Venlo and Horst) should be regarded as a green open space, or whether it is actually a long-term expansion area of the agrologistics complex around Venlo. And in the case of the latter, how to approach the residents of the village of Grubbenvorst on the plan damage that will occur as a result of this activity-allocation modification, or on the loss of their residential quality of life. This is a discussion, in which the individual, subjective, experience of the local inhabitants predominates.
- On a regional level: how to go about reconstructing not only the intensive livestock farming but also the other intensive forms of agroproduction. Since the scale will continue to increase, individually distributed businesses will continue to hark back to the 'mega-stall discussion'[280]. This will only be resolved when there is an actual decision by these businesses to opt for spatial concentration, as happened in the middle of the last century with broadly based industrial establishments in

[280] College van Rijksadviseurs (2007). *Advies megastallen*, College van Rijksadviseurs, The Hague, the Netherlands.

the cities. In this respect, spatial planning decisions are under discussion, which are ultimately legally anchored in activity allocation plans at the level of local policy. But in the example of New Mixed Farm, this discussion is being had at the intercommunal level between Horst aan de Maas and Venlo and actually on the higher scale level of the Peel region, that comprises North Limburg and East Brabant[281].

- On a national (EU-regional) level: the question as to whether industrial agroproduction, which is currently one of the most innovative sectors in the Dutch economy, will continue to be a spearhead in the long run, or whether this will end up being a small-scale development near the cities and otherwise huge open agricultural spaces situated mainly outside the Netherlands.

The decision-making in the Horst aan de Maas city council also reveals that such a policy body can transcend the level of extreme consensus leading to the escalated harmony mentioned by Termeer. With a close majority the council decided to press on with the innovation, despite the protests of many, thereby implementing the 'separate status' referred to by Minister Veerman. The original hypothesis, which on the basis of this example was tested not only on a virtual design but also on the passage of that design through the maze of rules and regulations which accompany the implementation of innovations in the Netherlands, should be reformulated in that sense:

> In all decision-making on the implementation of the integral agroparks design, with matterscape, powerscape and mindscape aspects, arguments from the world of justice and trustfulness take precedence over arguments from the world of truth.

> Social opposition to agroparks is directed primarily at intensive livestock farming.

This hypothesis can only be confirmed on the basis of the original design. Once glass horticulture and mushroom cultivation were removed from the plan, New Mixed Farm consisted exclusively of intensive livestock farming. What is remarkable in this example is that in a number of cases even the fellow entrepreneurs from glass horticulture adopted this argumentation in their decision not to participate.

[281] In the mid-1990s there were far-reaching plans to establish just one auction sale for the whole glass horticulture area on both sides of the Dutch-German border. This discussion came to nothing. But it does prove that the German area around Straelen should also be included in the plans for the Peel region.

6.4.7 Conclusion

New Mixed Farm is the most far-advanced agropark in the Netherlands in which intensive livestock farming is involved and in which all other sectors have withdrawn, partly for the image problem that intensive livestock farming has in this country. The long development process shows how difficult it is for small and medium-sized entrepreneurs to invest for the long term in complex system innovations and also how difficult it is for the government to facilitate such innovations with rules and regulations which are largely instrumentally focused on regulating what is already in existence and has been worked out in the minutest detail.

New Mixed Farm is an excellent example of a regional design. It concerns the spatial concentration and synergy development of existing businesses in the North Limburg and East Brabant Peel region. The Limburg provincial council and local city councillors play a dominant role, as does the KENGi broker KnowHouse, which has a regional focus.

And yet the appeal of the agropark concept in the Community of Practice, which grew up out of this project, continues to go transcend regional ambitions. It was the KENGi network around New Mixed Farm that in October 2006 laid the foundations for the design for Greenport Shanghai (Section 6.7).

6.5 WAZ-Holland Park

6.5.1 The design

Wujin Polder is part of the outer area of Changzhou in the Chinese province Jiangsu (Figure 29). Changzhou is one of the cities in the Southern Yangtze Delta Metropolis, with a population of 3.5 million. Wujin Polder was developed by the Wujin Agricultural Zone Authority (WAZ-A). The Master Plan for the polder focused on the parallel development of low density residential areas; recreational areas with a regional function; economic zones with offices and high-tech industry; and a high-tech agricultural development zone to act as a regional exemplar (Figure 30).

In September 2004 WAZ-A invited a consortium headed by Alterra, part of Wageningen-UR, to draw up a conceptual Master Plan for the development of the agricultural zone. That plan, WAZ-Holland Park, was delivered at the end of 2004. It comprised an agropark populated by Chinese and foreign companies; a Central Processing Unit (CPU) – for the joint management of waste and by-products; a recreational area with the theme of Chinese and Dutch agriculture; and a market square, for selling agropark products and as a meeting area (Figure 31).

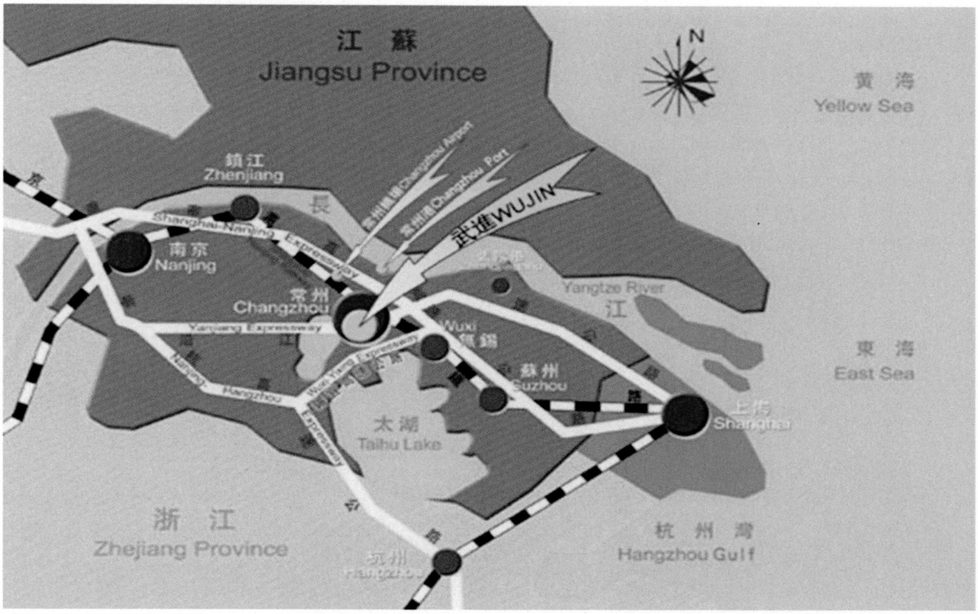

Figure 29. Changzhou in Jiangsu Province.
Smeets et al. (2004b).

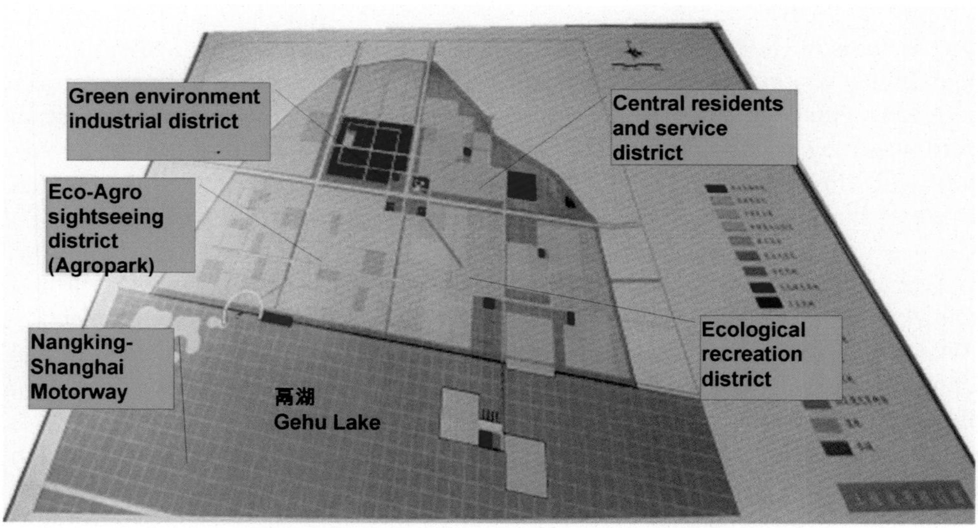

Figure 30. Wujin Polder.
Smeets et al. (2004b).

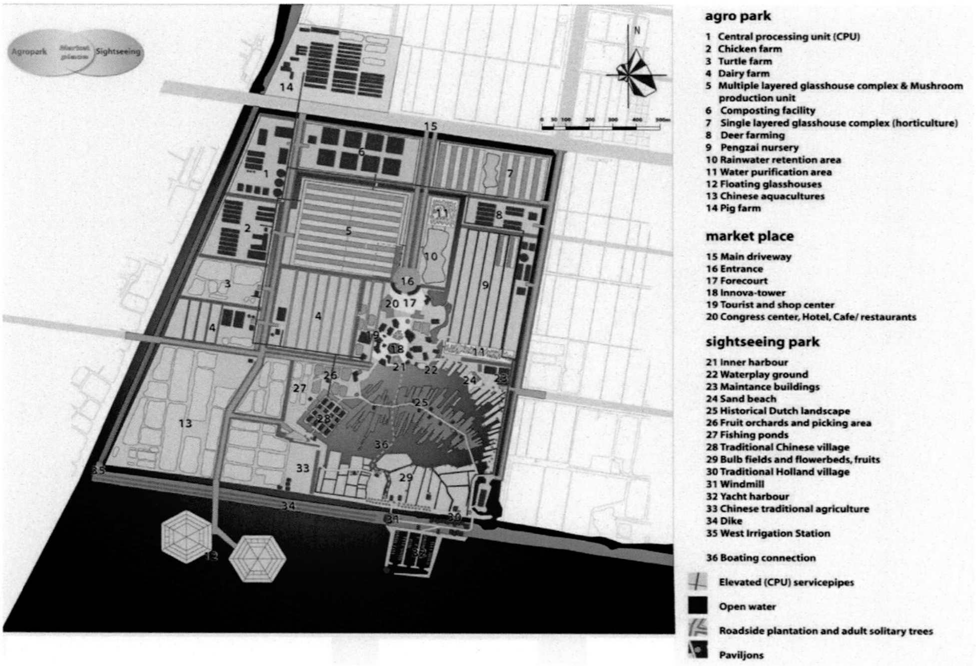

Figure 31. The Master Plan for WAZ-Holland Park.
Smeets et al. (2004b).

On the basis of the Master Plan, discussions were then held on implementing the plan. In that context, in April 2005 three WAZ-A employees took a look at examples of Dutch highly productive agriculture during a multi-day excursion; and a delegation from Wageningen-UR and KnowHouse returned to China for discussions in September 2005. But the talks ended in stalemate when the Chinese demanded that Dutch investors should contribute financially to the further development of the plans.

In February 2006 a meeting took place between WAZ-A, Wageningen-UR and KnowHouse, attended also by a representative from TransForum. At that meeting WAZ-A announced that a feasibility study on the CPU would be conducted by Tongji University in Shanghai. The Dutch organisations then offered to organise a delegation of potential investors on the basis of the feasibility study. But the feasibility study was not carried out. However, WAZ-A began implementing parts of the plan for which Chinese investors had been found (Figure 32). The principal victim of this threatens to be the CPU, for which neither Dutch nor Chinese investors have taken responsibility. There is still a possibility that the CPU will be taken up later as a joint project by entrepreneurs.

In the Netherlands, the WAZ-Holland Park design acted as a catalyst in mobilising the KENGi parties who were interested in the development of agroparks. For Wageningen-UR the design helped to support acquisition activities in the Netherlands, China and India from 2005 onwards.

Beginning in 2007 the infrastructure of the park for WAZ-A was put in place. A start was also made to the construction of the recreation area. The agricultural development area was further fleshed-out from the bottom-up, using the Master Plan as a rough guideline.

Two Chinese companies set up in the locations designed for them in the polder. One was a tree cultivator with a laboratory for tissue culturing, which has set up its operation on more than 20 ha of the polder, and has the option to expand to 50 ha (partly due to the lack of other investors). The other was a dairy company with 130 milkers on 36 ha and a dairy factory, which in addition to milk from the polder also planned to process milk from the surroundings on a large scale. A manure processing installation is linked up to the dairy business, producing electricity from the digestion of the dry fraction. The digestate is converted via vermiculture into organic fertiliser for the artificial grass field on which the liquid manure from the milkers would also be spread (Figure 32).

WAZ-Holland Park implementation

Google Earth image (taken on 11 June 2008) of Wujin Polder. The polder is framed in red. A recent satellite photo taken on 28 December 2006 (the grey strip in the left section of the image) shows the new dairy building of Mr. Wu, one of the investors in the Wujin polder. In the inset bottom left, a close-up of this enterprise. Below a photo of the dairy enterprise, taken on 23 March 2007, during a visit to Wujin Polder.

Figure 32. Implementation of WAZ-Holland Park.

A new neighbourhood was built in the central residential area of Wujin Polder for the subsistence farmers living in that area. They have effectively been able to hold on to their farmland. Some of them work in the new businesses.

6.5.2 Matterscape

The most striking feature of WAZ-Holland Park is the combination of the agropark with a market square and recreational area. Around the market square are the main entrance for visitors to the park and offices, horeca businesses and exhibition spaces (Figure 31). The Sightseeing Park has been designed as a recreational area for local inhabitants of Wujin Polder as well as for visitors to the agropark from elsewhere.

Animal production (dairy, chicken, aquaculture); vegetable production (glasshouse vegetables, tree cultivation); and mushroom production are planned for the agropark. Directly north of the park the large-scale intensive pig farm can link up to the CPU facilities. In the deer farm, blood from the deer horns is used to manufacture a product for the Chinese traditional medicine market.

The main processing of waste and by-products takes place in the CPU. Here, manure and biomass from the park and the wider surroundings are stored and co-digested. The CPU is linked up to a regional waste incinerator in Changzhou. The biogas from the digester is converted into electricity, CO_2 and heat to be used in the park. The CPU also has a composting installation.

No calculations were made in the Master Plan in relation to the resulting environmental emissions, energy savings and veterinary consequences.

6.5.3 Powerscape

The job of developing the Master Plan for WAZ-Holland Park was given to Wujin Agricultural Zone Authority (WAZ-A). This body acts as a project developer, using the plan to present the investment possibilities to external investors, and in case of interest, helping them to turn their investment wishes into reality. What is noticeable is that WAZ-A offers integral and turnkey support in the case of investments in the adjacent ICT park, but only the basic infrastructure in the agropark. Joint facilities like the CPU have to be set up by those external investors together.

Wageningen-UR invested its own money in the design of WAZ-Holland with Grontmij as the subcontractor. KnowHouse and Green Space and Agrocluster Innovation Network and a member of the Agrologistics Platform (all active participants in the Agrologistics Community of Practice) took part unpaid in the design process. On the Chinese side, in addition to the WAZ-A, staff from the Wujin district, an urban district of Changzhou, were also involved. One of the companies taking part in the New Mixed

Farm consortium, Christiaens bv, sent a member of staff who made an important contribution to the CPU design. There were no NGOs involved in the design phase.

The design did not include a profit forecast.

6.5.4 Mindscape

No attention was paid to animal welfare or working conditions in the design. A knowledge network, which included WAZ-Holland Park, was globally implemented.

The Chinese project developer seemed to be unfamiliar with the agricultural industry. WAZ-A works mainly on civil engineering and industrial project development. This relative unfamiliarity with the material made them reluctant to go as far with high-risk investments as they would have in the industrial, utility or residential buildings, where they often deliver turnkey projects and earn back their investments via direct resale or long-term lease contracts.

For the development of agroparks, the Chinese start with a consortium development similar to that for industrial investments. They expect a foreign partner to make a substantial investment in China and ignore the fact that those potential investors are small and medium-sized entreprises in China as well as in the Netherlands, and have neither the assumed desire nor the capacity to invest in this way.

A reinterpretation of the motives behind this assignment was formulated after the publication of the Master Plan by the Dutch parties, to explain the reluctance of WAZ-A to invest beyond the needs of basic infrastructure and recreation facilities. In order to develop Wujin Polder as an outer region of the city of Changzhou, the possibilities in terms of the project developer's interests are mainly the building of housing in higher price categories (with the accompanying recreation areas) and industry. WAZ-A has no problem removing the small-hold farmers from the polder. Much better housing has been created for them in the existing cores; and in the short term they can continue to use the land (in particular for growing rice and breeding fish). But there are also a number of big farmers in the polder: several tree cultivators, an intensive pig-breeding farm, a producer of chickens and a chicken manure processor. Sooner or later these entrepreneurs will have to move and the Master Plan will offer them an appealing vision, that corresponds with the ambitions for the rest of the polder. From this perspective, attracting foreign investors is not vital, rather 'window dressing', because the area serves primarily as a spatial overspill for the existing Chinese businesses. In 2007 WAZ-A indicated, when asked, that it still welcomed foreign investment but would continue with the development of the polder in its absence.

On 24 December 2004 the Master Plan was presented to WAZ-A at a meeting attended also by representatives of Wujin district, Changzhou city and Jiangsu Province.

6.5.5 Work process

The WAZ-Holland Master Plan consists of a collection of inventions. The two most important ones are the integration of agricultural production and processing via a CPU and the combination of intensive agricultural production with recreational and educational facilities. Only the latter invention has actually been implemented as an innovation with the partial execution of the Master Plan by the Chinese client. In addition, the design was carried out as a complete landscape plan, including a detailed water management plan and surface and underground infrastructure.

During the design process the KENGi network was not complete (Figure 33). No NGOs were involved and only one enterprise from the Netherlands. During the design workshop, however, a great many businesses in the region were visited and widely consulted. These businesses were also involved in the presentation of the design in China and commented on it at the site. Specific locations within the agropark were modified on the basis of these comments.

The design came about as a result of research by design, which took shape within one week in China, and was then developed in the Netherlands. During this design week in particular, when some of the participants held focused talks with experts in China, while others worked on the design, there was co-design, as described in

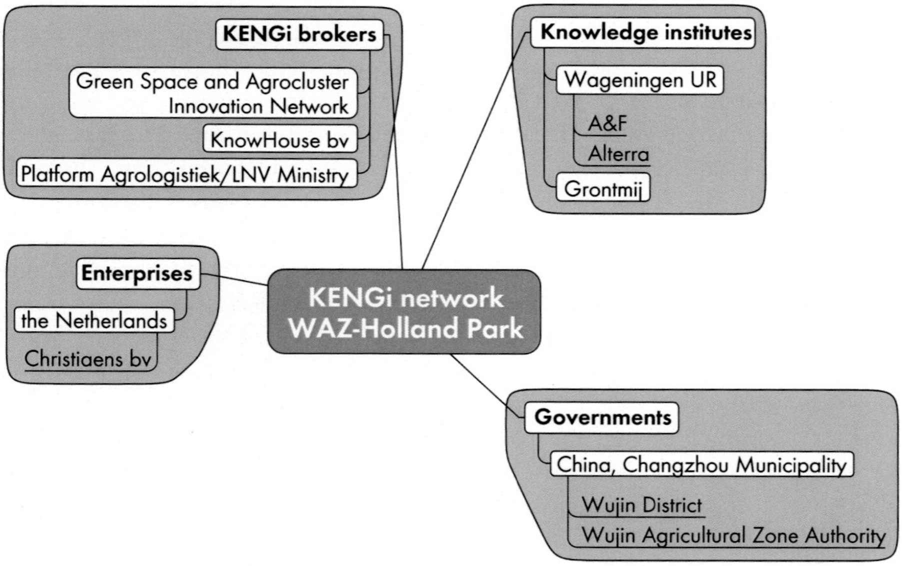

Figure 33. The KENGi network in the design process of WAZ-Holland Park.

Chapter 5. In this work process transdisciplinary work took shape between Dutch designers and scientists, representatives from the Dutch government and a company, and Chinese government staff. The emphasis during all this was on intercultural differences between China and the Netherlands. Most Chinese participants in the work process spoke no or limited English. The work discussions therefore had to be conducted through simultaneous interpreters.

Representatives from knowledge institutions and KENGi brokers, all participants in the Agrologistics Community of Practice, travelled to join the participants in the design workshop (November 2004) and conducted an analysis of the surroundings via their meetings with companies in the region and in Shanghai. Their findings were used to fine-tune the design and to get an idea of possible commitment from Chinese entrepreneurs to invest in Wujin Polder.

Key findings from this steering group were:
- high-tech companies only produce for export (low labour costs) or for hotels (high prices);
- much use is made of bypass flows, for which minerals and water are crucial, but with old technology, with combinations like worms and manure; pig liquid manure and fish; pig manure, plant fertilisation and biogas; chicken manure and composting;
- there are all kinds of systems of contract cultivation;
- there is a great need for recreational areas.

Back in the Netherlands, these people played an active role in the communication about the WAZ-Holland project to their own grassroots supporters. WAZ-Holland Park was also the first step by members of the Agrologistics Community of Practice at the international level.

6.5.6 Testing of the hypotheses

> **An agropark realises lower costs, greater added value and lower environmental pollution per unit of output and space.**

The WAZ-Holland Park design offers no new insight into this hypothesis. The design stands somewhere between the qualitative and hypothetical expectations pronounced in the Deltapark and the quantification thereof which first materialised around Agrocentrum Westpoort. The range of agroproducts was much broader than that in Deltapark. But a quantitative evaluation of the added yield from the use of industrial ecological integration was not carried out during the design phase.

An agropark can only come into being on the basis of an integral design of matterscape, powerscape and mindscape at both the global scale of Intelligent Agrologistics Networks and the local scale of a landscape

Much emphasis was placed in the design phase on the significance of WAZ-Holland Park for the region. The educational and recreational facilities which formed an integral part of the job were important for the inhabitants of the whole polder region and surrounding areas.

An agropark is a knowledge-driven system innovation and makes a significant contribution to sustainable development.

From the perspective of agroparks, WAZ-Holland Park got no further than being an invention, a design. It would have become an innovation had the entrepreneurs, who are at this moment getting established in the area on the basis of the spatial classification of the plan, actually worked with each other on the basis of industrial ecology.

An integral test on sustainable development did not take place in the design of WAZ-Holland Park.

The design and implementation of system innovations like agroparks necessitates the participation of knowledge institutions, enterprises, NGOs and government (or KENGi). It is a transdisciplinary process in which the explicit knowledge of knowledge institutions and the tacit knowledge of the other partners are developed in a process of continuous iteration. KENGi brokers act as the facilitators of this transdisciplinary process.

This hypothesis cannot yet be confirmed or rejected on the basis of the analysis of WAZ-Holland Park. The integral design did not leave the invention phase. In the Chinese planning tradition, the client can proceed undisturbed with the partial implementation of individual businesses and the recreational area. The NGOs, so pointedly present in the New Mixed Farm example, do not play a significant role in the undemocratic Chinese political system.

In all the decision-making concerning the realisation of the integral agroparks design, involving matterscape, powerscape and mindscape aspects, arguments from the world of justice and trustfulness take precedence over arguments from the world of truth.

This hypothesis is not rejected as far as the WAZ-Holland design is concerned. Parts of the design are being implemented because entrepreneurs want to invest in the area (an economic investment decision, in the area of powerscape), or because the presence of a recreational area is important for the other projects in the region.

But from the perspective of the setting in a different culture, it sheds new light on the way agricultural modernisation is regarded in North West Europe – especially concerning the position of intensive livestock farming. When the negative image of intensive livestock farming is not under discussion, an agropark appears to be a source of inspiration for the integration of recreation.

Social opposition to agroparks is directed primarily at intensive livestock farming.

WAZ-Holland Park adds to this hypothesis the observation that the resistance to further development in intensive livestock farming is strictly determined by North West European culture, in which the idyllic view of traditional farming and the emphasis on animal welfare issues dominate. In the formulation of the assignment by the Chinese, the agropark with its many forms of intensive livestock farming was expressly positioned as a modern agricultural-sightseeing park. Unlike the Dutch, the Chinese see modern agriculture in all its guises as a crowd-puller, strong enough to include a market square and recreation area, which they hope will attract a great many visitors. To this hypothesis too must be added that it only appears to apply in North West Europe:

In North West Europe social opposition to agroparks is directed primarily at intensive livestock farming.

6.5.7 Conclusion

Inspired by Deltapark and Agrocentrum Westpoort, the WAZ-Holland Park design came about in a co-design process with Dutch and Chinese participants. The inventions are the far-reaching integration of industrial ecology in a Central Processing Unit and the integration of agroproduction and processing park with recreation in an integral landscape design. Parts of the design are currently being implemented, and in that sense it can be described as a real innovation.

6.6 Biopark Terneuzen

6.6.1 The design

In the Terneuzen dockland area (Figure 34) an agropark is being built, in which an existing fertiliser manufacturer will join forces with glass horticulture businesses in a new cluster covering more than 200 ha. New industrial functionalities like biomass processing, bioethanol production and the purification and production of various grades of water will be added. The exchange of waste and by-products from these operations will offer the prospect of reduced costs, lower environmental emissions and less use of space.

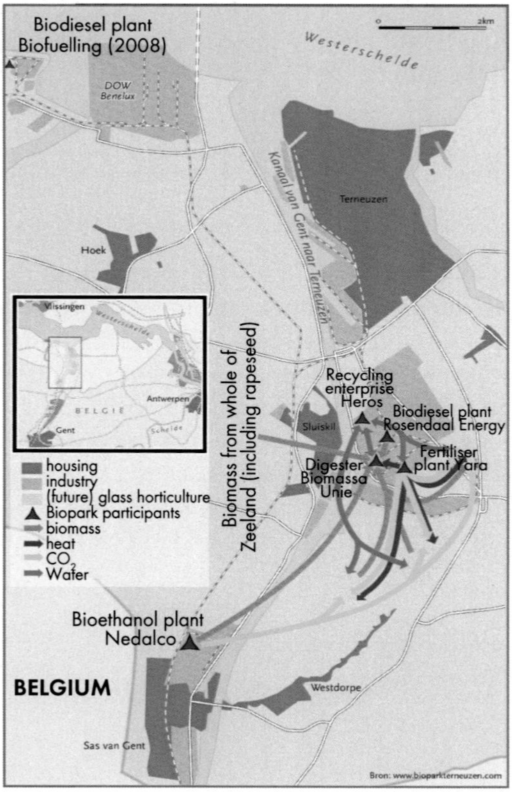

Figure 34. Biopark Terneuzen in the Terneuzen dockland area.
Boekema et al. (2008).

The first ideas for agropark development in the Zeeland docklands were launched in the *Meervoudig Ruimtegebruik* project in the South West Netherlands[282]. However, when the results of these design workshops were shown to the Zeeland provincial government, the latter rejected the idea of introducing forms of intensive livestock farming or glass horticulture to the region.

A few years later new staff at Zeeland Seaports took another look at the idea and this port authority has since played a key role in the development of the project. The essence of the idea was to link up the existing fertiliser business Yara via heat and CO_2 exchange with a new glass horticulture area and to add to that a few new matching industrial activities. Directed by Van de Bunt consultancy agency and with support from TransForum, the province of Zeeland and the municipality of Terneuzen, a coalition of knowledge institutions developed a trend scenario, in which existing development lines were extrapolated and three agropark scenarios were added. In these scenarios, activities were clustered on an ever-increasing scale and suitable new ones were added where necessary: the original plan, now called Biopark Terneuzen, the development of the Biopark within the Dutch land border and Bio-Valley Europa, for which a far-reaching collaboration between Ghent and Terneuzen is being discussed[283].

Between 2005 and 2007 a collaboration was set up between the existing fertiliser manufacturer and other large-scale industry developments in the docklands – a bioethanol factory, a biomass processor, a water purification company and a biodiesel factory. The development of a 240 ha glass horticulture area was also initiated. From 1 July 2007 onwards this joint venture was formalised under the name Biopark Terneuzen by all the stakeholders involved.

> 'The cluster can now evolve along two lines. Either the existing companies in the cluster expand, or new companies in the biotechnology sector are actively recruited to Terneuzen. These would have to be primarily high-value applications, such as the production of bioplastics, enzymes and vitamins. After all, these offer the most added value. (...) If Biopark Terneuzen gets off the ground, there will be exciting opportunities for further preservation of the economy of the planning area. (...) Clusters focusing on the bio-based economy are currently developing in the (....) canal zone. In a joint strategy Ghent Bio Energy Valley and Biopark Terneuzen should be able to grow into

[282] Goedman J., D. Langendijk, E. Opdam, S. Reinhard, I.d. Vries and M. Wijermans (2002). *Zee en land meervoudig benut; beknopt projectverslag*, Alterra, Wageningen, the Netherlands.

[283] Boekema F., M. Gijzen, F. Timmer and J. Dagevos (2008). Biopark Terneuzen. Een innovatief en duurzaam cluster. *Geografie* 17: 17-20; Timmer F., M. Gijzen, J. Dagevos and F. Boekema (2007). *Kanaalzone: Broedplaats van de biobased economy.* Radboud University, Nijmegen, the Netherlands.

> a cross-border (European) Bio-Valley, (...). The huge import and export flows will take place mainly by water.'[284]

By the end of 2008 a new project in the form of a further expansion of the original design had been approved, and was to be funded by the EU region, the Flanders Region and the Netherlands. It allowed for the establishment of a pilot plant for biofuel processing. Here the (mostly non-linear) up-scaling of the biofuel process from laboratory experiment to industrial scale will be simulated. In relation to the pilot plant, there will also be a new training programme for process engineers, provided by the University of Ghent in collaboration with HZ University of applied sciences in Middelburg, the capital of province of Zeeland.

6.6.2 Matterscape

Biopark Terneuzen covers the whole area of the Terneuzen canal zone. A new glass horticulture site of 250 ha net will be added to the existing and new industrial sites. The agropark contains the following functions and products[285]:

- Bioethanol factory: Koninklijke Nedalco is investigating the option of building a new bioethanol factory to the north of the production site of Cargill in Sas van Gent. Nedalco already has a factory at this site for alcohol for consumption.
- Biofuel additives factory: Rosendaal Energy is working on plans to build a biofuel additives factory on the Heros site east of the canal in Sluiskil. The capacity of this new production site is 250,000 tonnes per year. Natural oils and lipids will be used as raw material, sourced partly from Zeeland rape seed.
- Biomass plant: Biomassa Unie, EcoService Europe and Heros are working at the Heros site on a biomass plant for processing waste flows from the food industry and agricultural sector with a capacity of 135,000 tonnes biomass per year. The end products are green energy, NPK mineral concentrate, green coke and clean water.
- Waste water purification: Heros wants to upscale its existing wastewater purification installation in Sluiskil, in order to be able to facilitate waste flows from the planned glass horticulture, biofuel additives factory and biomass plant – with the aim of reusing the water.
- Glass horticulture: in 2007 work was started on setting up a large-scale glass horticulture area of net 250 ha on the east side of the Ghent canal to Terneuzen.

---

[284] Boekema F., M. Gijzen, F. Timmer and J. Dagevos (2008). Biopark Terneuzen. Een innovatief en duurzaam cluster. *Geografie* 17: 17-20.

[285] Projectbureau Biopark Terneuzen (2007). Biopark Terneuzen. Position paper. Available at: http://www.bioparkterneuzen.com/cms/publish/content/downloaddocument.asp?document_id=198. See also Gijzen M., F. Timmer, J. Dagevos and F. Boekema (2009). Biopark Terneuzen: Een duurzaam en innovatief voorbeeld voor zuidwest nederland. In: Smulders H., M. Gijzen and F. Boekema (eds.) *Agribusiness clusters: Bouwstenen van de regionale biobased economy*, Shaker Publishing, Maastricht, the Netherlands, pp. 37-48.

The project developer has made the purchase of CO_2 and heat compulsory. On 1 June 2008 80 ha gross of 300 ha gross was sold to Belgian vegetable cultivators.

- Biodiesel factory: In 2008 the Spanish company Biofueling will start up a factory at the Value Park Terneuzen. This factory will produce 200,000 tonnes biodiesel per year and has a strong synergy with the adjacent Dow complex.

The diagram in Figure 35 shows how industrial ecology is achieved between these elements of Biopark Terneuzen[286]. The biomass plant is central to this. It has been specifically designed for this agropark on the basis of work carried out by Broeze *et al.* (2007)[287], who developed the AF+ model for this purpose – building on the knowledge that was acquired while working on the Amsterdam Westpoort project. This model optimises the processing of waste and by-products from the various parts of an agropark with respect to each other, so that major savings can be made in existing product flows. Initially, the linkages are between the existing industries

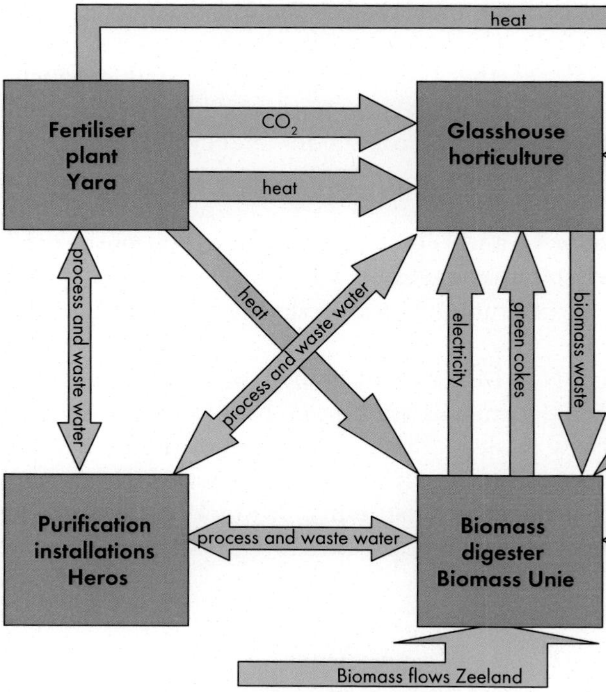

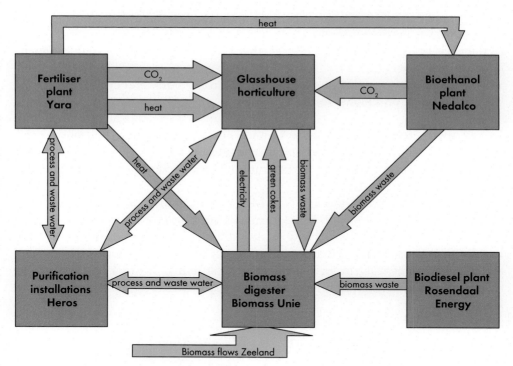

Figure 35. Industrial ecology in Biopark Terneuzen.
Projectbureau Biopark Terneuzen (2007).

[286] Timmer F., M. Gijzen, J. Dagevos and F. Boekema (2007). *Kanaalzone: Broedplaats van de biobased economy*. Radboud University, Nijmegen, the Netherlands: 18.
[287] Broeze J., E. Annevelink and M. Vollebregt (2007). *Onderzoek biomassa en energie biopark Terneuzen*, Report 848, Agrotechnology and Food Sciences Group, Wageningen, the Netherlands.

in the Sloe area. But the profit increases dramatically when primary agricultural production is added to the mix. So the possibilities offered by Biopark Terneuzen can only be realised by all parties and really implemented once the move is made towards combining industrial and agricultural production.

In addition to cost savings for the businesses, there is a considerable environmental gain: energy savings, CO_2 reduction, generation of green energy, water savings, less thermal pollution and a reduction in waste flows[288]. In the biomass plant 50% of the manure will be supplied from intensive livestock farming in the Zeeland province. Heros supplies water to the glass horticulture complex, which requires a maximum of 8 m^3/ha/hour. Yara and Nedalco supply 35 tonnes CO_2/hour to the glass horticulture area, which satisfies their total requirements. Regional agricultural and industrial (waste) flows from the agrosector form a link with Nedalco, as raw materials for bioethanol production. Rapeseed will be used as raw material for biodiesel production and the biodigester, and as raw material for the biodigestion process. The biodiesel factory supplies its waste product glycerine to the biodigester, which converts it into energy. A waste product from the biodigester is green cokes. Green cokes is supplied to the glass horticulture complex[289]. Further processing of the digestate produces a high-value, concentrated fertiliser. NPK minerals are not wasted but can be used worldwide in this form, making it possible to close the mineral cycle of intensive livestock farming on a global scale.

Energy savings are achieved because Yara will cover a minimum of 60% of the total heat requirements of the glass horticulture complex per year[290]. The biomass digester supplies green energy to the glass horticulture complex, producing 25,000 MWh/year.

As in Agrocentrum Westpoort and New Mixed Farm, manure processing does take on a massive scale in the park, but the animals are kept elsewhere. That results in substantially higher costs (especially in view of rising oil prices) in these livestock businesses. After all, many of the feed raw materials are shipped in reasonably cheaply and then have to be taken by lorry to the widely distributed livestock businesses. The manure from these businesses then has to be brought back to the docklands by lorry for processing.

[288] Boekema F., M. Gijzen, F. Timmer and J. Dagevos (2008). Biopark Terneuzen. Een innovatief en duurzaam cluster. *Geografie* 17: 17-20.
[289] Timmer F., M. Gijzen, J. Dagevos and F. Boekema (2007). *Kanaalzone: Broedplaats van de biobased economy*. Radboud University, Nijmegen, the Netherlands: 22.
[290] *Ibid.*: 22.

6.6.3 Powerscape

The assigner for the Biopark Terneuzen is Zeeland Seaports, with financial support from the province, the municipality of Terneuzen, Yara, Nedalco, TransForum and the knowledge institutions. The design project began in 2005 and was completed on 1 July 2007. Figure 36 shows the stakeholders in the process.

Biopark Terneuzen will create an estimated 2,350 new jobs. 80% of those will come from the siting of the new glass horticulture complex. The extra added value on the basis of extra employment is estimated at € 42 million/year[291].

> 'The firmer anchoring of industrial businesses by mutual dependency increases the chance of long-term business continuity in the canal zone. The extra investment for the physical connections are limited.
>
> - The starting up of alliances is a good strategy in an increasingly globalised economy.
> - The clustering of the businesses will attract other businesses and investors. The promotion of this unique concept may reinforce this autonomous

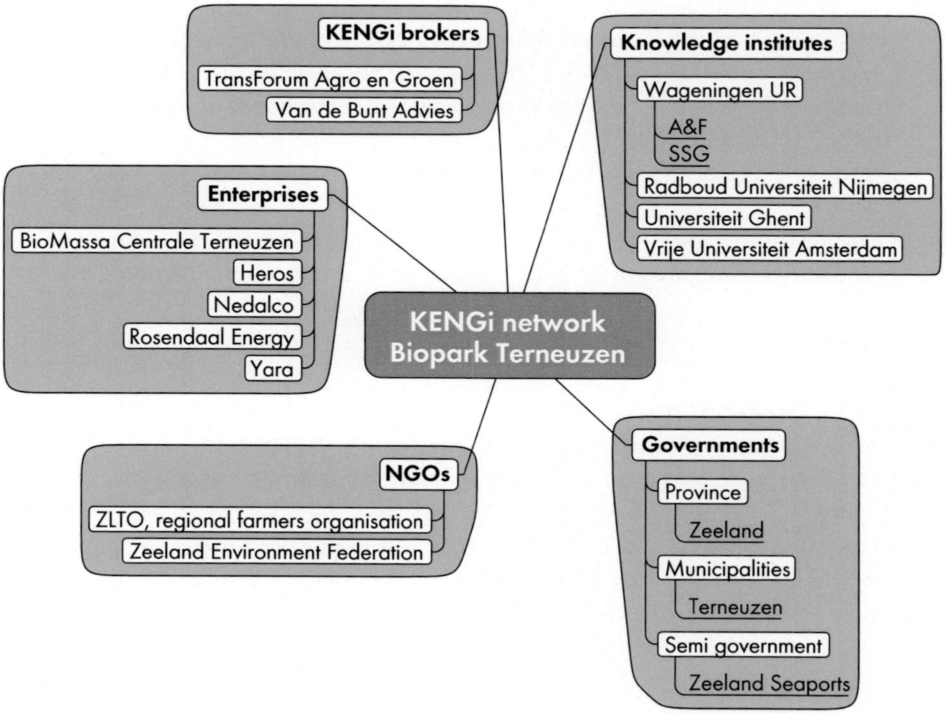

Figure 36. The KENGi network in Biopark Terneuzen.

---

291 *Ibid.*: 20.

movement; the canal zone is becoming more appealing for primary businesses connected to the Biopark, as well as for the general and technical sector, such as the processing industry and logistics.
- Waste flows which were traditionally regarded as costs on the balance sheet, are now a positive entry, since these can be profitably disposed of in the cluster. The foundations have been laid for expansion of the cluster, so that other industrial businesses will be interested in this cluster arrangement.'[292]

In the discussion about setting up in the new glass horticulture area, the advantages of scale take on a new light. Many Dutch glass horticulturalists gave as their main argument for not setting up in Biopark Terneuzen the fact that because of the mandatory purchase of heat and CO_2 from the industrial partners they could make only limited or no use of the benefits of electricity supply to the grid during peak hours, which would have earned them a lot of money.

> 'Collective energy supply and waste heat utilisation are being seen in a different light. With the higher energy prices (for gas and electricity) it may be very interesting for businesses bigger than 5 ha to generate energy themselves with Combined Heat and Power (CHPs) and to supply electricity back to the grid. In 2006 the glass horticulture sector even became a net electricity supplier as a result. In 2007 the interest in waste heat was minimal but collective energy generation remains interesting, provided that entrepreneurs who are participating in the "energy business" are in control.'[293]

This is one of the main reasons why Belgian glass horticulturalists, who don't use illuminated cultivation, are particularly interested in setting up in Terneuzen.

6.6.4 Mindscape

The Biopark Terneuzen project saw a new collaboration between the University of Ghent, whose expertise in the area of industrial ecology and process science has been developing for a long time, and Dutch knowledge institutions focusing on agropark development:

> 'The cluster formation also offers new opportunities to connect the knowledge infrastructure of Zeeland with universities and technical colleges in the rest of the Netherlands and Belgium (...). The exchange of knowledge

[292] *Ibid.*: 20.

[293] Ruijs M.N.A., A. Van der Knijff, J. Van der Lugt and C.E. Reijnders (2007). *Position paper glastuinbouw biopark Terneuzen. Deelrapport 2: Kansen voor glastuinbouw(complex) in biopark Terneuzen*, Projectcode 4057200, Landbouweconomisch Instituut, The Hague, the Netherlands: 13.

> and experience between companies and knowledge institutions can again lead to new innovations.'[294]

Individual decisions, which have clearly influenced the progress of the process around Biopark Terneuzen, are:

- The government of the province of Zeeland, which entrusted the Van de Bunt agency to look for interesting projects which would attract the support of TransForum, and following on from that, the decision to only elaborate the glass horticulture implementation from the resulting proposal, and to emphatically reject livestock farming in Biopark Terneuzen.
- Several new staff within Zeeland Seaport, who in contrast to a few years earlier, were given space to work on innovation trajectories. In addition, the other major project of Zeeland Seaports – a new container terminal at Vlissingen – threatened to run aground because of logistical problems in the surrounding region. The port industry was looking for other successful projects in addition to this container terminal.
- The reception by TransForum. The formula that this organisation uses to co-finance only on the basis of the joint input from business and knowledge institutions, forced these businesses to the table.
- The decision by project leaders and knowledge institutions to focus on the collaboration between Ghent and Terneuzen as a development perspective. As a result no party was selected in the dominant power struggle between Rotterdam, Vlissingen and Antwerp.

From the start, the Biopark Terneuzen project agency actively focused on the communication on the project. In this respect the existing expertise in this area at Zeeland Seaports was extremely important. Ambassadors were recruited for communication with the key stakeholders in the project – an executive councillor from the Zeeland province, an alderman from Terneuzen and the Director of Zeeland Seaports.

> 'Naturally, in the beginning not everyone was convinced. In the planning phase, however, the participating parties rapidly became aware of the major reciprocal advantages. Satisfactory solutions were soon found to difficult decisions. With the result that an initially difficult discussion on space gain and the negative environmental effects rapidly turned into a positive discussion about CO_2 and the strengthening of the regional-economic structure.'[295]

[294] Boekema F., M. Gijzen, F. Timmer and J. Dagevos (2008). Biopark Terneuzen. Een innovatief en duurzaam cluster. *Geografie* 17: 17-20. This collaboration has since resulted in a successful application for a € 21 million subsidy from the EU Region, Flanders and the Netherlands, to be used in setting up a test biorefinery and training for process engineers.
[295] *Ibid.*: 19.

As in the example of New Mixed Farm, the monitoring and evaluation of the work process in this project was carried out by the Athena Institute of the University of Amsterdam.

> 'The (...) vision on knowledge dissemination (...) which we use in this project, is based on interactive or contextual models of scientific communication. This is based on the principle that the content of the knowledge is linked to the context in which it is developed and the people who are involved in it. From this perspective of knowledge dissemination, it cannot unequivocally be established what the content of the knowledge is that has to be disseminated, for example in response to the research at Biopark Terneuzen into the coupling of waste flows.
> A second assumption of interactive or contextual communication models is that in the interaction with the possible recipient of the knowledge, the knowledge is not only customised, but new knowledge is also constructed and implicit knowledge is made explicit. The interaction, or communication, is part of the knowledge construction process, while in the transmission model the actual transfer of knowledge has no noticeable impact on the knowledge itself – which was already fixed.
> A third assumption, which touches on this, is that the potential target group of the knowledge is not only the recipient but the transmitter. It is likely that in an interaction between a scientist from Biopark Terneuzen and a project leader of another agropark, new knowledge and perspectives are introduced by the project leader, which the scientist was not aware of. The communication process then becomes an interactive and iterative process. In short, the discussion then centres not only around the form in which knowledge is disseminated, but also the content of this knowledge. Knowledge acquires form and content in communication between knowledge suppliers and knowledge demanders.'[296]

The monitoring and evaluating of the learning process across different comparable projects gives rise to the possibility of organising knowledge transfer from this project to other projects and others in the network. Regeer *et al.* use Learning histories here:

> 'In order to enable learning via "vicarious experience", we use the most important principles of "learning histories". A "learning history" is a way of establishing learning experiences and allowing people and organisations to learn from them. A "learning history" is characterised by the fact that it is not only the lesson taken from an experience that is told, but also the experience itself and the context in which it is acquired. Furthermore, different perspectives on the story are also reproduced. This story-form makes individual lessons accessible to others. People can recognise themselves in

[296] Regeer B. (2007). *Leren van biopark Terneuzen. Communicatie van kennis in context,* Afdeling Wetenschapscommunicatie, Athena Instituut, Amsterdam, the Netherlands: 8.

> a story, feel attached and thereby unconsciously make knowledge explicit. At the same time, another person's story offers access to new patterns and new dimensions of consciousness.'[297]

6.6.5 Work process

Industrial ecology, used successfully elsewhere in dockland areas or in chemical complexes between companies, was also suggested previously in Terneuzen. But it was the interposition of the agro-industry (glass horticulture and biomass processing) which gave wings to the process.

The work process in Biopark Terneuzen can certainly be seen as a form of co-design in connection with the Agrocentrum Westpoort and New Mixed Farm projects. Inventions that came about in these projects, were iteratively taken off the shelf in the design phase by the scientists involved, weighed up in a transdisciplinary environment with the KENGi parties concerned and converted into value propositions.

At the behest of KENGi broker TransForum, Regeer began explicitly describing the procedures in this transdisciplinary innovation laboratory via Learning Histories. This gave rise to mode 2 learning, in which, aside from the content of the learning process, the learning to learn also becomes a key component.

> 'The relationship between the research (the developed scientific knowledge) and the system (an agropark) was also discussed on a more scientific-sociological basis. On the basis of the available knowledge, to what extent can it be established which are the best decisions to be made about the system? What are the technological possibilities and what was taken into account when making the decisions? The Biopark Terneuzen case seems to show that a lot of generic knowledge was missing and that trials should be conducted in agroparks to obtain the required installations. In addition, the Biopark Terneuzen case also reveals that decisions were made on the basis of different considerations (e.g. high-risk investments versus cheaper option).'[298]
>
> 'The "research" concerns the Biopark Terneuzen "system", which is again part of local and rural "social" contexts. When planning, it is important that:
>
> - there are all kinds of intrinsic interactions between the three levels – overlap therefore between research and system, and between system and social context as well (....);
> - the traditional division of roles between knowledge supplier and knowledge demander is blurred. This report shows that scientists also ask questions

[297] *Ibid.*: 11.
[298] *Ibid.*: 26.

and project leaders also import knowledge. The various intrinsic items can therefore come from different sources;

- knowledge has a cognitive and social dimension and is always identifiable in the different versions of the story (interplay between presumptions informed by social priorities, choices, finances, phase in the process, etc.).'[299]

There is no doubt that the most important intrinsic learning experience of Biopark Terneuzen is that the non-inclusion of animal production (not advisable from a technical perspective) helped win over both the government and industrial parties. As a consequence Biopark Terneuzen avoided becoming a target of the environmental movement, as happened in the case of Agrocentrum Westpoort and New Mixed Farm. The implementation time of the project is considerably shorter as a result of the absence of intensive livestock farming. Not only are the procedures simpler, because it is just a matter of adding glass horticulture to a dockland complex, which is set up to facilitate heavy industry in the highest category of the Nuisance Act. But because it concerns a simple addition of one agro-function to industry, there was no question of an additional sustainability test, which the Horst aan de Maas city council deemed necessary for New Mixed Farm. A second advantage of not including animal production in the first phase, is that the manure processing has not been designed to connect to intensive livestock farming in the park. In this situation the scale of the intensive livestock business is the starting point, and these are often relatively small businesses. In Biopark Terneuzen a large-scale installation was designed, on a par with the other existing industries, enabling the use of more efficient technology.

The Community of Practice (CoP) which arose around Biopark Terneuzen consists mainly of parties other than the group which emerged from the projects described earlier. Only the presence of Wageningen-UR, Athena Institute and TransForum is a constant in both CoP lines. The absence of interference by the national government is striking.

6.6.6 Testing of the hypotheses

An agropark realises lower costs, greater added value and lower environmental pollution per unit of output and space.

The hypothesis is supported again. Simultaneous improvement of economic and ecological efficiency also seems entirely possible – by linking primary agricultural to industrial agroproduction – and creates major advantages. The insight gained into

[299] *Ibid.*: 31-32.

the industrial ecology of agricultural primary production, as well as the linkage with other, non-agricultural industrial businesses, has been modelled by Broeze *et al.* (2007) in the AF+ model[300].

The economic benefits are not confined to cost savings: 20% of the new jobs that have emerged in the region are in the industrial parts of Biopark Terneuzen and create extra added value.

However the linking of major industrial companies to (relatively) small glass horticultural businesses creates a disadvantage for the latter that has so far not been examined in the examples. The industrial ecology link does offer the prospect of lower costs, but in this case glass horticulture cannot generate the extras, which fellow entrepreneurs elsewhere earn by supplying energy to the grid from their own CHP.

An agropark can only come into being on the basis of an integral design of matterscape, powerscape and mindscape at both the global scale of Intelligent Agrologistics Networks and the local scale of a landscape.

As with Deltapark and Amsterdam Westpoort, Biopark Terneuzen was designed in a dockland area. It is therefore connected by definition to the global space of flows, in which the big multinationals in Terneuzen operate. But later in the design process, it was not just the matterscape aspects (such as the discussion on the CHP link in the glass horticulture) which played an important role. The effective exclusion of intensive livestock farming for mindscape reasons was decisive for rapid implementation.

An agropark is a knowledge-driven system innovation and makes a significant contribution to sustainable development.

Now that Biopark Terneuzen has been completed, it is more than a fully-fledged innovation. As a result of the new cross-border collaboration between Belgium and the Netherlands – both between knowledge institutions, and between entrepreneurs and businesses – the project can be called a system innovation.

The various publications that have appeared on the Biopark Terneuzen, show that the project contributes on many fronts to sustainable development, by reducing environmental pollution and energy consumption, via economic cost benefits, by a positive contribution to job creation and by new contributions to the knowledge infrastructure.

---

300 Broeze J., E. Annevelink and M. Vollebregt (2007). *Onderzoek biomassa en energie biopark Terneuzen*, Report 848, Agrotechnology and Food Sciences Group, Wageningen, the Netherlands.

The design and implementation of system innovations like agroparks necessitates the participation of knowledge institutions, enterprises, NGOs and government (or KENGi). It is a transdisciplinary process in which the explicit knowledge of knowledge institutions and the tacit knowledge of the other partners are developed in a process of continuous iteration. KENGi brokers are the facilitators of this transdisciplinary process.

Biopark Terneuzen has no implicit opponents, like those that appear in the other Dutch agropark projects. Parties from all KENGi sections were involved during the design process and the process shows at least that this involvement doesn't have to hinder fast and successful implementation. The work of Regeer *et al.* convincingly demonstrates that the collaboration between knowledge institutions and other KENGi partners was transdisciplinary.

In all the decision-making concerning the realisation of the integral agroparks design, involving matterscape, powerscape and mindscape aspects, arguments from the world of justice and trustfulness take precedence over arguments from the world of truth.

In his observations on the background to this project, Mark van Waes, the director of Van de Bunt and project leader of Biopark Terneuzen, uses the diagram in Figure 37, throwing new light on this hypothesis:

In this diagram we see again the three domains of matterscape (technology and logistics), powerscape (organisation, legal, permits but also power) and mindscape (ambition, strategy and culture). In his interpretation of this figure Van Waes confirms my hypothesis as regards attention in the design. All five domains named by him are important as success factors, but only the technological-economic and structure factors receive sufficient attention. This gives rise to an important interpretation of the powerscape domain, which seems to be justifiably called the 'power' domain by Jacobs[301]: Endless discussions can be had about the legal structure, permits and internal organisation but the essential powerscape questions are: Who gets what? and Who is allowed what?.

Also important in the Biopark Terneuzen powerscape is that the need for economic development in the Terneuzen dockland area was supported by government and population. Support from the state-level Agrologistics Platform appeared not to be necessary in this project. Furthermore, this project never came into the firing line of

[301] Jacobs M. (2004). Metropolitan matterscape, powerscape and mindscape. In: Tress G., B. Tress, W.B. Harms, P.J.A.M. Smeets and A. Van der Valk (eds.), *Planning metropolitan landscapes. Concepts, demands, approaches, Delta series*, Wageningen University, Wageningen, the Netherlands, pp. 26-39.

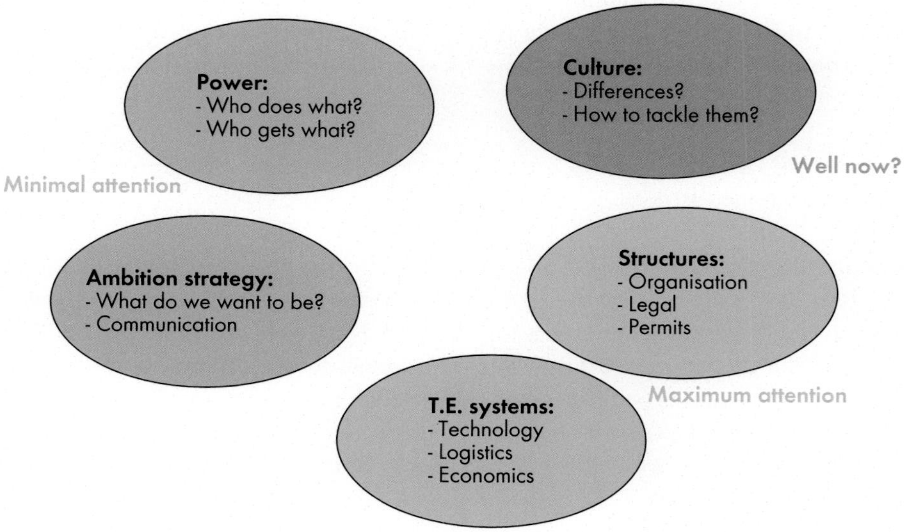

Figure 37. Success factors for consortia and clusters.
Diagram courtesy of Mark van Waes, project leader for Biopark Terneuzen.

environment and animal welfare groups because of the absence of intensive livestock farming in the design. From the beginning, the Zeeland Environment Federation supported the plan, because of the many environmental benefits arising from the industrial ecology.

The designers were able to achieve great synergy in the collaboration between the Netherlands and Belgium. Not only regarding the collaboration between knowledge institutions, but also in terms of the collaboration possibilities of the European Biovalley between Ghent and Terneuzen and by opening up the Belgian potential to glass horticulturalists. Poorly developed spatial planning in Belgium meant that, among other things, space for large-scale expansion of glass horticulture, as undertaken in the Netherlands in the project locations, was virtually non-existent[302].

In the Biopark Terneuzen mindscape the conscious decision to exclude intensive livestock farming and the extensively developed collaboration between Terneuzen and Ghent were especially decisive factors.

[302] De Geyter X., G. Bekaert, L. de Boeck and V. Patteeuw (2002). *After-sprawl; onderzoek naar de hedendaagse stad*. NAi Uitgevers, Rotterdam, the Netherlands.

The hypothesis is therefore confirmed again. While the technological and economic benefits of industrial ecology were repeatedly highlighted, the arguments from powerscape and mindscape were decisive for successful implementation of the design.

In North West Europe social opposition to agroparks is directed primarily at intensive livestock farming.

This hypothesis is explicitly and once again confirmed in the example of Biopark Terneuzen. The inclusion of the processing of manure from intensive livestock farming in the design, but the exclusion of animals on site, proved explicitly that it is primarily the inclusion of animals (i.e. the social debate on animal welfare) and less the environmental aspect of intensive livestock farming that is important here. Biopark Terneuzen takes on a major part of the (environmental) burden via manure processing but earns no money from the synergy that intensive livestock farming in an agropark would create.

6.6.7 Conclusion

After the first dismissive reactions to the Zeeland Voedt plan, Biopark Terneuzen became a success story. With the dynamic approach and the ever increasing urgency of the energy problem, the focus chosen also delivers an attractive prospect for the future, which can be further developed. It also appears possible in the chosen layout, to concentrate the widely distributed intensive livestock sector in Zeeland Flanders in Biopark Terneuzen instead of letting them expand further at their dispersed sites in the rural areas. But the decision to develop large-scale manure processing first and only later add intensive livestock farming on site, appears, at least in the Netherlands, as a much more intelligent strategy.

6.7 Greenport Shanghai

6.7.1 The design

Chongming Dao is an island north of the city of Shanghai, in the estuary of the river Yangtze. In 2009 it was connected via a bridge-tunnel to Shanghai on the mainland of China (Figure 38). The island is growing several hundred metres a year on the eastern side because of river deposits, and every 10 years the newly deposited land is diked. The most recent diking is called Dongtan, and it is here that Shanghai city is working on a plan for Dongtan Ecocity, intended to be an example of sustainable civil engineering. The project has been entrusted to the Shanghai Industrial Investment Company (SIIC), a project developer whose shares are owned by the city of Shanghai. Dongtan Ecocity is divided into four zones, to be developed into a garden city; an area

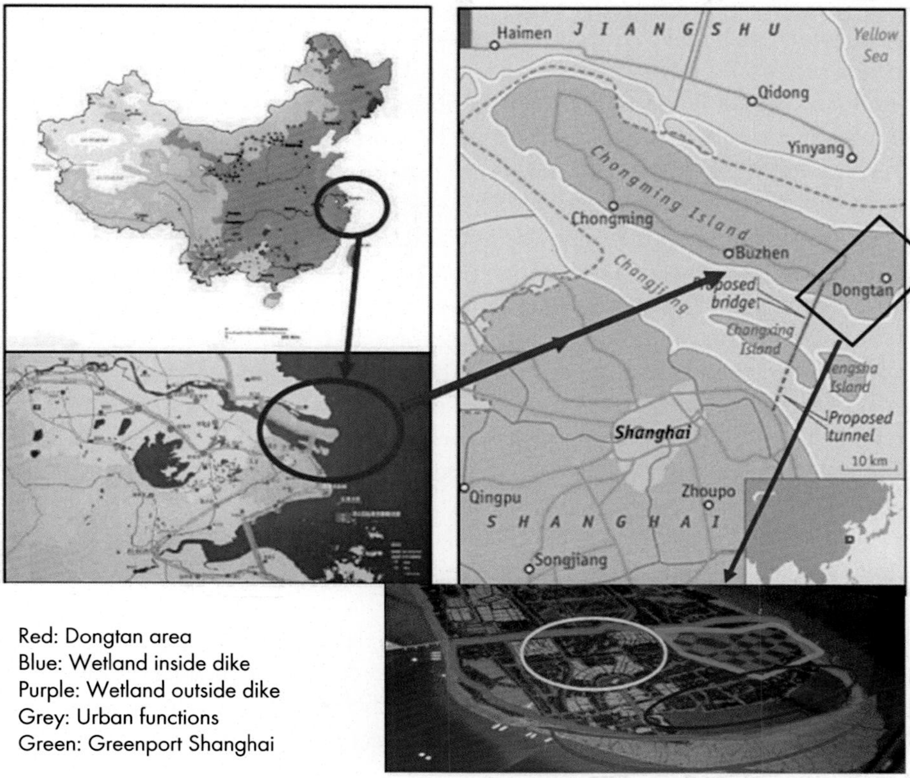

Figure 38. Location of Dongtan.
Smeets et al. *(2007).*

with offices and educational functions; a nature conservation zone with wetlands; and a 27 km² area for modern agriculture.

For the development of this agricultural area, SIIC sought to collaborate with Tongji University and via Chonghua Zhang, an associate at this university who also had connections with Wageningen-UR, the Master Plan for WAZ-Holland Park caught the attention of SIIC, just like a design produced earlier by Alterra/Wageningen-UR for nature development in the Dongtan area[303]. This gave rise to a collaboration between SIIC and Wageningen-UR, which became a TransForum project: Greenport Shanghai.

[303] Blom G., M. Brinkhuijsen, W.B. Harms, M. van Mansfeld, I. Mulders, P.J.A.M. Smeets, L. Xiuzhen and E. Zuidema (2002). Mianzi for all: Shanghai international wetland park 2002. Landscape ecological design, arranging concepts and principles for the chongming east headland. Alterra 2002, WUR Wageningen-Bureau Zuidema Rotterdam, the Netherlands / Shanghai Industrial Investment Holdings Co; Shanghai, China PR.

In October 2006 TransForum and KnowHouse jointly organised the first delegation to Shanghai, which included all the entrepreneurs who took part in New Mixed Farm and colleagues active in the North Limburg network of KnowHouse. Other participants included two regional Ministers from Limburg province, staff from the Limburg Development and Investment Company, the Economic Affairs alderman from Venray and the director of the provincial farmers organisation. During this mission, the group devised a number of principles on which the design for Greenport Shanghai was to be based. They proposed no longer working with a blueprint principle, as in previous designs, but first setting up a demonstration park to display the various aspects of industrial agriculture in practice. This demonstration park will enable research to be carried out into market and production conditions, and, by adding a trade park, it will be possible to bring products quickly on to the market and monitor the actual size of the market for these products. Only if there appears to be a big enough market for a specific product and only if it can also be produced optimally in Shanghai, will this chain be integrated as fully as possible in Greenport Shanghai.

This Demo>Trade>Processing>Production plan signalled a radical change in the design principles applied up to that point, from a supply-driven to a demand-driven design.

But the SIIC continued to ask for a very detailed Master Plan for Greenport Shanghai in order to initiate permit and formal planning procedures. The promise made by SIIC to have Dongtan Ecocity and all its components ready for visitors on the occasion of the opening of the World Exhibition in 2010 placed great urgency on the whole process.

A consortium of designers from Wageningen-UR, knowledge brokers from TransForum and KnowHouse, entrepreneurs and several independent advisors, began work on this task under the leadership of Alterra/Wageningen-UR. The Master Plan Greenport Shanghai was developed between November 2006 and June 2007 during several workshops[304].

A new planning methodology was developed for Chinese concepts, positioning the Master Plan between the two other plans which had been requested in this phase[305]. The Chinese authorities involved accepted this construction. At the same time the principle devised by the entrepreneurs took shape: the plan contains demo, trade, production and processing. No-regrets specifications for ecology, water, zoning and

304 Smeets P.J.A.M., M.J.M. Van Mansfeld, C. Zhang, R. Olde Loohuis, J. Broeze, S. Buijs, E. Moens, H. Van Latesteijn, M. Van Steekelenburg, L. Stumpel, W. Bruinsma, T. Van Megen, S. Mager, P. Christiaens and H. Heijer (2007). *Master plan greenport Shanghai agropark*, Report 1391, Alterra, Wageningen, the Netherlands.

305 Buijs S., P.J.A.M. Smeets, H. Guozheng, B. Xinmin and M. Van Mansfeld (2007). *Masterplan greenport Shanghai agropark. Knowledge report 1. Planning methodology*. Report 1391 sub. 2, Alterra, Wageningen, the Netherlands.

main infrastructure are laid down in the design according to the 'layer approach'. In this no-regrets plan, four scenarios for production and processing have been developed, firstly to clarify from the inside out which requirements the no-regrets plan must satisfy; and secondly, to delineate in later phases the space in which Greenport Shanghai can evolve – in relation to the debates that are still due to take place on trade, production and processing.

On 1 July 2007 this phase was completed with a presentation of the Master Plan in Shanghai. Since then implementation has proceeded along three lines. SIIC began elaborating the no-regrets plan, the Demopark and all other legal and administrative procedures. Wageningen-UR and Grontmij, who both helped draft the Master Plan, were invited to participate in an advisory capacity.

TransForum organised an international knowledge network to bring about the collaboration between knowledge institutions and Greenports in the Netherlands (Greenport Venlo first, then Westland at a later stage) and Greenports in the rest of the world. This Platform on Innovation of Metropolitan Agriculture was set up in October 2007 in Beijing and has since come under the umbrella of the International Food and Agribusiness Management Association (IAMA).

At the request of Limburg province KnowHouse, began organising and recruiting entrepreneurs with an interest in investing in Greenport Shanghai. At first, these were participants from the previous delegation, but the group has rapidly grown. Organisation models for Greenport Shanghai are being further developed with these entrepreneurs.

The Dutch implementation activities are being coordinated by a Steering Group with directors from the organisations involved under the leadership of former DG Oostra from the Ministry of Agriculture and the Dutch Agricultural Attaché in Beijing.

The hope is that by 2010 there will be a Demopark in place at Greenport Shanghai. Which production, processing and trade activities will follow will become apparent from the experience acquired on the basis of the Demo>Trade>Processing>Production principle. Until there is more clarity on that subject, SIIC is taking small steps – for example, by improving the existing aquaculture cultivation, already present in the Dongtan region.

6.7.2 Matterscape

The Greenport Shanghai area is 2,700 ha. At the basis of the Master Plan is the no-regrets plan, which includes designs for zoning, water management, ecology, transport infrastructure and the infrastructure of the Central Processing Unit (CPU), in which the industrial ecology will take shape (Figure 39).

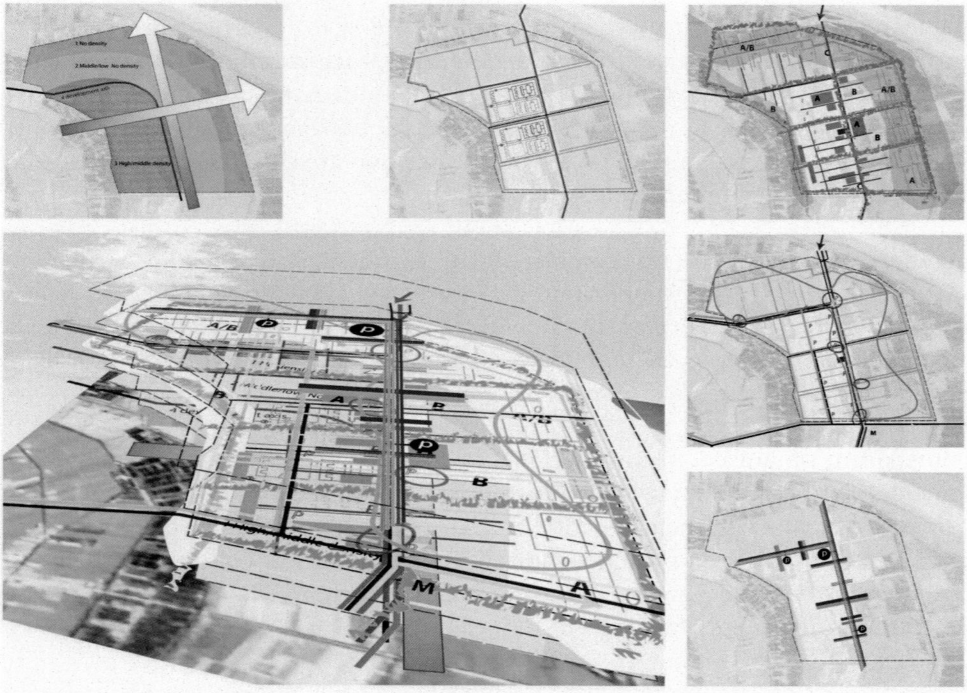

Figure 39. Design of the no-regrets options for Greenport Shanghai. Clockwise from top left: the zoning principle, the water system, the ecological system, the infrastructure and the CPU pipelines. Bottom left the layered integration map.
Smeets et al. *(2007).*

On the basis of the no-regrets plan, four scenarios have been developed, each of which presents a different direction for the development of the agropark (Figure 40). The components are balanced via their contribution to and withdrawal from the CPU, whereby sufficient scale for an economical operation and minimum environmental pollution are always the most important criteria.

The Master Plan therefore includes four possible implementations. The basis is the above-mentioned Demo>Trade>Processing>Production principle, whereby the entrepreneurs decide at a much later stage whether or not to set up in the park.

6.7.3 Powerscape

The assigners for Greenport Shanghai were the Shanghai Industrial Development Company (SIIC), TransForum and Wageningen-UR. The participants in the KENGi network are listed in Figure 41.

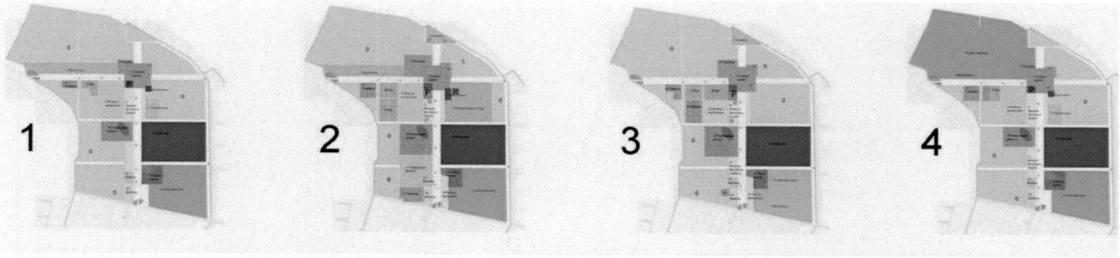

	1: Basic mushroom	2: Large scale	3: Pig top breeding	4: Dairy included
Open area	2,542 ha (94.1%)	1,908 ha (70.7%)	2,356 ha (87.3%)	2,522 ha (93.4%)
Built-up facility agriculture	93 ha (3.4%)	608 ha (22.5%)	233 ha (8.6%)	107 ha (4.0%)
Built-up non- agriculture	65 ha (2.4%)	184 ha (6.8%)	111 ha (4.1%)	71 ha (2.6%)
Internal use of power (kWh/y)	29 mln	625 mln	115 mln	120 mln
Surplus power (kWh/y)	-0.9 mln	260 mln	52 mln	80 mln
Internal use of heat (MY/y)	119 mln	2,490 mln	459 mln	477 mln
No. of pigs	50,000	1 mln	1 mln (piglets)	50,000
No. of chicken	2 mln	7 mln	3 mln	2 mln
No. of cows	0	0	0	8,000
Vegetables (tonnes/y)	9,000	240,000	50,000	24,000
Mushrooms (tonnes/y)	6,000	30,000	0	6,000
Inhabitants	4,000	25,000	12,000	6,000

Figure 40. Four scenarios for implementation of trade, processing and production in Greenport Shanghai.
Smeets et al. *(2007).*

Non-governmental organisations do not appear in this network. In China they do not function in the areas relevant to the development of agroparks and certainly not at the local level. Dutch non-governmental organisations, especially environmental and animal welfare organisations functioning at the national level, either do not want to participate in the development of agroparks at all, or effectively only want to do so in relation to the Dutch situation.

Many of the environmental aspects of Chinese agriculture which are discussed in the context of sustainable development (water contamination, water consumption, soil and air pollution) are explicitly addressed at the national level by the government authorities. The problem is often that the aims are communicated to the local level

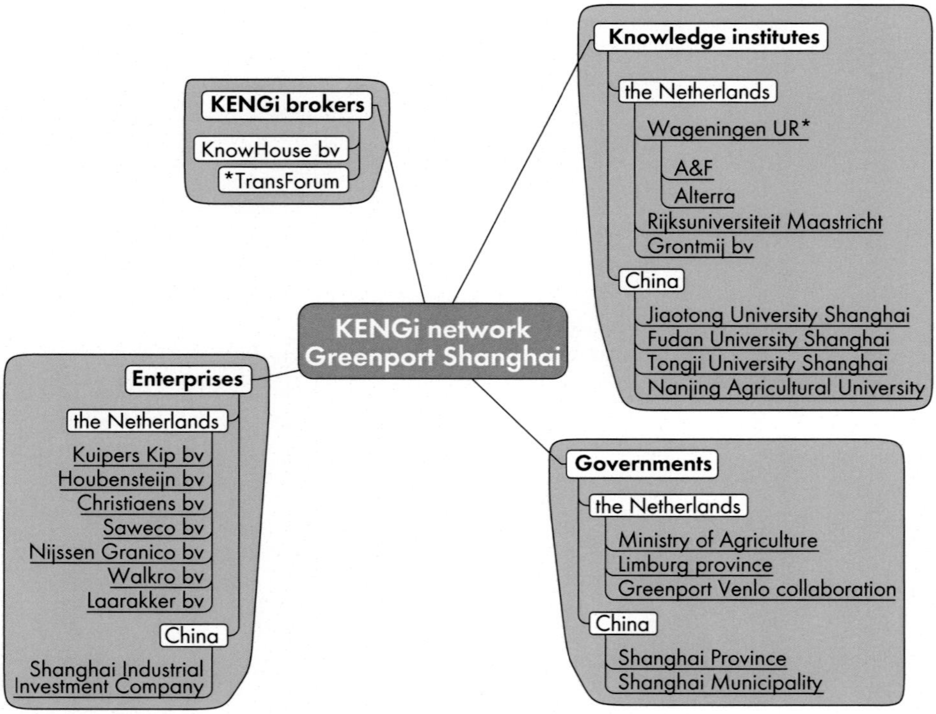

Figure 41. The KENGi network of Greenport Shanghai.

but the accompanying funds and power are not transferred. This results in less transparency and increases the risk of local corruption[306].

The national government also places great emphasis on aspects like food security and food safety.

The issue of biodiversity is high on the agenda throughout the Dongtan region, because the entire area outside the dike is a designated RAMSAR site (The Ramsar convention is an intergovernmental treaty that provides the framework for national action and international cooperation for the conservation and wise use of wetlands and their resources). Huge areas within the dike are also designated conservation areas, and much attention is focused on recreational facilities here. The plan for Greenport Shanghai foresees ecological connection zones between these areas within the agropark region.

306 OECD (2007). *Oecd review of agricultural policies China*. OECD, Paris, France: 26.

Once the Shanghai city council had accepted the Master Plan on 1 July 2007, both the Chinese and Dutch parties continued with their work on the implementation.

6.7.4 Mindscape

The social debate being held in North West Europe on animal welfare in intensive livestock farming is virtually non-existent in China. The current Dutch norms for animal welfare have been adhered to in the design for Greenport Shanghai and actively brought to the attention of Chinese assigners, wherever a connection could be established between productivity and animal comfort[307]. For the rest, the position has consistently been taken that the Chinese assigner explicitly asked for the deployment of state-of-the-art technology from the Dutch agrocomplex, where the far-reaching norms concerning animal welfare are fully integrated. The participating Dutch authorities and entrepreneurs also made reference to the indirect controls to which they are subject by Dutch community organisations. As soon as it appeared that the Dutch norms were not being applied in the design, they would be accused of using taxpayers' money to support the export of animal suffering or environmental pollution.

No HRM aspects were developed at the Master Plan level. In the context of project development, the problem of the forced relocation of farmers, who live and work on the location, often to worse living or working conditions, is not an issue in the Dongtan area. Since the draining of the region in the mid-1990s, inhabitants (legally or illegally) have been discouraged from living there. A few farmers work for SIIC in agriculture or fish breeding, and have been assured that they will be enabled to participate in the development of the area.

The formation of a so-called Knowledge Value Chain has already been instigated on the basis of Greenport Shanghai in the Master Plan and thereafter during the conference 'Innovating Agriculture in metropolitan Areas', held in Beijing in October 2007.

> 'The implementation of an agropark requires transfer of knowledge. Learning and education in the countries and regions where an agropark is being implemented must deliver the know-how to use and develop them in a sustainable way, adapted to the local conditions. The continuous participation of knowledge institutes like Wageningen UR and Chinese universities in a joint venture will be an asset, not only for them to earn revenues from the invested knowledge but also to develop this knowledge on the basis of scientific monitoring and evaluation of the agropark practice that, used in

---

307 Examples of this are the connection between temperature and the welfare of pigs, which means higher productivity and the prevention of animal transportation where possible, to drastically reduce stress and death.

> this way turn into laboratories and is the basis for new innovations in the agropark production and processing but also in the way the government is dealing with these systems. This new knowledge in turn will be transferred to students, some of whom will be the future managers and workers in the Agropark. In doing so the knowledge institutes implement the whole knowledge value chain. The transdisciplinary approach that is characteristic of the agropark knowledge, expands the knowledge value chain to the integrated level of the other stakeholders involved (entrepreneurs and government employees and NGOs) that put in their tacit knowledge.'[308]

The knowledge value chain is being further developed from the Greenport Shanghai project by TransForum on an international level under the auspices of the International Food and Agribusiness Management Association (IAMA).

Individual decisions, which have clearly influenced the process of Greenport Shanghai are:

- The decision by participants in the network around WAZ-Holland Park (Wageningen-UR and KnowHouse) to start on the acquisition phase at SIIC together with TransForum and Chonghua Zhang as broker.
- The decision by KnowHouse, TransForum and Wageningen-UR to persuade the KENGi network around New Mixed Farm to go on a working visit to China. This working visit ended up not only being a positive step for Greenport Shanghai, but conversely also substantially reinforced the network around New Mixed Farm in North Limburg.
- The decision by Limburg province and the LNV Ministry to actively support the implementation of the Master Plan.

After the Master Plan had been completed, a brief audiovisual presentation with English and Chinese text was produced on the basis of a three-dimensional virtual simulation of the design[309]. The film has since been used on many occasions by participants in the KENGi network as a source of information on the project, but also as a means of attracting new participants to that network. In this film 'Greenport Shanghai' is allocated as the new brand name, to help promote the idea of agroparks outside the Netherlands. The SIIC supports the idea not only of using Greenport as the designation for new agroparks, but also of using it as a brand name guaranteeing top quality. From that perspective Greenport Shanghai would have to acquire the

---

[308] Smeets P.J.A.M., M.J.M. Van Mansfeld, C. Zhang, R. Olde Loohuis, J. Broeze, S. Buijs, E. Moens, H. Van Latesteijn, M. Van Steekelenburg, L. Stumpel, W. Bruinsma, T. Van Megen, S. Mager, P. Christiaens and H. Heijer (2007). *Master plan greenport Shanghai agropark*, Report 1391, Alterra, Wageningen, the Netherlands: 60.

[309] This film can be seen on the website http://www.Greenportshanghai.com and the Master Plan can be downloaded.

role of quality controller, to be assumed later in the Greenport network which is still to be built.

6.7.5 Work process

The Greenport Shanghai Master Plan is an invention. To date only SIIC has made steps towards actual investment and has started on implementation. In this sense, it can be called an innovation, even though these preparations are limited to the no-regrets options in the Master Plan. The value of this invention is also being increased further afield, because new design assignments are being created in other places in China and India by the Dutch KENGi network on the basis of the report and particularly as a result of the film about Greenport Shanghai.

During the design process for Greenport Shanghai, the KENGi network came up with a new planning methodology, greatly inspired by the Dutch concept of spatial development policy, as advanced by the WRR (Scientific Counsel for Government Policy)[310].

> 'The first characteristic of the (...) Agropark is the way it represents a new type of area development: neither fully urban nor fully rural but a balanced mix of mutually dependent built and non-built functions, linked by infrastructure enabling the implementation of the closed cycle principle. Because of the novelty of this intermediate position between traditional urban and rural planning, it is difficult to apply existing planning methods, regulations and procedures to the planning and development of the Agropark. For the next steps on the way to a fully operational status it might be helpful to designate the Agropark as an experiment in "integrated urban-rural planning", aiming at realising the concept of metropolitan agriculture. (...) The second characteristic of the planning for the Agropark is its development orientation. This new planning paradigm is necessary to ensure successful implementation, for which the old control oriented planning is not sufficient. This is because implementation depends in the first place on entrepreneurs willing to invest in productive activities. They will assess the possibilities to maximize returns, minimize risks, and guarantee continuity and growth. For them freedom to organize production processes as efficiently as possible, to decide on where to purchase their inputs, and which types of products to bring to the market, is crucial. Traditional "blue print" land-use planning seldom offers sufficient freedom for entrepreneurs to make investments attractive. (...) [A] third characteristic, that needs to be adopted is a process planning method. This is for two reasons: not only because of the unpredictability of actual implementation that makes it necessary to offer

310 Wetenschappelijke Raad voor het Regeringsbeleid (1998). *Ruimtelijke ontwikkelingspolitiek.* Wetenschappelijke raad voor het Regeringsbeleid, The Hague, the Netherlands.

> maximum flexibility while at the same time safeguarding essential qualities, but also because – as stated above – the concept of urban-rural integrated planning (or planning for metropolitan agriculture) is new, doesn't fit in existing legal frameworks, and asks for a "learning-by-doing" approach. With regard to this new concept of urban-rural integrated planning, it cannot be expected that the Chinese legal context will be changed just to accommodate a single project that falls outside normal practices. What is needed is recognition as an experiment, in a way that offers opportunities to learn lessons from practical action. If successful, the experience from this project may well lead to a future review of present formal regulations.'[311]

The Master Plan played its intended role in the formal Chinese planning system between Strategic Plan and Industrial Plan, despite the fact that because of the scenario methodology used it did not use the blueprint approach that would normally be expected of such a plan in China. The city and province of Shanghai therefore requested and obtained experimental status for this departure from normal practice. Meanwhile, the methodology used has been drawn to the attention of the Chinese National Development and Planning Committee. In this sense there has been an innovation in the domain of the Chinese government.

Following the approval of the Master Plan, the various parties involved proceeded to implement the plan further. The SIIC concentrated on administrative and permit procedures and on the no-regrets components. On the Dutch side, TransForum and KnowHouse proceeded to organise and further expand the KENGi network. The first entrepreneur's delegation, which laid the groundwork for the Demo>Trade>Processing>Production principle, consisted of entrepreneurs from Limburg, from the KnowHouse network in Greenport Venlo, and has now been joined by the LNV Ministry, Greenport Netherlands, and the provinces of North Brabant and South Holland. All manner of exploratory talks are being held with other governments and potential investors in the Netherlands as well as in India. In China too, it was only when the conceptual Master Plan was finished that work began on creating support in China's national government.

What is relevant here is that the target group of Dutch businesses was small and medium-sized entreprises for the most part, each often too small individually to take the step to invest in China. A support network will have to be organised from the Netherlands in which government authorities, enterprises, financiers, knowledge institutions and other interested parties share the risks and potential benefits of the initial steps being taken at Greenport Shanghai. In a far-reaching integration

[311] Buijs S., S. P.J.A.M., H. Guozheng, B. Xinmin and M. Van Mansfeld (2007). *Masterplan greenport Shanghai agropark. Knowledge report 1. Planning methodology,* Report 1391 sub. 2, Alterra, Wageningen, the Netherlands: 10-11.

like an agropark, this can only happen in an open innovation process, in which the parties involved share extensively the ins and outs of their own production processes within their own sectors and across the different sector boundaries. The basis for collaboration is that the Dutch invest primarily in terms of know-how and time (and to a limited extent with money) and that the major financial investments are made by SIIC and via joint ventures with other Chinese businesses.

In the Greenport Shanghai example, the organisation of this open innovation network is being facilitated by KnowHouse and TransForum, who have both received subsidies for this purpose from the Limburg and Dutch governments respectively, and who are also financing their own efforts and additional efforts of knowledge institutions and entrepreneurs. In the Shanghai Greenport work process, two disadvantages were highlighted by this construction:

- The contributions from these knowledge brokers are regarded by the Chinese partners as subsidies for the project. That creates expectations for the next phase. The Dutch government appears to have an interest in stimulating its companies to do business in China. In the negotiations about the steps to follow the Master Plan, the Chinese partners again appear to expect the Dutch parties to pay part of their costs themselves. Where that is no longer the case, because, for example, there is a ceiling on the innovation subsidy from TransForum, the negotiations have run aground and the development process has come to a standstill.
- Innovation subsidies, such as that from TransForum, come with their own additional, often complex financial parameters. These are co-funding projects, in which inputs from parties in cash and in kind, are weighed up against each other and have to be in proportion[312]. In the Greenport Shanghai international collaboration the input of Chinese partners, such as SIIC and universities appointed by SIIC, is included in this co-funding. So these too must comply with the Dutch norms of the underlying BSIK (Investment in Knowledge Infrastructure) directive, which for this purpose must not only be translated in their entirety into Chinese, but must also be integrated into the Chinese system of auditor's reports and time-accounting systems.

This phase is, in fact, still only the second step in the design process, whereby after the design of the Matterscape in the Master Plan, the design of the Powerscape takes centre stage. This process is being seriously hindered by the urgent need for implementation, because the SIIC is required by its own clients to be able to show several elements of Ecocity Dongtan at the opening ceremony of the World Exhibition in 2010. This is putting particular pressure on the entrepreneurs in the KENGi network taking part in the design process to make a rush decision about investing in Greenport Shanghai. The three lines of design, implementation and decision-making, as interpreted on the

[312] For TransForum these parameters are laid down in the general terms and conditions for subsidies of the BSIK programme.

basis of the Regional Dialogue described in Chapter 5, are being mixed up again in this procedure.

Furthermore, the experience from the regional dialogue also teaches us that the KENGi network of organisations in the different phases of the three lines should be made up of people with various skills. That is a problem with participants from small and medium-sized entreprises. In a one-man business the entrepreneur must simultaneously devise a strategy, drum up support, take decisions and organise funding. He/she is rarely a master of all four. Selecting on these qualities was not properly carried out in the design of matterscape Greenport Shanghai. The design skills of the entrepreneurs involved in this phase were not sufficiently assessed.

There is therefore a risk of the design phase being completed prematurely. This is not possible with 4 scenarios, because then the design is only half-finished. Since there is still a final phase of design to go (making decisions in the playing field defined by the scenarios, decisions that have to be made by entrepreneurs in a design mode), implementation cannot be started yet.

As a result, efforts are effectively being focused on setting up the *in-situ* demonstration facility only for 2010. As a result there are not likely to be any companies undertaking trade, production or processing activities at that stage. There is still too much uncertainty for them in the area of logistics and sales opportunities.

The most important learning experience to emerge from the various evaluation and reflection consultations with participants, is that to an even greater extent than in the Dutch situation where KENGi networks are trying to bring about an innovation leap, building up trust is crucial in an international, multicultural situation[313]. This takes time. In the complexity of a multicultural setting, where the three lines of design, support and decision-making should not be mixed, this means continuously and gradually enticing new participants, by showing them appealing images of what has already been achieved.

313 See also Rotmans J. (2003). *Transitiemanagement: Sleutel voor een duurzame samenleving.* Koninklijke van Gorcum, Assen, the Netherlands: 74 ff.

6.7.6 Testing of the hypotheses

An agropark realises lower costs, greater added value and lower environmental pollution per unit of output and space.

As far as the hypothesis in production ecology is concerned, the Greenport Shanghai offers no new insights. The design stands somewhere between the qualitative and hypothetical expectations pronounced in the Deltapark and the quantification thereof which first materialised around Agrocentrum Westpoort. The AF+ model[314] derived from the New Mixed Farm and Biopark Terneuzen is also used in the designs for Greenport Shanghai. But a quantitative evaluation of the added yield from the use of industrial ecological integration was not carried out during the design phase.

An agropark can only come into being on the basis of an integral design of matterscape, powerscape and mindscape both on the global scale of Intelligent Agrologistics Networks and on the local scale of a landscape.

In the first design round of Greenport Shanghai, in which the accent lay on the design of matterscape, this hypothesis is not only supported but has been translated into a new design method. In this method, the matterscape elements, on which there is social consensus in the world of justice and trustfulness, are developed as no-regrets options; and important decisions, which in the case of Greenport Shanghai have to be made primarily by the entrepreneurs, are developed into different scenarios.

An agropark is a knowledge-driven system innovation and makes a significant contribution to sustainable development.

Greenport Shanghai is still in the design phase and must therefore be referred to as an invention. The scenario planning method used did, however, lead to the creation of an experimental status in Chinese government planning, which can be regarded as an innovation.

[314] Broeze J., E. Annevelink and M. Vollebregt (2007). *Onderzoek biomassa en energie biopark Terneuzen*, Report 848, Agrotechnology and Food Sciences Group, Wageningen, the Netherlands.

The design and implementation of system innovations like agroparks necessitates the participation of knowledge institutions, enterprises, NGOs and government (or KENGi). It is a transdisciplinary process in which the explicit knowledge of knowledge institutions and the tacit knowledge of the other partners are developed in a process of continuous iteration. KENGi brokers are the facilitators of this transdisciplinary process.

To date, no community organisations have played a significant role in the design for Greenport Shanghai. Knowledge institutions, KENGi brokers and government authorities have introduced the various norms and values of NGOs as a design requirement. But the absent non-governmental organisations do not share in the transdisciplinary knowledge development. The KENGi brokers' participation seems to have advantages and disadvantages. In the first phase of this design process, which is more vulnerable in its international and multicultural setting than similar processes in the Netherlands, they are a decisive element because of their input in the area of know-how and experience, and their contribution by way of major subsidies. But when they bring heavily rigged accountability to the process from the Netherlands, their presence has a delaying effect. Their subsidies appear to put foreign clients on the wrong footing because of the impression they give that the Dutch government will continue to pay out money in all phases of the innovation process.

In all decision-making on the implementation of the integral agroparks design, with matterscape, powerscape and mindscape aspects, arguments from the world of justice and trustfulness take precedence over arguments from the world of truth.

A confirmation of this hypothesis can be found in this project in the newly developed planning method (powerscape) which works on the basis of scenarios. What is interesting from the perspective of transdisciplinary knowledge development is that the entrepreneurs instigated it by devising the Demo>Trade>Processing>Production principle. The decisions within the playing field of these scenarios are made via the Demo>Trade>Processing>Production principle on the basis of existing market preferences or on the basis of new market introductions (powerscape), but different political options or options in the area of mindscape may also be decisive.

The reality around the forced implementation of the design by SIIC is an example of mindscape dominance. The pressure on SIIC to show something at the opening of the World Exhibition in 2010 is so great that they are effectively choosing a design in which a wide range of production activities will be minimally developed and the focus will be on the establishment of the Demopark.

But the key motives behind the innovation leap, which the Chinese authorities and the city of Shanghai want to achieve with Greenport Shanghai, are the guarantee of sufficient, healthy food and the reduction of environmental pollution, both of which traditional agriculture is struggling to cope with. These are arguments that belong to the world of Powerscape and Matterscape, and so this hypothesis is still on shaky ground against a global background.

> In North West Europe social opposition to agroparks is directed primarily at intensive livestock farming.

SIIC and the city of Shanghai would very much like to implement the agropark, preferably with a major contribution from foreign investors. The national government is giving its full support to these ambitions. In the design process for Greenport Shanghai, the Chinese client made no objections to the integration of sizeable intensive livestock farming operations in the agropark. This hypothesis is therefore confirmed.

A new hypothesis about the design process, which can be deduced from the method which was invented for the Master Plan and from the state of affairs after it was completed, states that:

> Design of the agropark orgware with KENGi parties in a multicultural setting can only take place once sufficient trust has been built up between stakeholders on the basis of an appealing matterscape design, in which there are still many open options.

A subsequent hypothesis, which largely results from the attempts to forcefully set in motion the actual implementation and investment decisions on the basis of the Matterscape Master Plan, states that:

> Only in a broadly composed KENGi network, focusing on open innovation, can the Dutch SMEs in the agrosector create system innovations on an international level and participate in them.

The experiences with the use of the Greenport logo, and the ensuing support for that from the various Greenports in the Netherlands, also leads to a hypothesis concerning the Dutch Greenport strategy. The hypothesis states that:

> **Greenport seems to be an attractive international logo, under which the Dutch agrosector can propagate system innovations and quality management in its global network. Extensive collaboration and synergy between the Dutch Greenports is vital for this purpose.**

6.7.7 Conclusion

Greenport Shanghai is an appealing invention, tremendously charismatic because of its scope and its location near one of the most dynamic metropolises in the world. However, the speed with which the client as well as the designers thought they could implement a large part of their design, became a serious problem, even in this environment. Fundamental factors in the powerscape and mindscape, such as mutual trust and a working multicultural KENGi network, can apparently only be forced to a limited extent. They take time to be built up, but can quickly fall back down again.

6.8 IFFCO-Greenport Nellore

6.8.1 The design

The development of Agropark IFFCO-Greenports India Kisan Special Economic Zone Nellore, hereinafter referred to as IFFCO-Greenport Nellore, is in full flow at the time of writing. This example has not been included because of the design of the agropark. This design has not yet been finished. But the powerscape development which led to this first agropark design in India, can be regarded as a synthesis of the experience acquired of working on this area, particularly in Wageningen-UR, in the last few years. The crucial aspect in this project is the strategic collaboration from September 2005 onwards between Wageningen-UR and the Indian Yes Bank, which lies at the heart of the various agropark projects in India.

The work in India also made us aware of the importance of the logistical infrastructure which has emerged in North West Europe in the last few centuries and which is an essential component of this system. It explains the increased public interest in agrologistics in recent years. Most of the food produced in India is taken straight from the manufacturers by middle men to the consumer markets and sold fresh to customers. Only 2% of all food produced in India is processed. That explains the huge losses experienced in this country, where temperatures can rise to above 40 °C and where the poorly developed infrastructure is seen as the biggest problem facing the economy.

At the same time India is, broadly speaking, self-sufficient as regards food. But that food is produced by 60% of the population, mainly subsistence farmers, who, in a democracy like India, constitute a significant political power.

The problems that arise when an agropark has to be inserted in the Indian situation therefore require much more consideration than in the previous designs, both in terms of the relationship with the existing farmers and in relation to agrologistics and the market where the products made at the agropark will be sold. I will pay particular attention to these aspects in my discussion of this example.

The agropark designs in India therefore begin as Intelligent Agrologistics Networks (IAN), which incorporate agroparks as well as distribution and consolidation centres. An IAN ensures that large-scale and industrially produced products from the agrosector around the world reach the consumers via the close-knit network of supermarkets and specialist stores in the metropolises in as fresh a state as possible. IANs (long-standing in Western metropolises but still in the design stage in emerging economies) are the response of agroproduction to the emergence of a global network society, which was examined in Chapter 2 of this work. As with all networks, they function on different spatial levels of scale, from regional to global.

The concept of Rural Transformation Centres (RTC) was developed with an eye on the producers, who want to bring their products to market via an IAN. RTCs include not only the collection of agricultural products, but also activities focusing on rural development. RTCs are also included in the IAN, and are the link between the agropark and the local communities. They work as a collection centre for local produce but can also be used for developments in agriculture and those involved in agriculture on site.

At the request of INDU Projects Ltd, an Indian project developer, feasibility studies were conducted in 2007 on agroparks in South East India in the Bangalore, Hyderabad and Chennai triangle. One of the feasibility studies concerns an area of more than 1,100 ha in Nellore in Andhra Pradesh. This was the place chosen to develop a Master Plan. In the summer of 2008 work got underway on the Matterscape component of the IFFCO-Greenport Nellore Master Plan on the three levels of scale of the IAN, the agropark and the RTCs.

6.8.2 Matterscape

The main problem for the agrosector in India is the virtual absence of processing. As a consequence, the agropark designs in India focus firstly on creating processing capacity and the necessary storage space. But the whole logistic chain from producer to consumer will also have to be tackled from this perspective, because that is where the biggest problems lie. Secondly, efforts are being made to capitalise on the new market demands of the fast-growing middle classes and the 'out-of-home market' in India. The Indian middle classes want a broader food package, more processed products and products with higher safety guarantees, delivered via supermarkets, which have incidentally only been allowed in India in 2005. They would agree to pay a higher price for food as a result. This too provides one of the solutions that need to be found, primarily among new forms of processing of existing products and different logistics.

From the moment food processing begins to become more important, it is no longer a way of getting rid of surpluses, as is the case now. Processing does place greater and different demands on the quality of the products to be processed. This transition can therefore not occur without the introduction of other forms of production, many of which can be added to this agroprocessing in agroparks.

IFFCO-Greenport Nellore has therefore been set up as part of an IAN, which includes other agroparks and in which consolidation centres are the link between the large-scale production in the agroparks and the demand for agroproducts from the metropolis (Figure 42).

From the perspective of the producer, an agropark supplies his products throughout the year as efficiently as possible, without depending on land or the seasons. The consolidation operates between the agropark and the buyers in the metropolis, serving a metropolitan market and responding to changing consumer preferences throughout the year. In theory it is located less than an hour from the centre of a metropolis. The essence of a consolidation centre is that it redistributes the thick homogeneous flows from producers and processors into small compound flows, as they are demanded by the supermarkets and other buyers in the metropolises. The IAN is crucial so that

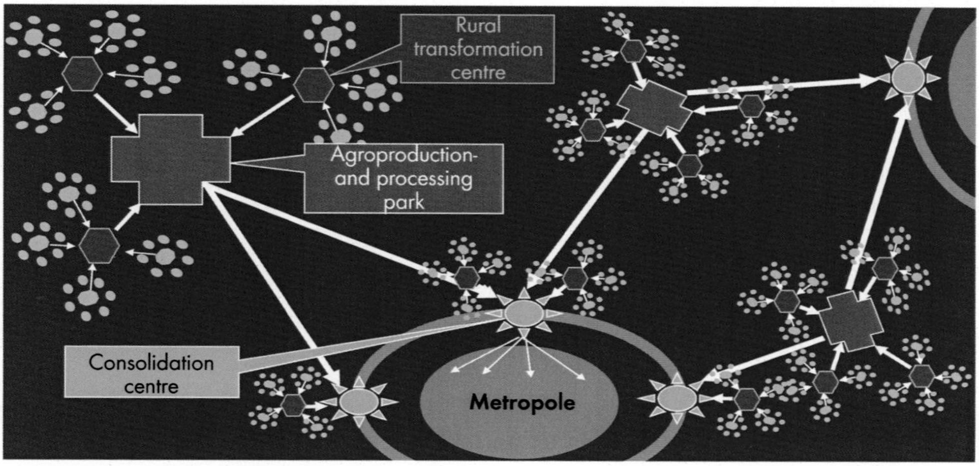

Figure 42. Consolidation centres, agroparks and rural transformation centres in an IAN.
Smeets et al. *(in press).*

production, processing and consumers can be linked up in this way: these networks link regional producers and global flows of basic products to modern wholesalers, supermarkets and via these to consumers in the ever-increasing middle class in the metropolises. Freshpark Venlo is a prototype of a consolidation centre, which supplies a sixth of the 20 million consumers in the South East of the Netherlands and the German Ruhr area with vegetables, fruit and flowers. In addition to the consumer-oriented consolidation parks, there are also export-oriented centres operating in a similar way.

In the Indian situation, the agropark is primarily focusing, as part of the IAN, on processing products from the existing agricultural sector. It will also produce intensively and land independent. In view of the increasing demand for convenience products, an agropark will also always have to be able to process significant trade flows of products, which are not produced in the vicinity or in the park itself. As always, these agroparks are also places where research and development can take place, and from where know-how and experience can be shared with the producers (Figure 43).

In the network the products from existing primary producers are collected in Rural Transformation Centres. These act as knowledge centres in their local environment for the associated farmers and as service centres (health care, credit facilities, ITC,

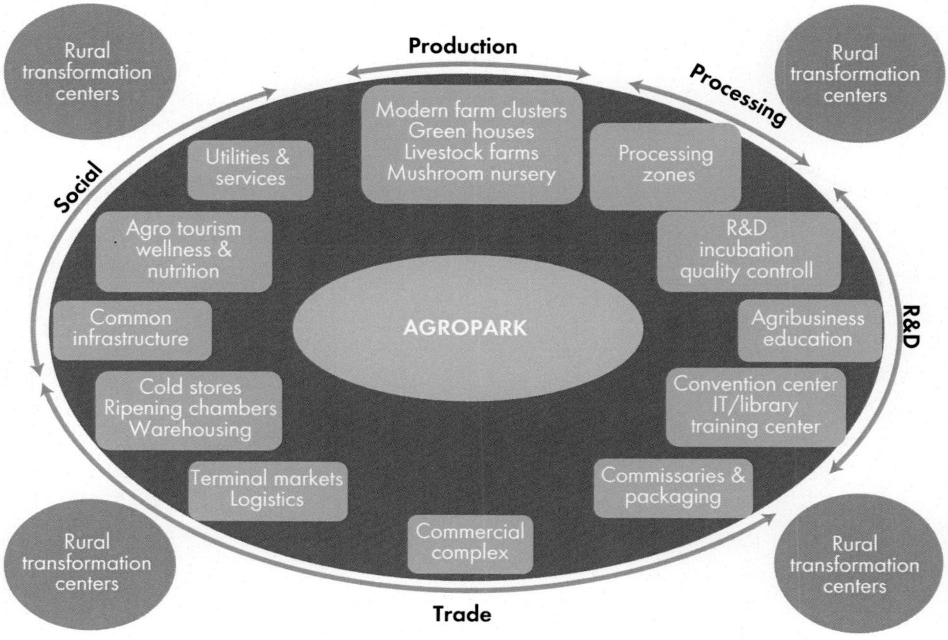

Figure 43. Functions of an agropark as part of an IAN.
Smeets et al. (in press).

etc.). Products go from the RTCs to agroparks to be processed, along with products from the agropark itself, or products that are imported into the agropark (Figure 44).

RTCs, agroparks and consolidation centres are differentiated here so as to clarify their function. In practice, they can be combined in various ways, depending on their specific location:

- agroparks with an RTC at the heart;
- agroparks with a consolidation centre at the heart;
- agroparks with an export centre at the heart.

Agroparks with an RTC at the heart are the centre of a highly productive agricultural area. The size of the area depends on the degree of specialisation in the region. The more specialised, the easier it is to reach the required critical mass to make the logistic process efficient. From this point of view, it is worth promoting specialisation in the intake area from the agropark.

For agroparks of the consolidation centre type, the number and proximity of metropolitan consumers is crucial. In reality Dutch consolidation centres need at least 3 million consumers for a consolidation centre to operate efficiently. The supermarkets used by these consumers must be serviceable within an hour.

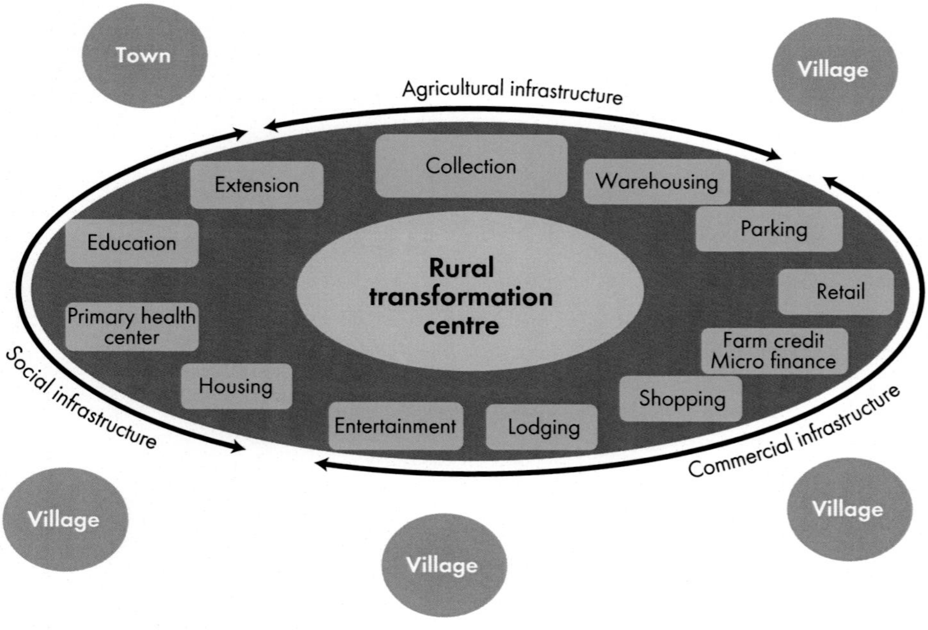

Figure 44. Functions of a rural transformation centre as part of an IAN.
Smeets et al. *(in press).*

For agroparks of the export centre kind, the proximity of a seaport with regular container connections with other key ports around the world, is vital. As far as high added-value perishable produce is concerned, air and rail may also be an option.

IFFCO-Greenport Nellore (conceptual Master Plan shown in Figure 45) will be set up as an agropark with an RTC at the heart, but will also be served by several other RTCs in the area. The draft design to be submitted is based on the production of meat (goats, sheep and chickens), milk, and vegetables in glass houses. Half of the park will be used for processing internal and imported meat, milk, vegetables and rice.

The calculations, made using the AF+ model for designing a Central Processing Unit, highlight the fact that cooling of the various components is critical to success. Efficient glass horticulture in particular cannot take place unless there is a relatively cheap cooling source. The relatively high humidity makes it difficult to use some of the cooling techniques here that are currently deployed in the Middle East.

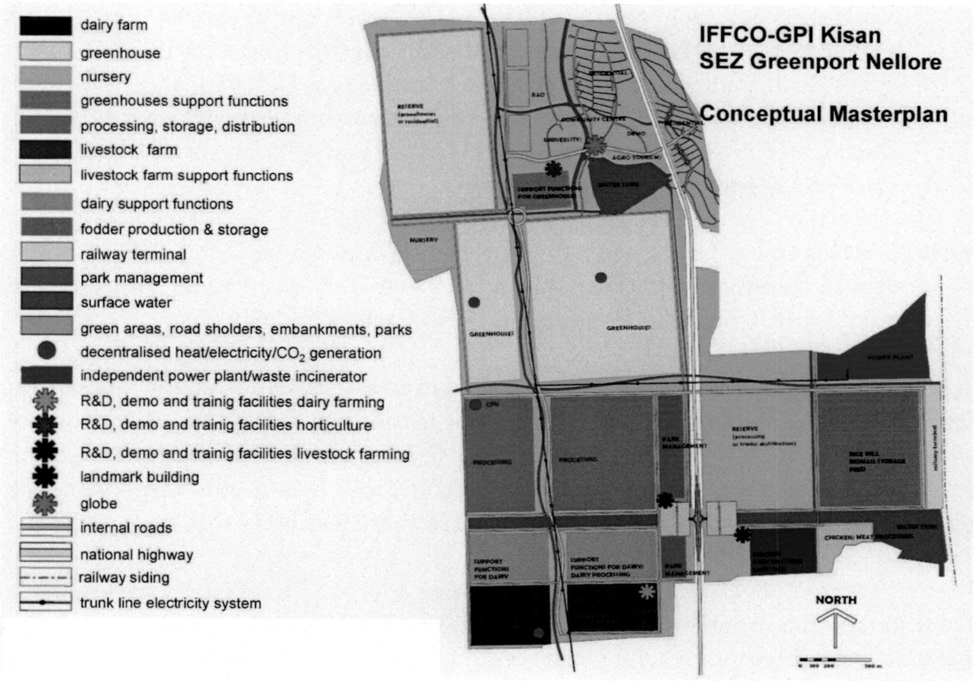

Figure 45. Conceptual Master Plan IFFCO-Greenport Nellore, June 2009.
Smeets et al. (*in press*).

6.8.3 Powerscape

With the help of experiences gained in the WAZ-Holland Park project, the collaborating knowledge institutions of Wageningen-UR concluded that the best approach to working in a multicultural setting such as China in the long term would be to form a collaboration with strategic partners so that several projects could be tackled simultaneously. This option was first discussed in 2004 when Wageningen-UR also decided to investigate the possibilities of agropark development in India.

This idea became more concrete during the first exploratory visit to India in 2005 when a suggestion for strategic collaboration was made by the Yes Bank. Yes Bank – the knowledge bank – was set up in 2004 by former employees of Rabobank India. The name of the bank was partly the result of a desire to radiate a positive image to customers. On its website Yes Bank portrays itself as a 'knowledge based industry (...) helping our clients to develop great ideas and nurture them to fruition.' The website describes the collaboration with Wageningen-UR as follows:

> 'YES Bank has secured a strategic alliance with Alterra, a leading international knowledge institution and part of Wageningen University Research Centre, Netherlands. The key objective behind this partnership is to achieve successful implementation of the unique Integrated Agri Food Park concept in India wherein high-tech Agri production, processing and logistics would be integrated in a manner that ensures the creation of efficient, effective and environment friendly business models.'[315]

Since 2005 this consortium has developed various propositions along different lines, via tenders to commercial parties and via talks with various regional authorities in the states of West Bengal, Andrha Pradesh, Maharashtra and Gujarat.

In July 2007 these acquisition activities resulted in an assignment from INDU-projects Ltd, an Indian project developer, to carry out feasibility studies into the possibility of developing an agropark in the Bangalore, Chennai and Hyderabad triangle in South India. The former owner of Christiaens bv, who is currently responsible for acquisitions in his company, also participated in these feasibility studies.

The motive for INDU-projects is an innovative approach to project development. Traditionally the money is earned by purchasing land, building housing or utility operations on it and then selling the resulting project to the end-users. The profit comes mainly from the added value of the land. INDU began developing thematic projects. Its first development was Health City near Bangalore, a project integrating a hospital, wellness facilities, hotels, a medical university and all the necessary housing

[315] Http://www.yesbank.in/food_agribusiness.htm.

and utilities. In order to assure the intrinsic quality of the project, the development was carried out in close cooperation with one of the top medical universities in the USA. After the project has been implemented, both partners remain owners of the park management; only some parts are outsourced to specialist companies, while the housing and some of the business buildings are sold. These theme projects thus deliver more added value than traditional project developments.

INDU also sees agroparks as thematic projects and wants to make money in a similar way; on the one hand from the sale of part of the utility buildings and housing connected to the agropark, but also by assuming a long-term role in the park management. They regard Yes Bank and Wageningen-UR as the intrinsic quality controllers of the process.

INDU and Yes Bank also participated in the symposium on Metropolitan Agriculture that took place in October 2007 in Beijing, and in the establishment of the Platform on Innovation of Metropolitan Agriculture. INDU-projects adopted the Greenport concept which became Greenport Shanghai and has since set up Greenports India.

One of the locations in the feasibility study in the Bangalore – Chennai – Hyderabad triangle (BCH-triangle) was an area of 1,100 ha in Nellore, which was offered to INDU as a site for an agropark development by the Indian Farmers Fertilisers Co-operation (IFFCO), a cooperative with more than 50 million members. The state authorities are playing a key role in the development of the project, both in the release of the land in question and the granting of the requisite permits. The initiative received much approbation from Andhra Pradesh's Chief Minister Reddy, and at an event attended by 50,000 farmers from the surroundings, he laid the first stone on 21 March 2008 for what is now called Agropark IFFCO-Greenport India Kisan Special Economic Zone Nellore (kisan means farmer).

Greenports India and IFFCO have set up a joint venture for Greenport IFFCO Nellore. This acts as project developer and assigner for Yes Bank and Wageningen-UR. New in the job formulation is the option taken by Yes Bank and Wageningen-UR to convert some of the revenues from this job into shares in the enterprise that will be responsible for the park management.

The Yes Bank and Wageningen-UR consortium in collaboration with INDU is meanwhile developing other agroparks in the IAN which also includes Nellore. The collaboration between Yes Bank and Wageningen-UR has become an Agrologistics Intelligence Unit, which gives advice on development problems in the Indian agrosector from a broad base of Dutch know-how. As far as the issue of cooling of greenhouses, stalls and processing areas is concerned, this collaboration has extended to include Dutch companies who are dealing with these problems around the world. Some of these companies have also indicated a desire to search in a pre-competitive collaboration for new solutions to the cooling problem in the exploration of foreign markets, such as India. If that works, the transdisciplinary collaboration would evolve into a new system innovation, whereby SMEs from the Dutch agrosector would innovate on a customised basis for the international market.

In the period since November 2008 Yes Bank and Wageningen-UR have organised a number of business delegations from the Netherlands and Israel. A start has been made during these delegations to organise joint ventures with Indian companies, who will undertake activities in the agropark, in RTCs and/or in the IAN.

The KENGi network, active in mid-2009 in India, is shown in Figure 46.

Non-governmental organisations, not including political parties, have not yet become involved in the social debate around Nellore agropark. The first stone laying ceremony did receive plenty of media attention. This event was explicitly seized on by the sitting coalition of Chief Minister Reddy, in which the Congress Party has the upper head, as proof of its connection with the farmers of Andhra Pradesh, because the agropark intends to create more value for these farmers by processing existing produce.

The absence of specialist KENGi brokers in this network is noticeable. This role has been assumed by the collaborating knowledge parties, whereby Wageningen-UR on the Dutch side and Yes Bank for India, are mobilising their networks and then

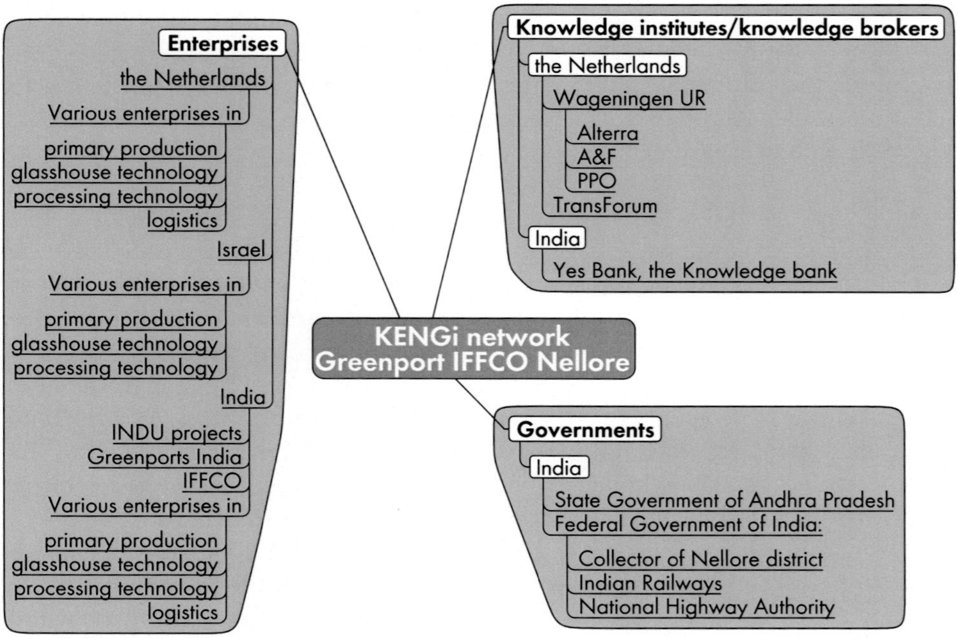

Figure 46. The KENGi network around IFFCO-Greenport Nellore.

linking them up. Partly as a result of experiences of Wageningen-UR in India and also because of the delayed effect that Dutch subsidy had on the second phase of Greenport Shanghai, Wageningen-UR and Yes Bank decided to relinquish initially at least Dutch subsidies when creating a new innovative position in the Indian market. The extra administrative obligations that go with international co-funding set-ups were also a reason for rejecting support from KENGi brokers.

A global cost-benefit analysis was conducted in the feasibility study for potential agroparks in the BCH-triangle, for agropark sites in this part of India. The results were positive. In the accompanying risk assessment, the biggest risks were as follows:

- syndication risk: the risk that, particularly in the initial phases of the project, the limited number of clients at that moment would not be able to produce the necessary funds (on time);
- interest rate fluctuation: leading to higher costs for loans;
- foreign exchange fluctuation: leading to higher costs for capital and foreign investments.

In the summer of 2008 the finishing touches were made to the Master Plan with the accent on the Matterscape elements. Bearing in mind the experiences in China on this point, Yes Bank and Wageningen-UR both spent a lot of time and energy building up a network of Dutch businesses, who undertook a first mission to India in November 2008.

6.8.4 Mindscape

Meat consumption in India consists mainly of chicken, sheep and goats' meat. Beef is not eaten because most of the population are Hindu. Pork is not consumed by the Muslim population and not eaten much by Hindus. Chickens are produced mainly in large-scale concerns for meat and egg production. Sheep and goats are bred almost exclusively in extensive grazing systems. Chicken, sheep and goats are slaughtered on the 'wet markets' in cities in front of the customers, so that the latter can be certain that the animal is healthy.

Since cows are sacred in Indian culture, special facilities will need to be provided, if milk production is to be included in agroparks. Beef consumption is out of the question in India, because of the Hindu population. But the slaughter of dairy cows that are no longer productive, is not accepted in this culture and is a taboo. Explorations have so far highlighted three solutions to this problem:

- Near the borders with Pakistan and Bangladesh, cows taken out of production are exported to these countries and slaughtered there. This partly explains why states like Gujarat and West Bengal have relatively high dairy production.
- In other places cows that are taken out of production are taken by Muslims and slaughtered in their abattoirs.

- But there are also several facilities, under community control, where decommissioned cows can stay until they die a natural death. This solution is a real possibility for under-productive cows in India. But highly productive cows would have to be detrained before they could continue living without the high milk production to which they are accustomed.

The knowledge value chain is developed in the Master Plan for IFFCO-Greenport Nellore from various angles. In the Rural Transformation Centres several facilities are included which are explicitly aimed at improving the situation of the participating farmers, such as a health care centre, a facility for education, information and training and a financial services branch.

In the joint venture, which is now shaping the Master Plan and which will soon act as Park Manager, long-term arrangements are being made about the form of association.

The contracts that are being set up between the collaboration Wageningen-UR – Yes Bank and the Nellore joint venture make provisions for Wageningen-UR and Yes Bank to have shares in the final agropark. Via success fees, linked to reaching certain milestones, extra bonuses on top of the normal consultancy fees have been agreed on. These can be converted into shares if so desired. This creates a much more long-term relationship between the knowledge institutions and the commercial enterprises that go to make up the Intelligent Agrologistics Network. This relationship is a sustainable starting point from which to develop the knowledge value chain.

Different subjective decisions have been decisive throughout the process:

- Staff at Yes Bank and Wageningen-UR, who decided to set up a strategic collaboration between both institutions with a wide range of collaborative forms as a perspective, ranging from an Intelligence Unit on Indian Agriculture to commercial participation in agropark projects.
- The directors of INDU who decided to apply the theme park innovation to the agroproduction sector.
- Staff at Yes Bank and Wageningen-UR who decided to decline the subsidised participation of KENGi brokers.
- Chief Ministers from Andhra Pradesh, Maharashtra en Gujarat who decided to give the go-ahead for government support to the development of agroparks. It is worth noting that the Chief Ministers from Andhra Pradesh and Maharashtra, both at the head of a centre-right coalition, both use the creation of more added value by processing and the reduction of losses in the logistics system as the most important argument here. Both may lead to more income for farmers. In Gujarat Chief Minister Modi of the centre right BJP sees the agropark as a reinforcement of the industrial corridor between New Delhi and Mumbai, most of which lies in this state.

From the beginning the Yes Bank-Wageningen-UR consortium has placed a lot of emphasis on targeted communication about the agropark proposition to different groups. In September 2005 workshops with representatives from regional governments and businesses were organised in New Delhi, Chennai and Calcutta, to discuss the concept of agroparks. The results of these workshops were presented to high-ranking government officials, in the presence of specific members of the press.

In the period from September 2005 onwards, Yes Bank (sometimes together with Wageningen-UR) marketed the concept further during many bilateral acquisition talks. This resulted in various tenders being issued, whereby those for INDU and IFFCO in 2008 could be converted into a collaboration, and in the summer of 2008 pre-tenders were also issued in Gujarat and Maharashtra.

After INDU and Yes Bank had both taken part in the symposium on Metropolitan Agriculture in Beijing, INDU set up Greenports India, as a subsidiary. Greenports India then took over the existing contracts for the development of agroparks with Yes Bank and Wageningen-UR.

6.8.5 Work process

As a result of the invention of the IAN (with consolidation centres, agroparks and RTCs) the ground became available for the IFFCO-Greenport Nellore and work got underway on ground preparation and infrastructure. This part of the project is therefore an innovation, and in view of the new collaborations that IFFCO, Greenports India, Yes Bank and Wageningen-UR have agreed in this context, can even be viewed as a system innovation. The federal government will be involved in this network via India Railways, which is planning to set up a terminal in IFFCO-Greenport Nellore.

As with the Chinese examples, NGOs are also absent from the design process in India. In the case of IAN Nellore, the interests of poor farmers are being protected via political coalitions which are currently being formed by the government both at the federal and state level, and which have a large part of their electorate among the small and landless farmers. IFFCO, as a cooperation of more than 50 million farmers, is also particularly sensitive when it comes to their interests, even though IFFCO has the more wealthy farmers among its grassroots supporters, who are in a position to buy fertiliser.

When forming the KENGi network, Yes Bank and Wageningen-UR deliberately opted for a dual role as knowledge supplier and broker between the other KENGi parties. This collaboration is a transdisciplinary process, whereby the scientific work performed by Wageningen-UR in the Dutch situation via the formation of a joint venture with Yes Bank and later other organisations is linked to the tacit knowledge of these organisations. For the knowledge institutions the IAN, the agroparks and

RTCs will function as practical laboratories and a practical training facility, generating revenue to fund strategic and fundamental research. The Yes Bank – Wageningen-UR consortium operates internally as a Community of Practice. In addition, the collaboration which was originally set up by a few of the staff at Alterra and from the Food and Agribusiness Advisory and Research department (FASAR) of Yes Bank, is slowly expanding to the whole of Wageningen-UR as well as within the Yes Bank.

The work process in India began with the creation of support in the federal government and various state authorities. This can justifiably be compared with the role that the discussion around Deltapark played in the Netherlands in creating support and even opposition to agroparks. As with Greenport Shanghai, the design process of the actual IFFCO-Greenport Nellore project was set up as a landscape dialogue which was extended to include various scenarios. However, the scenarios were not developed in parallel but in an iterative process in which different entrepreneurs got the chance to include their wishes.

In 2008 TransForum approved a project application specifically designed to monitor and evaluate the learning experiences in the Indian project. This facilitated the connections with the other agropark projects supported by TransForum, where the learning experiences have been systematically established.

To date the important learning experiences from the IFFCO-Greenport Nellore include the following:

- Working with a partner strong in international collaboration (Yes Bank) has major advantages, particularly if this partner is also intent on generating system innovations rather than on the successful implementation of projects, as was the case in both Chinese examples.
- A top-down approach, whereby the system innovation is first anchored via communication campaigns to governments, the business world and media, incurs higher running costs but creates a good breeding ground in which generic problems concerning government regulation and planning can be resolved and concrete projects can then be implemented.
- Dutch government subsidies or KENGi brokers funded by the government are not necessary if the institutions who want to implement the system innovation are prepared for this long incubation process.

6.8.6 Testing of the hypotheses

An agropark realises lower costs, greater added value and lower environmental pollution per unit of output and space.

The hypothesis cannot be confirmed or rejected on the basis of the existing agropark designs. It is clear, however, that the cooling system required for the use of (semi) closed greenhouses in particular, will require a great deal of energy. If this is demand is met directly by fossil fuel, then the cost in terms of environmental pollution will be higher. But this does not damage the hypothesis. On the contrary. The cooling problem can be resolved, for example, by clustering companies that produce high-temperature waste heat as a by-product such as electricity plants or waste incinerators. This further implementation of industrial ecological clustering would then lead to a confirmation of De Wit's hypothesis outside the expertise of agroproduction, for which the hypothesis was originally developed.

An agropark can only come into being on the basis of an integral design of matterscape, powerscape and mindscape both on the global scale of Intelligent Agrologistics Networks and on the local scale of a landscape.

This hypothesis is confirmed in India by the various motives used by the governments to approve the projects. These also put a different slant on the design of the agroparks to be implemented. In the Indian situation, the new element of the Rural Transformation Centres is also added to the Intelligent Agrologistics Network, acting on the one hand as a collection centre for primary products from existing farmers to the agro(processing) park, and on the other hand as an intermediary for all kinds of information flows to small farmers in their environment. Work was carried out simultaneously on the designs for the three parallel spatial levels (network, agropark, RTCs).

An agropark is a knowledge-driven system innovation and makes a significant contribution to sustainable development.

Thanks to the enthusiastic way in which IFFCO and the state authorities of Andhra Pradesh have gone about implementing this project, the incomplete Master Plan for IFFCO-Greenport Nellore can already be seen as an innovation. In this respect, the new collaboration between IFFCO, Greenport India, Yes Bank and Wageningen-UR can be viewed as a system innovation.

The agropark designs in India are still in the design phase. It is too early to assess their performance in the area of sustainable development. An *ex-ante* evaluation has not yet taken place either.

The design and implementation of system innovations like agroparks necessitates the participation of knowledge institutions, enterprises, NGOs and government (or KENGi). It is a transdisciplinary process in which the explicit knowledge of knowledge institutions and the tacit knowledge of the other partners are developed in a process of continuous iteration. KENGi brokers are the facilitators of this transdisciplinary process.

In India the debate about agroparks is taking place in the political arena. Here there is a more integral weighing up of the pros and cons than in the Netherlands, where the discussion is dominated by 'one-issue' movements. In states with different political coalitions in power, the expected benefits are explained in a different way. The necessary participation of KENGi parties is not confirmed by the Indian situation. The collaboration on the system innovation of IFFCO-Greenport Nellore is a transdisciplinary process, whereby the matterscape design is iteratively effected via the feasibility study and, thereafter, different versions of the implementation in the Master Plan. The role of KENGi brokers is adopted in the collaboration between Wageningen-UR and Yes Bank.

In all decision-making on the implementation of the integral agroparks design, with matterscape, powerscape and mindscape aspects, arguments from the world of justice and trustfulness take precedence over arguments from the world of truth.

The motives for designing and implementing agroparks, in this example too, have their origins primarily in the domain of powerscape, such as more added value for farmers or a better economic structure. Matterscape arguments are cited but take second place. Mindscape arguments play a role where there are religious objections to the slaughter of cows or the consumption of beef or pork.

In North West Europe social opposition against agroparks is directed primarily at intensive livestock farming.

The hypothesis is confirmed in the Indian example. Although only the chicken and dairy production exist in India as intensive non-land-dependent systems, there is no broadbased social opposition to it. From an animal welfare standpoint, there is no resistance to agroparks in India. There are, however, various objections within population groups to the slaughter or consumption of cow meat or pig meat, but these are religion-specific and are not directed at the livestock-holding systems or the welfare of animals.

Design of the agropark orgware with KENGi parties in a multicultural setting can take place once sufficient trust has been built up between stakeholders on the basis of an appealing matterscape design, in which there are still many open options.

The agropark designs in India are not yet advanced enough to test this hypothesis.

Only in a broadly composed KENGi network, focusing on open innovation, can the Dutch SMEs in the agrosector create system innovations on an international level and participate in them.

The agropark designs in India are not yet advanced enough to test this hypothesis. But a start has been made to building up this network.

Greenport seems to be an attractive international logo, under which the Dutch agrosector can propagate system innovations and quality management in its global network. Extensive collaboration and synergy between the Dutch Greenports is vital for this purpose.

The first part of the hypothesis is confirmed in the Indian situation. The clients who have made a substantial commitment to the development of agroparks immediately adopted the Greenport label and established themselves as Greenports India. The agropark designs in India are not yet advanced enough to test the second part of this hypothesis.

6.8.7 Conclusion

The work on agroparks in India created several inventions, which seen retrospectively are also important for the designs in China and the Netherlands. The realisation of the importance of the Intelligent Agrologistics Network (assumed in the Netherlands and China because it was already present) is the most important of these. A different work process was also used in India. As with the Dutch designs, there was no KENGi broker, and unlike in China, the projects were undertaken together with a strong Indian partner, with a well-developed network at the national and state level. In just three years the designers were thus able to make the journey from initial introduction to concrete projects. Building on the know-how and experience acquired previously in the Netherlands and China, the preinvestments necessary for this could be brought in by Wageningen-UR and Yes Bank without recourse to government subsidies. This again was an advantage in terms of speed and readiness for business.

6.9 Iterative testing of the resulting hypotheses

In this section, the resulting hypotheses, as synthesised from the seven examples via the research by design, are once more held next to all the examples, insofar as these hypotheses have been newly formulated or reformulated in relation to the starting point in Chapter 5.

6.9.1 The first three hypotheses

The first three hypotheses formulated in Chapter 5 are confirmed without reformulation in the examples in which they were tested. They state that:

> An agropark realises lower costs, greater added value and lower environmental pollution per unit of output and space.

> An agropark can only come into being on the basis of an integral design of matterscape, powerscape and mindscape at both the global scale of Intelligent Agrologistics Networks and the local scale of a landscape.

> An agropark is a knowledge-driven system innovation and makes a significant contribution to sustainable development.

6.9.2 The fourth hypothesis

The fourth hypothesis is more difficult to establish. It states that:

> The design and implementation of system innovations like agroparks necessitates the participation of knowledge institutions, enterprises, NGOs and government (or KENGi). It is a transdisciplinary process in which the explicit knowledge of knowledge institutions and the tacit knowledge of the other partners are developed in a process of continuous iteration. KENGi brokers are the facilitators of this transdisciplinary process.

The participation of NGOs in the design process seems to be particularly problematic in various examples. I will come back to this in the last section of this chapter.

6.9.3 The fifth hypothesis

In its original formulation based on the theoretical principles in Chapter 5, this hypothesis states that:

> In all the decision-making concerning the realisation of the integral agroparks design, involving matterscape, powerscape and mindscape aspects, subjective choices of individuals are ultimately decisive. The world of trustfulness (mindscape) is therefore dominant in the last instance.

On the basis of the New Mixed Farm example, in which politicians had the final word and brushed aside the mindscape arguments of many opponents, the hypothesis was reformulated to state:

> In all the decision-making concerning the realisation of the integral agroparks design, involving matterscape, powerscape and mindscape aspects, arguments from the world of justice and trustfulness prevail over the arguments from the world of truth.

The hypothesis was more broadly formulated as a result, but what remains is that the matterscape arguments from the world of truth (for example, energy and environmental aspects) seem to be subordinated to economics, social issues and cultural arguments. In this sense the examples of Deltapark and Agrocentrum Westpoort are also confirmed, where the arguments used are mainly in the domain of mindscape. The Agrocentrum Westpoort example seems to validate the fact that when weighing up the economic argument (powerscape) of lower costs and thus more added value, against the negative image of intensive livestock farming (mindscape), the latter argument was decisive.

6.9.4 The sixth hypothesis

This hypothesis was formulated in response to the first example (Deltapark) and originally stated that:

> Social opposition to agroparks is directed primarily at intensive livestock farming.

In response to the examples in China and India, it can be concluded that this opposition is applicable in North West Europe but does not play a significant role in China and India. Therefore, this hypothesis was modified:

> In North West Europe social opposition to agroparks is directed primarily at intensive livestock farming.

This hypothesis is confirmed in all examples.

6.9.5 The seventh hypothesis

In response to the Greenport Shanghai example, three new hypotheses were formulated. The first of those stated that:

> **Design of the agropark orgware with KENGi parties in a multicultural setting can take place once sufficient trust has been built up between stakeholders on the basis of an appealing matterscape design, in which there are still many open options.**

On the basis of the first and second examples, respectively Deltapark and Agrocentrum Westpoort, this hypothesis can neither be rejected nor supported, because both examples got no further than a matterscape design and because both projects were discontinued precisely in that trust-building phase between the KENGi parties.

In the work process applied in the New Mixed Farm example, this hypothesis is confirmed. What's more, the original matterscape design is repeatedly modified, because there was not enough trust between potential partners. First the glass horticulturalists withdrew because they no longer wished to be associated with the negative image of intensive livestock farming. Then the mushroom business went bankrupt. The location of New Mixed Farm was also modified in response to the debate that emerged in political circles about the project. The question is whether in the current situation, trust can be rebuilt. New Mixed Farm is now being implemented on the basis of a political decision, whereby various representatives from the opposition groups have announced that they are not yet ready to withdraw their opposition. There are now two possibilities: either the designers of New Mixed Farm try to win over the trust of the original opponents and include them in the design process of the powerscape and mindscape, which is still ongoing. Or New Mixed Farm is implemented, without the support of all KENGi parties. This would disprove the fourth hypothesis, which states that only if all KENGi parties are involved, will it be possible to implement system innovations like agroparks.

In the example of WAZ-Holland Park, the hypothesis is confirmed in the sense that after the completion of the matterscape design, it was not possible to create a basis for trust between the participating KENGi partners in the multicultural setting of Chinese-Dutch negotiations. Parts of the matterscape design are now being implemented by parties each working independently and not in the integral collaboration necessary in an agropark.

Finally, in the example of Biopark Terneuzen, the hypothesis is completely confirmed. A robust basis for trust was built up between the parties due to the absence of intensive livestock farming and as a result too of the consensus which existed between the KENGi parties involved on the economic significance of the agropark and on the industrial-ecological progress which could be made in the Terneuzen docklands. The parties involved are now working together on implementing the orgware and software for Biopark Terneuzen.

In hindsight, the conclusion from this testing is that the hypothesis is not refuted by any of the examples and is confirmed by many of the examples. Only in the example of New Mixed Farm is an alternative development line envisaged, which if successful, will falsify the entire reasoning behind the necessary and positive involvement of KENGi parties. I will come back to this in the last section of this chapter. The addition 'in a multicultural setting', made in response to the multicultural design in Greenport Shanghai, can reasonably be omitted. The reformulated hypothesis therefore states that:

> **Design of the agropark orgware with KENGi parties can only take place once sufficient trust has been built up between stakeholders on the basis of an appealing matterscape design, in which there are still many open options.**

6.9.6 The eighth hypothesis

This hypothesis was also formulated in response to the Greenport Shanghai example. It states that:

> **Only in a broadly composed KENGi network, focusing on open innovation, can the Dutch SMEs in the agrosector create system innovations at an international level and participate in them.**

As in the previous hypothesis, the starting point here is also a fully participating KENGi network and the same remarks made in relation to the previous hypothesis also apply here. The rest of the hypothesis concerns the international operations of Dutch industries. This was only discussed in the WAZ-Holland Park case, and the hypothesis is neither confirmed nor rejected in this project. The Chinese client in the WAZ-Holland Park example explicitly asked for Dutch investment in the implementation of the core of the agropark – the CPU. But there were no individual Dutch companies willing to take on this investment, much less a KENGi network that could provide the financing. Work on building up this network only began in the context of Greenport Shanghai, and in the IFFCO-Greenport Nellore project the network is beginning to bear fruit, particularly in the collaboration between enterprises and knowledge institutions.

In conclusion, the hypothesis is not falsified in the other examples, but it is still too early to confirm it on the basis of the examples presented here.

6.9.7 The ninth hypothesis

The last hypothesis, developed in response to Greenport Shanghai, must also be evaluated in its international setting. It states that:

> **Greenport seems to be an attractive international logo, under which the Dutch agrosector can propagate system innovations and quality management in its global network. Extensive collaboration and synergy between the Dutch Greenports is vital for this purpose.**

This hypothesis cannot be tested in the Dutch examples. This is primarily because in the Netherlands, glass horticulture is predominant in the further development of the Greenport logo, now that it has been declared a policy category in the most recent National Policy Document on Spatial Planning. Only the development of New Mixed Farm will actually materialise as a Greenport, but it is concentrating initially on intensive livestock farming. Nonetheless, an agenda will arise from this. If the Greenport logo is going to be used on an international scale to promote agropark development, the concept will have to be expanded in the Netherlands to the full range of agroproduction and not limited to what is after all a relatively random selection of vegetable products.

Likewise, in the WAZ-Holland Park example, the addition of 'Holland' as a motto and as an inspiration for the design, builds on the quality image that Dutch agroproducts also had in China, but the Greenport logo is only being used in Greenport Shanghai.

In conclusion, the hypothesis is not falsified in the other examples, and in the light of the above it should be reformulated and added to the eighth hypothesis, as follows:

> **Greenport seems to be an attractive international logo, under which the KENGi parties, focused on open innovation in the agrosector, can propagate system innovations and quality management in the global network. For this purpose Greenport must embrace the entire Dutch agrosector, and extensive collaboration and synergy between the existing and future Greenports in the Netherlands is essential.**

6.10 Conclusions from research by design

6.10.1 The first hypothesis

An agropark realises lower costs, greater added value and lower environmental pollution per unit of output and space.

This hypothesis, derived from De Wit's resource use efficiency theory, is confirmed by the model-based calculations carried out in the form of *ex-ante* analyses in the examples of Deltapark, Agrocentrum Westpoort, New Mixed Farm and Biopark Terneuzen. But it has not been demonstrated empirically, because none of the agropark in production have been monitored and evaluated on this aspect.

There are two main reasons for this improved efficiency. The first is that an increase in the intensity of the production process (in terms of more capital input per ha and capital input per man hour) goes hand in hand with increases in scale, in turn due to fixed costs associated with production fall per unit of output. The second is that the spatial clustering of different sectors results in the low-(transport)cost exchange of waste and by-products. This is the basic principle of industrial ecology, and it applies to different agrosectors but also to the exchange between agro- and non-agroproduction, as demonstrated in the Biopark Terneuzen example.

The first reason requires further comment. In recent decades agroproduction has gone from being supply-driven to demand-driven, and this demand is particularly defined in the economic centres of the world. Agroproduct prices are calculated in the trade-off between demand and supply. Resource use efficiency is a dominant factor here, but not the only one. Biopark Terneuzen proves that the inclusion of a CHP installation for glass horticulturalists offers an economic advantage, not because they can produce cheaper electricity than the industrial partners in the park, but because they can respond more flexibly to market demand, only supplying electricity at those times of day when demand is greatest and therefore the price is highest. This flexibility makes these horticulturalists more efficient, not in (ecological) terms of kWHs supplied per unit of energy used, but in terms of creating a higher price for supplying on time.

It is for these sorts of reasons, which also apply to consumer responsiveness of all agroproducts, that Greenport Shanghai was set up according to the Demo>Trade>Processing>Production principle, while the earlier agropark designs were devised as more supply-driven stand-alones. This was developed further into Intelligent Agrologistics Networks in India, where consolidation centres organise the harmonisation with demand markets, that an agropark cannot access precisely because of this resource use efficiency. All of this has already been stated in the original hypothesis.

The first hypothesis therefore supports the third hypothesis:

An agropark is a knowledge-driven system innovation and makes a significant contribution to sustainable development.

This hypothesis is confirmed in the development of the subsequent examples. Deltapark, Agrocentrum Westpoort and WAZ-Holland Park are inventions, which could not be practically implemented either as a value proposition, and therefore not as an innovation, or a system innovation. But the work put into them did provide a mass of valuable data pointing the way to sustainable development. They also instigated a social debate, which tested the perspectives of agroparks for sustainable development. New Mixed Farm and Biopark Terneuzen are innovations and they were tested in their separate development process on their potential contribution. The results were positive in both cases, and decisive for various permit procedures and for the decision by the entrepreneurs involved to start with the actual implementation. Biopark Terneuzen has advanced to such an extent in this process, and has also created new relationships in collaborations between Dutch and Belgian entrepreneurs, that it could be labelled as a genuine system innovation.

Greenport Shanghai, like the first three examples mentioned above, has not yet developed further than an invention, of which the contribution to sustainable development can only be assumed because the design builds on the know-how of knowledge institutions, governments and entreprises that are involved in Dutch projects. The images of the matterscape design now to hand, and the experiences in the design of the orgware has been extremely important in the rapid development of several projects which began in India. IFFCO-Greenport Nellore is one of those that can be regarded as a system innovation, because the project is being implemented, and because it arose from a coalition of Dutch and Indian knowledge institutions, governments and businesses, that are working together on the basis of newly developed relationships.

Finally, all seven examples are knowledge-driven in the sense that the most constant long line that can be drawn through the seven examples is the participation of Wageningen-UR, whereby researchers built up a body of knowledge on the basis of the successes and failures in these examples, and created a network of government organisations and companies, which is working on an increasingly successful and efficient basis.

In conclusion, the first and third hypothesis can be summarised as follows:

An agropark as knowledge-driven system innovation realises lower costs, more added value and reduced environmental pollution per unit of output and surface area and in so doing makes a significant contribution to sustainable development.

6.10.2 The second hypothesis

An agropark can only come into being on the basis of an integral design of matterscape, powerscape and mindscape at both the global scale of Intelligent Agrologistics Networks and the local scale of a landscape.

A global consensus is emerging on quality standards that on many fronts can be better assured by industrial production methods than by traditional agriculture. But as soon as mindscape arguments begin to play a key role in determining the quality, a relapse might occur because, for example, consumers choose non-industrial products rather than industrial products. We observe this in the dominant role played by the opposition – sometimes actually present but sometimes assumed by decision-makers – of consumers or citizens to intensive livestock farming in the discussion on agroparks in the Netherlands.

The design of an agropark should therefore include powerscape and mindscape alongside matterscape, as stated in the second hypothesis, without pretending to be able to have complete control over the world. Basically, if emotions in the real world seem to be the decisive argument in determining whether or not to oppose agroparks or buying produce from them later, while these carry so many promises in terms of resource use efficiency, a story needs to be developed that touches these emotions instead of endlessly repeating matterscape arguments that bypass them. In other words, there is a strong link between the above-mentioned hypothesis on production-ecology and the hypothesis on landscape theory.

As far as the advantages of spatial clustering are concerned, the issue here is optimisation and not maximisation. Different components of agroparks can also be too close together. This can again relate to matterscape arguments, for example, the risk of diseases spreading from chickens to pigs, so that there is a need to consider placing buffers between both forms of production. This has to be tailored to the local situation.

But mindscape arguments can also play an important role. Significant in this respect is the withdrawal of glass horticulturalists from New Mixed Farm because they did not want to be associated with the – in their eyes – negative image of intensive livestock farming. This is similar to the deliberate exclusion of the same intensive livestock farming by the designers of Biopark Terneuzen. These mindscape arguments are by definition culturally determined and so may differ on a regional and even local level. The hypothesis is therefore confirmed as a conclusion:

An agropark can only come into being on the basis of an integral design of matterscape, powerscape and mindscape at both the global scale of Intelligent Agrologistics Networks and the local scale of a landscape.

6.10.3 The fourth hypothesis

The design and implementation of system innovations like agroparks necessitates the participation of knowledge institutions, enterprises, NGOs and governmental organisations (or KENGi). It is a transdisciplinary process in which the explicit knowledge of knowledge institutions and the tacit knowledge of the other partners are developed in a process of continuous iteration. KENGi brokers are the facilitators of this transdisciplinary process.

The starting point of the hypothesis – the collaboration between KENGi parties, which is a necessary prerequisite for generating system innovations – is not unequivocally highlighted by the examples. It is confirmed by the break-down of the Deltapark and Agrocentrum Westpoort projects, which ran aground because of express opposition from environmental and animal welfare organisations in the case of Deltapark and because of the suspected acceptance problems in the case of Agrocentrum Westpoort, which were conveniently confirmed by the rejection of those same organisations. Biopark Terneuzen demonstrates that when all KENGi parties are of one mind, the system innovation can actually be accomplished – and on time. But New Mixed Farm shows that there is another way, whereby after careful consideration a political decision rejects the protests of many but not all environmental protection and animal welfare organisations, and still gives the project the green light. Here the local authorities in particular seem to have the objections of these organisations tested and, as it were, take on their function. In this sense, these organisations are helping to improve the design despite themselves. However New Mixed Farm is still in the invention phase, thus creating doubt about the hypothesis but not falsifying it.

The non-governmental organisations are completely absent from the foreign examples. Despite that, IFFCO-Greenport Nellore developed into a system innovation. But in that case, knowledge institutions, governments and businesses included many of the arguments usually put forward by environmental protection and animal welfare organisations in the design. NGOs participated from a distance. The state of affairs at New Mixed Farm shows that an integral test on sustainable development, like the one carried out by Blonk Milieu Advies[316], could compensate for the unwillingness of NGOs to take part in the experiments with agricultural system innovations.

This part of the hypothesis is therefore confirmed.

[316] Kool A., I. Eijck and H. Blonk (2008). *Nieuw gemengd bedrijf. Duurzaam en innovatief?* Blonk Milieu Advies, SPF Gezonde Varkens, Gouda, the Netherlands. See Section 6.3 for a desciption of this test on sustainable development.

The second part of the hypothesis – 'It is a transdisciplinary process in which the explicit knowledge of knowledge institutions and the tacit knowledge of the other partners are developed in a process of continuous iteration' – follows on logically from the fact that there is a collaboration between knowledge institutions with explicit knowledge on the one hand and governments and companies with tacit knowledge on the other hand.

The third part of the hypothesis is not indisputably confirmed by the examples either. In the Netherlands, the KENGi brokers (Innonet, KnowHouse, TransForum, Van de Bunt) seemed to play an important, sometimes crucial, role in the development of the projects. But their role was less evident in the foreign examples. In India, Yes Bank and Wageningen-UR were jointly the KENGi broker.

In summary, the conclusion concerning this hypothesis states:

> The design and implementation of system innovations like agroparks necessitates the participation of knowledge institutions, enterprises, NGOs and governmental organisations and a positive outcome from an integral test on sustainable development. It is a transdisciplinary process in which the explicit knowledge of knowledge institutions and the tacit knowledge of the other partners are developed in a process of continuous iteration.

6.10.4 The fifth hypothesis

> In all the decision-making on the realisation of the integral agroparks design, involving matterscape, powerscape and mindscape aspects, arguments from the world of justice and trustfulness take precedence over arguments from the world of truth.

This hypothesis was inspired by the experience with the Dutch projects. In the case of Deltapark and Agrocentrum Westpoort in particular, the negative image of intensive livestock farming, behind which are the arguments from the world of trustfulness (mindscape), was the decisive reason behind the discontinuation of their development. By excluding intensive livestock farming, but including manure processing, Biopark Terneuzen demonstrated, with its rapid implementation, that the issue centres mainly around animal welfare. Cornelis gives the hypothesis a philosophical basis in his original formulation, because in his vision the citizen in the future society will manage the domains of powerscape and matterscape by communicative self-steering (by definition mindscape).

However, the example of New Mixed Farm shows that these superior citizens (assuming that the opponents of this project are already communicative self-steering citizens of the 21st century) do not come to unequivocal conclusions.

On the one hand the opponents – environmental and animal welfare organisations – whose arguments are largely dispelled by the various tests on sustainability. Future neighbours of the project, whose fears about more particulate matter and heavy traffic are justified, but whose local council, looking on from above, decide that these objections do not outweigh the advantages (not only in terms of particulate matter and heavy traffic but also in many other areas) to be enjoyed from the spatial concentration in this project. Some representatives from knowledge institutions put forward arguments to substantiate those of the opponents.

On the other hand the supporters of the project: entrepreneurs, governments and other citizens, also supported by knowledge institutions, all backed by a close majority in the local council.

Effectively the only decision-making process left is democracy, in which the majority decides, once the arguments of the minority have been carefully considered and weighed up. Powerscape takes precedence over mindscape.

How does this work in China? The Chinese do use arguments from the various domains when making decisions on agroparks and these vary depending on the different examples. But the arguments used in the Netherlands about animal welfare are met with polite surprise in China, and are only understood if a link can be made between animal welfare and higher productivity. In other words, it is the difference between animal welfare and animal comfort, where animal comfort belongs to the domain of matterscape and animal welfare arises from a cultural view about how we as people think we should treat animals. The Chinese clearly have different views on this matter: 'Do you mind if we solve our human welfare problems first?'

In India there is yet another view. There, groups which adhere to a specific belief, are strict about their views on certain animals. Hindus do not kill cows and Muslims do not eat pork, but will slaughter cows from the Hindus. The essential arguments for the system innovation agroparks are sought primarily in powerscape and matterscape. The above-mentioned religious aspects are considered in the detailed implementation.

The hypothesis therefore applies to North West Europe, part of Sloterdijk's 'Crystal Palace', and there it is actually the expression of a discussion which has not yet reached a conclusion, according to many supporters of industrial agriculture. Are animal welfare and animal comfort only negatively affected by the scale on which animal production is being set up? In the discussion recently held in the Netherlands on mega-stalls in the landscape, this argument was sidelined.

This all leads to the conclusion that on a global scale there is no clear prevalence of one or two of the three spheres in the decision-making process. Matterscape, powerscape and mindscape are tools for the designer and all need to be weighed up in the political decision-making process. The hypothesis is not tenable.

6.10.5 The sixth hypothesis

In North West Europe social opposition to agroparks is directed primarily at intensive livestock farming.

This hypothesis has been convincingly demonstrated in all examples, whereby the contrast between New Mixed Farm – with its purely intensive livestock farming – and Biopark Terneuzen – without intensive livestock farming – emphasises how sharp the distinctions are. The New Mixed Farm project was started in 2003 and in mid-2008 is still embroiled in endless permit procedures, notably in a city and province which are certainly not the greatest opponents of intensive livestock farming and were allocated 'separate status' by a Minister. Biopark Terneuzen, which was started in 2005, is being implemented mid-2008 in a province where the council opposed an earlier design for large-scale glass horticulture. It is distressing to see how the pig and chicken farmers keep running up against a brick wall in their attempts to innovate, whereas in China and India, where they are participating in the Greenport Shanghai and IFFCO-Greenport Nellore projects, they are regarded as the best in the world with the highest scores for environmental protection, animal health, economic yield, animal comfort and food safety.

In the radical rejection of modern livestock housing systems, animal protection societies are exporting animal suffering and the environmental movement is exporting environmental problems. The consequence of this rejection is that, in the Netherlands, standards are often unilaterally and prematurely tightened, while intensive livestock farming is competing in a European or global market. As a result the Dutch entrepreneur is put at a disadvantage and because consumers in Europe and around the world, despite the pleas of the same groups, are eating not less but more meat, animal welfare will deteriorate on balance and the environmental problems created by old-fashioned farming systems will increase on balance. This is not the wisdom we expect from the guardians of the glacial time perspective.

The hypothesis must therefore be tightened in the following conclusion:

The organised campaigns against agroparks as a system innovation by organisations such as Friends of the Earth Netherlands, the Socialist Party and the Animal Protection Foundation, all of which are concerned with environmental and animal protection, totally ignore the demonstrable improvements to the environment and animal welfare that these agroparks can bring both in the Netherlands and elsewhere in the world.

6.10.6 The seventh hypothesis

Design of the agropark orgware with KENGi parties can only take place once sufficient trust has been built up between stakeholders on the basis of an appealing matterscape design, in which there are still many open options.

In the iterative testing which was carried out in Section 6.9, the second part of this hypothesis was confirmed: trust is a key factor for the system innovation, where the emphasis is not only on doing different things, but also on the different relationships between the stakeholders who do different things. Trust is the connective tissue in the orgware. But the question that has to be formulated in response to the discussion of the fourth and sixth hypotheses is, who precisely belongs to this inner circle of stakeholders. In the opinion in response to hypothesis 4, I argued that the role of NGOs here was a matter for debate. In response to hypothesis 6 it becomes evident that some of these groups would have a counter-productive effect if they achieved their goals. In addition, it is also the case that the issues which these groups claim to represent are often seen differently in a multicultural setting. The most clear example of this is that in the complex called animal welfare in the Netherlands, in China it is not recognised as animal welfare and is rejected as animal comfort in a cost-benefit analysis, while in India it is defined from a religious point of view. My conclusion is therefore that NGOs no longer need to be involved per se (as part of the nice abbreviation 'KENGi parties') in every system innovation process. Only Knowledge institutions, governments and entrepreneurs should take on this role. The primary task in every concrete project is to look for citizens who want to be involved constructively in discussing the design of this system innovation from a communicative self-steering position and from the local setting in which the agropark has to be designed as an 'inhabited expanse' and as a landscape. Whether they are organised groups or just individuals from communicative self-steering is secondary. What is interesting is that in various projects discussed here, individual citizens have played such a role. In two of the Dutch examples they were from a NGO, but acted on their own initiative, sometimes against the wishes of these groups. The following conclusion therefore arises from the hypothesis:

Design of the agropark orgware with knowledge institutions, enterprises, governments and citizens from the local area, where the park is to be implemented, can only take place once sufficient trust has been built up between these parties on the basis of an appealing matterscape design, in which there are still many open options.

6.10.7 The ninth hypothesis

Greenport seems to be an attractive international logo, under which the KENGi parties, focusing on open innovation in the agrosector, can propagate system innovations and quality management in the global network. For this purpose Greenport must embrace the entire Dutch agrosector, and extensive collaboration and synergy between the existing and future Greenports in Dutch is essential.

The hypothesis expresses the learning experience of Greenport Shanghai and has already been included in the organisational development in the work on the Indian projects. But it needs to be amended in response to the reformulation of the fourth and seventh hypotheses. Since the issue here is working in a global network, in the space of flows, the citizens that are the link with the local environment of each previous design, are less important in this network than the collaboration between the other three parties: knowledge institutions, governments and entrepreneurs. In conclusion, therefore:

Greenport seems to be an attractive international logo, under which a broadly formed network of knowledge institutions, governments and entrepreneurs, focusing on open innovation in the agrosector, can propagate system innovations and quality management in the global network. For this purpose Greenport must embrace the entire Dutch agrosector, and extensive collaboration and synergy between the existing and future Greenports in Dutch is essential.

7. Discussion

In Chapters 2, 3 and 4 urbanisation, network society and the spatial development policy are described as well as the specific form of agriculture, referred to as metropolitan foodclusters. As formulated in Chapter 1, the scientific aim of this publication is to find answers to the questions whether agroparks contribute to sustainable development in metropolises, how an agropark is developed and how the design should be realised.

The scientific method I have used to examine this aim is research by design. As far as the results of the design are concerned, research by design was based on two theories: the resource use efficiency theory and the theory of the three dimensions of landscape. To study the design process, I have used the co-design theory.

In this final chapter, I place the conclusions from the research by design back in the context of urbanisation, network society, spatial development policy and metropolitan foodclusters, and I will attempt to give them meaning in the light of these social issues and challenges. This will be done in Section 7.1 with the resource use efficiency theory and in 7.2 with the theory of the three dimensions of landscape. In Section 7.3 I develop a methodical feedback system by combining and anchoring the findings from the design process into the method of co-design.

In addition to this theoretical and methodical elaboration, I discuss some of the social consequences on the basis of two more practical conclusions that I have drawn from this work. Section 7.4 contains a spatial elaboration of the policy that the national government should in my view implement, as an elaboration of spatial development policy in the area of agriculture in the network society, intelligent agrologistics networks and agroparks. Finally, Section 7.5 contains a discussion of the knowledge infrastructure which will have to be adapted as a consequence.

7.1 Resource use efficiency of metropolitan foodclusters

7.1.1 Resource Use Efficiency applies at ever higher integration levels

The first conclusion of the previous chapter states that an agropark as a knowledge-driven system innovation realises lower costs, more added value and reduced environmental pollution per unit of output and surface area and in so doing makes a significant contribution to sustainable development.

This conclusion paves the way for a further expansion of the working domain of the resource use efficiency theory. The theory was originally formulated at the level of crop parcels, such as grain, and made judgements on the combined use of nutrients, processing, biocides, etc. But the number of resources had already been expanded in the area of glass horticulture. In a closed glasshouse it is cost-effective to finely regulate the various components of the glasshouse climate (temperature and humidity at various levels, CO_2, light interception and light use, etc.) at the same time as monitoring traditional inputs such as nitrate, phosphate and water. Each of these fine controls in turn contributes to an even better fine-tuning of each of the other single factors, and also to increased productivity of the whole system. Production ecologists have already extended the working domain of the theory to include the crop systems level and the combination of arable and livestock farming systems[317]. The WRR used the resource use efficiency theory for a Europe-wide discussion on optimising land use. In Chapter 5, I postulated that the domain of the theory up to and including these developments concerned mainly the sphere of matterscape and to some extent the domain of powerscape.

The *ex-ante* analyses of various agroparks in Chapter 6 show that the theory will also apply to the complete industrial ecological complex that is at work in agroparks. The higher the level of integration, and the more completely it is implemented, the greater the increase in resource use efficiency of the park. This applies to the agrosystems themselves, if they are mutually integrated, for example the pig and chicken breeding in the example of New Mixed Farm, which will earn money from the sale of electricity, heat and CO_2; or the intensive livestock farming and glass horticulture in Agrocentrum Westpoort, which should jointly accomplish the same thing. This also applies to the integration of chain components, for example the meat chicken chain in New Mixed Farm, in which production and processing are combined in one company; or the combination of production, processing and trade in Greenport Shanghai and IFFCO-Greenport Nellore. It is even applicable to the combination of integral chains and the accompanying logistics in the example of IFFCO-Greenport Nellore. The case of Biopark Terneuzen highlights the fact that the working domain of the theory crosses agrosystems borders, because Biopark Terneuzen is the integration of the chemical industry and primary agricultural production. This extends the integration level of the theory from crop parcels via crop systems and land use to complex industrial ecological systems, or better, industrial ecological networks, some of which are spatially clustered. The examples from Chapter 6 show that, if more physical factors

[317] Glendining M.J., A.G. Dailey, A.G. Williams, F.K. Van Evert, K.W.T. Goulding and A.P. Whitmore (2009). Is it possible to increase the sustainability of arable and ruminant agriculture by reducing inputs? *Agricultural Systems* 99: 117-125.

are integrated within the agropark system, the efficiency of the system as a whole will increase[318].

In the final example – IFFCO-Greenport Nellore – there is an even more far-reaching prospect. Logistics and consumer responsiveness – properties of the intelligent agrologistics network – are becoming new and different characteristics of resource use efficiency. The traditional resource use efficiency, aimed at crop parcels and crop systems, attempts to find answers to the question how the supply of resources can best be combined with the aim of reducing costs. In the WRR calculations the discussion is transferred to the domain of social choices (powerscape), when discussing the optimisation of land use. But when applied in the intelligent agrologistics system, the focus turns to the integrated use of resources in a demand-driven market[319]. After all, ultimately it is not only the costs that determine the profitability of agriculture, but the price of the products too. As a result, both the ability to know the demand and dynamics of the market (a knowledge question, see below under knowledge as a resource) as well as the capacity to respond to it quickly, are becoming crucial features of the system. Consolidation centres, as specific parts of Intelligent Agrologistics Networks, are a special form of agroparks which enable precisely this fast response. They are the most far-reaching specification of logistics as a resource.

---

[318] None of this is a plea purely for spatial clustering. There's an optimal point there too, which is determined by external circumstances. Yet a lot can be expected from bringing together glass horticulture and intensive livestock farming around logistic hubs. Intensive livestock farming also comprises dairy farms which have their cows standing in stalls the whole year round. But that is primarily because the spatial spread of these primary production businesses is so extreme in the former rural areas (partly as a result of decades of successful land reallotment, in which increasing the farmhouse plot was the predominant starting point, partly also because these farms originated on the site where much more land-dependent agriculture was once practised). And so in many cases any clustering of all these businesses, in which product processing from domestic products plays a less dominant role than the import and export from and to the external market, offers advantages. But the boundaries are determined by veterinary and phytosanitary factors. Directly because diseases can be transmitted and indirectly because of legislation that may impose a transportation ban in the event of an outbreak of a certain disease, which would then affect all parts of an agropark.

[319] In the WAZ-Holland Park and Greenport Shanghai designs this was first developed by including demonstration and education as crucial parts of the agropark, with the possibility of direct contact with consumers for the entrepreneurs. In IFFCO-Greenport Nellore a much more radical approach was taken, by making agroparks part of the intelligent agrologistics network around metropolises. In the Biopark Terneuzen example this aspect was highlighted in the dilemma of individual vs. collective electricity production. The solution that was designed (electricity generated on a large scale) was, in itself, the best one. But in the current electricity market in the Netherlands, it would appear to be more profitable for Dutch horticulturalists, who benefit from integrated electricity production (for assimilation lighting in the winter) and CO_2, to work with individual, smaller scale (and thus less efficient) electricity generation. Which is why the glass horticulture in Biopark Terneuzen is appealing to Belgian horticulturalists, who do not work with assimilation lighting.

7.1.2 Space as a resource

In the resource use efficiency theory, as originally formulated, space is the common denominator in all the comparisons that are made, and is therefore a neutral factor in the original theory. The efficiency of the use of production means per hectare increases proportionately if more production means are used in tune with one another. Therefore, within the limits of the resource use efficiency theory, as originally formulated, it was also possible to opt for more or less use of space[320]. If the environment requires it (for example, because the flushing away of nitrate in the vicinity of a water-collection area is unacceptable), the use of production means is reduced. The consequence of this, i.e. that the efficiency of the use of all production means also diminishes, was accepted in the discussion held at the time. However, this is becoming increasingly debatable, given the new insights into the finiteness of phosphate stocks and also as a consequence of the massive input of fossil fuels which is effectively needed for the production of nitrate fertiliser. The WRR report, 'Ground for Choices', also showed that extreme reduction in land use in land-dependent agriculture was possible and, given the reduction in fertiliser and biocide use, was even desirable, without affecting the level or quality of production[321].

In the designs of industrial ecological networks, spatial clustering in agroparks seems to provide great advantages; starting with the greatest possible reduction in transportation with its direct savings, and the indirect benefits such as reducing veterinary risks and animal suffering. But in this situation space becomes a scarce element in the metropolitan environment where agroparks are being developed. After all, not every site is suitable.

Conversely, clustering also has its limits: too many functions in too confined a space cause logistical problems. Veterinary precautions also often require distances and isolation. But solving this problem is nothing more than a qualitative addition to the quantitative restrictions of the availability of space. A site suitable for an agropark must be big enough right from the word 'go'. Space is seen as a limited resource that should be added to the traditional list of resources such as nitrogen, phosphate, biocides, etc.

As a result, the freedom to opt for extensification, which existed previously, is further curtailed. In the metropolitan environment, where the available space is scarce, and

---

[320] De Wit C.T. (1993). Tussen twee vuren. In: Themagroep Landbouw-Milieu (ed.) *Intensivering of extensivering*. Studium Generale, Agricultural University Wageningen, Wageningen, the Netherlands: 16. 'so "intensification or extensification" cannot be fulfilled by one or the other. Intensive agriculture should be applied wherever possible and extensive agriculture wherever necessary'.

[321] In the most extreme scenario which focuses on maximum reduction of land use, the area for cultivation with constant production falls from 127 mn to 26 mn ha (80% reduction), nitrogen surplus from 11 to 2.1 bn kg (81% reduction) and pesticide use from 400 to 21 mn kg active ingredient (95% reduction).

where there are also different limits set on the use of the available space, this choice is non-existent or at least drastically curtailed. If the aims, which for agriculture can be formulated from the perspective of sustainable development, are to be taken seriously, extensification of land use is not an option. This means that productivity increases, which can be achieved with agroparks, are no longer an option, and that space must be reserved for this purpose whenever governments try to plan metropolitan space. As with 'compact urbanisation', productivity increases in agricultural production are becoming a guiding principle for the spatial development policy.

7.1.3 The resource use efficiency theory and mindscape

The higher the level of integration, and the more completely it is implemented, the greater the resource use efficiency. The theory also seems to work for software resources such as knowledge and creativity[322].

The more knowledge is injected, the more chance there is of a successful and complete design. The development line running through the seven examples of the previous chapter, reveals that the earlier and more fully entrepreneurs, government staff and citizens are involved in the design, the more robust these agroparks, or at least the design of them, will be, and the more successfully they are likely to be implemented.

The success of the knowledge resource can undoubtedly permeate the operational phase of agroparks, in which knowledge of the market and the ability to capitalise on that, results in even better consumer responsiveness and a better anticipation of the rapid changes which are occurring among consumers in the metropolises. And this applies in the wider sense to all monitoring and evaluation of the processes in the park and the ensuing management decisions.

In its traditional formulation, resource use efficiency theory was a plea for multidisciplinary collaboration between the natural sciences, agricultural sciences and economics. Its use in the Ground for Choices[323] report served as an invitation to politicians, policy-makers and scientists to collaborate[324]. From the examples described here, and on the basis of the knowledge resource in the collaboration of the KENGi partners, it now invites transdisciplinary collaboration between scientists in the area of explicit knowledge and entrepreneurs and other stakeholders in the

---

[322] Van Ittersum M.K. and R. Rabbinge (1997). Concepts in production ecology for analysis and quantification of agricultural input-output combinations. *Field Crops Research* 52: 197-208, also introduce knowledge-intensive and knowledge-extensive labour as a production factor.

[323] Wetenschappelijke Raad voor het Regeringsbeleid (1992). *Ground for choices; four perspectives for the rural areas in the european community.* Wetenschappelijke raad voor het Regeringsbeleid, The Hague, the Netherlands.

[324] Van Ittersum M.K., R. Rabbinge and H.C. Van Latesteijn (1998). Exploratory land use studies and their role in strategic policy making *Agricultural Systems* 58: 309-330..

area of tacit knowledge. Using the success of the transdisciplinary collaboration between stakeholders, the resource use efficiency theory thereby becomes a plea for the participative approach in powerscape.

It seems from the reflections of Regeer and Hoes[325] on some of the agropark projects discussed, that when knowledge is seen as an input, it is primarily the emergence of a mode 2 learning position that applies as a proliferation factor in the efficacy of knowledge input. In short, it is not just the quality of the knowledge itself that ensures a smarter and more productive system, but chiefly the reinforcement of the learning capacity of the network (of people), that jointly designs the proposition, brings value to it and allows it to work, and with it the ability to be able to respond quickly within that system to external changes[326].

Resource use efficiency therefore works as a positive feedback mechanism on the meta level, whereby a system will work more efficiently – if the mutually tuned use of various resources in the spheres of matterscape, powerscape and mindscape are on a higher level, i.e. the integration level is higher.

7.1.4 Resource use efficiency in metropolitan foodclusters

In more general terms, the above is a plea for far-reaching clustering of industrial agriculture in agroparks and for the integral design of agroparks and the accompanying intelligent agrologistics network, in response to the global process of the emergence of the network society and urbanization, which plays such a key role in it. The examples and the generic conclusions that were formulated via research by design in Section 6.10, and the resulting generalisation formulated above, show that they must be designed simultaneously as space of place and as space of flows.

In the space of place, the actual agropark is designed in the three dimensions of the landscape: matterscape, powerscape and mindscape. In this 3D landscape resource use efficiency keeps its increasingly radical promises of higher productivity with the most efficient input of production means possible. This reduces economic and ecological costs as well as the amount of space that agroproduction takes from its metropolitan environment. With the input of the transdisciplinary knowledge resource, not only are

---

325 Regeer B. (2007). *Leren van biopark Terneuzen. Communicatie van kennis in context,* Afdeling Wetenschapscommunicatie, Athena Instituut, Amsterdam, the Netherlands; Hoes A., B. Regeer and J. Bunders (2008). Transformers in knowledge production. Building science-practice collaborations. *Action Learning: Research and Practice* 5: 207-220..

326 The latter was in fact predicted by the success of the traditional study clubs of horticulturalists, who as competitors made use of their mutual proximity and who shared their knowledge, albeit selectively, with each other, and thereby instigated innovations.

the initial designs better, but a learning system emerges that can adapt, and continue to increase its responsiveness to its direct environment.

In the space of flows the agropark is part of an intelligent agrologistics network that, in addition to agroparks, contains sourcing areas with land-dependent agriculture and consolidation centres (see Figure 39 in Section 6.8). This network must be defined on the scale level of the metropolis (and in the case of India, for example, must also be partially designed). It then becomes a sophisticated spatial-temporal cohesion, in which logistics play a central role. Together these metropolitan networks form the global network society, in which they each hold a unique position; on the one hand in relation to important production centres for primary raw materials, energy, feed, raw materials from bio-based economy and leftovers and by-products from agroproduction; and on the other hand in relation to the other important metropolises, inhabited by the consumers who generate demand for the products from the park.

The metropolis is the most significant form of urbanisation in the network society. It creates a new demand for agroproducts: not only for different foods but also different products ranging from biofuels to pharmaceutical products. It forces agroproduction, which supplies these products, to assume a new form – agroparks in intelligent agrologistics networks. Resource use efficiency is the driving theory pointing the way to more far-reaching integration of these systems as a solution to the problem of sustainable development.

7.2 Landscape theory

Two conclusions in Section 6.10 are the basis for a general view of the connection between the landscape theory formulated in Section 5.3 and its use in the spatial development policy discussed in Chapter 2. They state that:

- An agropark can only come into being on the basis of an integral design of matterscape, powerscape and mindscape at both the global scale of Intelligent Agrologistics Networks and the local scale of a landscape.
- The design and implementation of system innovations like agroparks necessitates the participation of knowledge institutions, enterprises, NGOs and governmental organisations and the positive outcome of an integral test on the characteristics of sustainable development.

In Chapter 6, in the discussion on the fifth hypothesis, it also became clear that when creating agroparks no hierarchy emerged between matterscape, powerscape and mindscape, in the sense that truth, justice and trustfulness dominate at different times in the development process and any one of them can be decisive at any given moment.

But the creation of a hardware design in matterscape was a logical first step in each of the examples, resulting each time in a conceptual master plan. Thereafter, however, the discussion focused on powerscape and mindscape.

7.2.1 Agroparks and spatial development policy

In Section 5.3 it was postulated that the landscape is the concrete form of Sloterdijk's as well as Castells' 'non-compressible' spaces of place. The landscape is also the spatial integration framework of the agropark, which is anchored via various network constructions (logistics, knowledge, water, etc.) in the regional and global space of flows. The above conclusions drawn from this, summarise how the spatial integrity of that landscape must be dealt with in the design. In reality, that process starts with a conceptual master plan of the hardware. In the Netherlands, the process of spatial development policy then begins.

In the spatial development policy, the relevant level of scale of the landscape and therefore of the agropark is most synonymous with the level of the city. It is probably no coincidence that in Dutch spatial planning law, one of the forerunners in the world, the legal binding of spatial planning is laid precisely on that local level of scale of the local spatial plan. If public involvement has to be organised – and this involvement typifies precisely the quality of the Dutch spatial planning armamentarium – then it will be most successful if it occurs at the level of the citizens space of place, of his landscape.

The different examples of agroparks in the Netherlands show that it is possible to implement this spatial planning from the bottom up in this country, without there being a spatial vision at the national level of the matter in question, i.e. the spatial planning of the agroproduction and logistics. There was certainly never any vision on the national level concerning the spatial planning of agriculture[327]. Land consolidation and thereafter land development were instruments used at the regional level and no higher. The most recent National Policy Document on Spatial Planning refers to five Greenports. The rest of the spatial policy in the rural areas is left to the province and city authorities. Nevertheless, there are still a few principles that form the backbone of the spatial policy of recent decades. The WRR listed these principles in its publication on spatial development policy[328].

'...

- Concentration of urbanisation, compact urbanisation in order to preserve the spatial quality of the green areas.

[327] Possibly with the exception of the *Koersbepaling Landelijk Gebied* in the Fourth National Policy Document on Spatial Planning (Extra), but see Section 4.3 on this matter.
[328] Wetenschappelijke Raad voor het Regeringsbeleid (1998). *Ruimtelijke ontwikkelingspolitiek.* Wetenschappelijke raad voor het Regeringsbeleid, The Hague, the Netherlands: 195-200.

- Spatial cohesion, holding together of urban functions, which appears increasingly difficult and so causes congestion.
- Spatial differentiation, which is disappearing fast as a result of stealthy urbanisation.
- Spatial hierarchy, which is declining fast as a result of the dispersal of high-value urban functions, and
- Spatial justice, which is being undermined by extensive suburban development.'

The design of agroparks primarily fulfils the first three of these principles.

- Concentration of urbanisation: agroparks redefine highly productive forms of agriculture as an urban activity (metropolitan foodclusters) and spatially reorganise this into as compact an area as possible.
- Spatial cohesion: agroparks localise the industrial agricultural activities into the urban area, near multimodal logistical hubs, to reduce congestion.
- Spatial differentiation: agroparks create a 'space pump', whereby industrial agricultural activities with a small direct but a big indirect use of space (odour circles, etc.) disappear from the green areas and thus create more space for other functions.

In the previous section I also concluded that 'productivity increases in agriculture' should be added as a principle of spatial development policy.

It therefore seems possible in bottom-up designs to preserve the principles of spatial planning without having to establish it in a spatial main structure. It could be described as the power of the KENGi approach, which builds on the collective intelligence of the participants, so that even if the great visions remain in the background, the basic principles are very much present and can be translated into practical designs. The KENGi approach seems therefore to be an excellent basis for innovative landscape designs, especially now that they too are substantiated by the resource use efficiency theory, where the highest possible integration level of knowledge resources is concerned.

The example of New Mixed Farm in particular shows that the government must redouble its efforts in the area of system innovations: it cannot just deploy its existing armamentarium of laws, plans and permits, because that by definition trails behind the system innovations being discussed. They must also offer space for experimentation, in which they then valorise the innovation as one of the KENGi partners, on an equal footing with the others.

The researchers in the KENGi must not only acknowledge the full value of the tacit knowledge of the other partners. They are also the guardians of the transdisciplinarity:

it is their job to test this tacit knowledge on quality and to turn it ultimately into explicit knowledge.

Some institutionalised NGOs appear to play an obscure role in blocking the system innovation of agroparks. And in some of the examples they only participated indirectly because their arguments were introduced by other parties. But it has also become clear how important the involvement of citizens is, especially at the local level, where the agropark has to be designed as part of the landscape. Of course it is only logical that these citizens unite in a local action group for that purpose and therefore form an NGO.

As far as the enterprises in the KENGi collaboration are concerned, the backing of the SMEs deserves most attention. Primary production in the Netherlands certainly consists mainly of SMEs, who are responsible for many innovations in the agrosector. Only if they organise themselves into networks, as happened in the case of New Mixed Farm, will they be able to get system innovations like agroparks off the ground.

This generic conclusion on the development process of agropark designs is specific for the Netherlands. It would be overstepping the mark to draw detailed conclusions from the three foreign examples on the development of these projects in their own environments, with the exception of the conclusion on the KENGi principle: without structural cooperation between, at least, knowledge institutions, government and enterprises, these projects will not last long either.

7.2.2 The aesthetics of agropark designs

The aesthetics of design, the final aspect of landscape theory, developed on the basis of the agropark designs, deserves a separate mention. The first two designs discussed, Deltapark and Westpoort-Agrocentrum, did not get beyond the stage of purely functional working drawings. The agropark design in Biopark Terneuzen was also limited to the functional cohesion of industrial ecology. In the process of designing New Mixed Farm aesthetics only crept in during the design. When the discussion with citizens in Horst reached its climax, the entrepreneurs realised that the aesthetics of the design was going to play a major role in the discussion. So they called in a landscape architect, who first redesigned the functionally-designed buildings and paid particular attention to the aesthetics of their position in the landscape of LOG-Witveldweg. These designs for intensive livestock farming buildings are therefore among the first in the Netherlands in which aesthetics does not consist merely of the addition of planting schemes, such as those hiding modern mega-stalls to make them less visible in the landscape. The New Mixed Farm buildings are very visible. Attention has been paid to their design.

In the foreign examples, a completely opposite approach was taken right from the beginning. Greenport Shanghai and IFFCO-Greenport Nellore are designed as living-working landscapes, in which the agropark is a centre for education and demonstration in addition to being a site for production, processing and trade. Attention to aesthetics is self-evident in this approach: it will be difficult to get people to come and find out about the latest developments in agriculture if the landscape is unappealing. The most radical design is that for WAZ-Holland Park, which is both an agropark and a recreational area.

The aesthetics of this design are an aspect of the mindscape par excellence. The examples from the previous chapter clearly show that much of the social debate, which precedes the political decision to set up an agropark and the entrepreneurs' decision to invest, is held in this mindscape domain. Failing to pay any attention to the aesthetics of the design, puts that design at a disadvantage in this discussion.

7.3 Methodical elaboration

The aspects described in the previous section can be used conversely as a generic methodical basis for the design of system innovations as agroparks. We are concerned here mainly with the generic conclusions on the third theory from Chapter 5 – that of the process. The most important generic conclusions concerning the process in Section 6.10 state:

- From the fourth hypothesis: the design and implementation of system innovations like agroparks necessitates the participation of knowledge institutions, enterprises, NGOs and governmental organisations and a positive outcome from an integral test on sustainable development. It is a transdisciplinary process in which the explicit knowledge of knowledge institutions and the tacit knowledge of the other partners are developed in a process of continuous iteration.
- From the seventh hypothesis: design of the agropark orgware with KENGi parties can only take place once sufficient trust has been built up between stakeholders on the basis of an appealing matterscape design, in which there are still many options open.

7.3.1 Working on system innovations

In Chapter 5 I postulated that agroparks are knowledge-driven system innovations.

System innovations are aimed at renewing the system itself. That has been made clear in Section 7.1. The resource use efficiency theory invites us to look for far-reaching system integrations in every design process for future agropark designs. In the area of matterscape it is primarily a matter of integrating different agroproduction sectors and chain component, as for example in the New Mixed Farm project. But other examples also reveal that good combinations can be made between agrosectors and other industrial sectors, such as the chemical industry, the energy sector, and waste processing companies. The theory therefore continues to offer the prospect of increasing economic and ecological productivity, the higher the integration level. Whereas, at the time of formulating the resource use efficiency theory, the agro-ecosystem was still defined as a crop parcel, now it is an agropark, incorporated in an intelligent agrologistics network which is simultaneously a space of place – a landscape – and is part of the space of flows. In that sense the system is constantly being renewed.

In Section 7.2 I concluded that earlier spatial planning had been shifted from the national policy which implements the dominant lines of spatial planning via a spatial vision from the top down, to a system in which agroparks are bottom-up regional designs, with which to implement the still applicable principles of this national policy. This redirected the responsibilities in spatial planning as referred to in the report by the WRR on spatial development policy. The principles play a background role in the process but with the early involvement of all KENGi partners in the design, the focus is much more on the collaboration between these partners than on the handing down of a National Vision from top to bottom in an activity allocation model.

In system innovations the interest groups involved form fundamentally different relationships. The previous section highlighted the fact that the KENGi partners are important here, whereby the role of knowledge institutions, government and enterprises has been clarified in any case. The different examples demonstrate how other relationships among these parties have taken shape. The knowledge institutions make the step from inter- to transdisciplinarity; governments no longer sit alone at the table handing out directional plans and testing the resulting designs afterwards, but instead create experimentation space in which they themselves become co-designers. Enterprises set up joint ventures with knowledge institutions and also sell their knowledge. Working on a system innovation means that existing mutual relationships must be explicitly visualised for the process, and the change in that relationship must be part of the design. A different role played by NGOs is a lot less evident from the examples.

The KENGi parties should not only be involved in the matterscape design, but should also focus on powerscape: what is the innovation ambition in the process? How are the experimentation spaces organised, so that these innovations can evolve without being immediately regulated to death? Where and which parties have legal, political and/or advertising power, and where are they prepared not to relinquish the power game in order to give the innovation a chance?

The parties should also be aware of their joint mindscape: how do they deal with the various types of knowledge (explicit and tacit) not only among those directly involved but also among the grassroots supporters of the participants in the design process?

And how will they tackle the different dimensions of the mindscape (aside from knowledge, emotion, aesthetics and education)?

7.3.2 The test on characteristics of sustainable development in the process

During the process, the resource use efficiency theory also gives a few handles for the ex-ante testing of, in particular, profit and planet aspects of sustainable development. In terms of the input of traditional resources such as nitrogen, phosphate, biocides and fossil fuel, it can simply be stated that efforts must be made to minimise the use of each of these per unit of product. The theory demonstrates first and foremost that this trade-off must occur in mutual cohesion.

For a generic method of testing sustainable development, the report by Blonk Milieuadvies[329] is a good guideline. In this report the performances of intensive livestock farming in New Mixed Farm are compared to the average performances of intensive livestock farming in the Netherlands and the spatial level of scale on which they score is also indicated from the various aspects. Figure 47 summarises the results of the test carried out on the New Mixed Farm design[330], which was discussed in detail in Section 6.4.

The test confirms the predictions of resource use efficiency: over a broad spectrum of aspects of sustainable development, New Mixed Farm shows improvements compared to the average scores of intensive livestock farming in the Netherlands. The only negative aspect is that the few lower scores, which occur exactly at the local level of scale, are a consequence of the extreme spatial concentration of production. Pollution from ammonia, odour, and particulate matter, increases locally. At the regional level, this pollution decreases over a much greater surface area.

It becomes more of a problem with the power- and mindscape resources such as space and knowledge. These can be partly expressed in terms of money (land prices, man-hours for researchers) but this value doesn't represent everything. How can an odour circle be dealt with when it affects land prices? Even more difficult is the testing of aspects such as animal welfare or aesthetics, using objective standards. In a generic sense, it can be concluded that the KENGi parties will have to make a judgement about this during the design process itself and later in the form of a regularly recurring discussion on these aspects.

---

329 Kool A., I. Eijck and H. Blonk (2008). *Nieuw gemengd bedrijf. Duurzaam en innovatief?* Blonk Milieu Advies, SPF Gezonde Varkens, Gouda, the Netherlands.
330 *Ibid.*: 65.

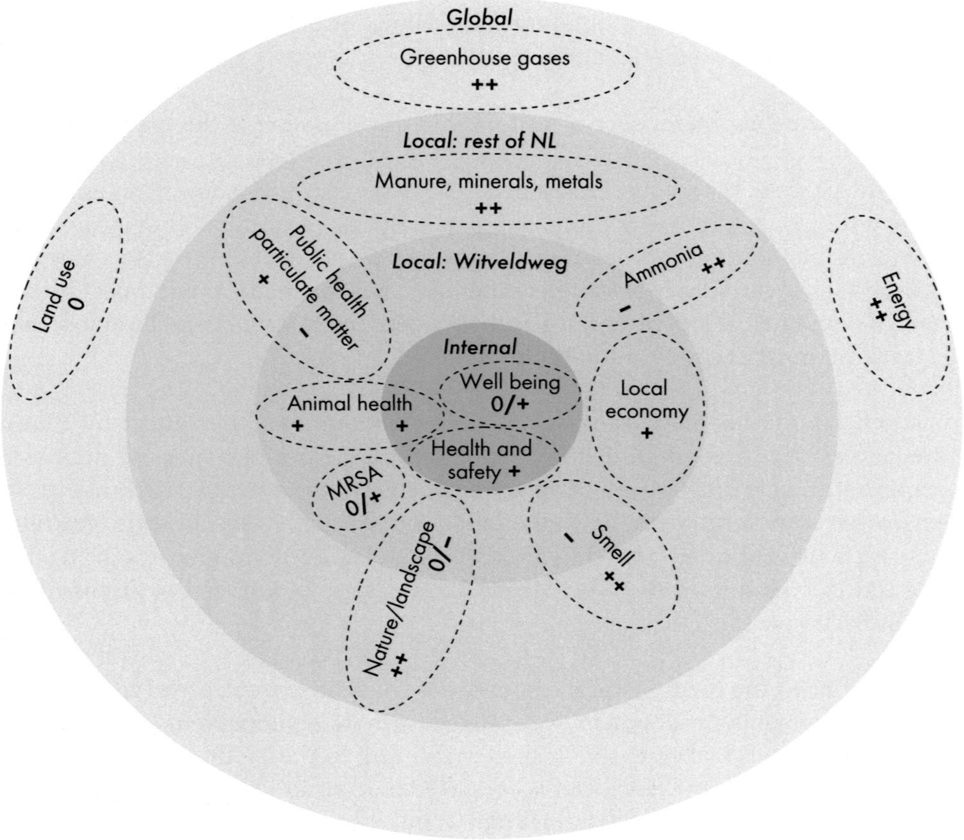

Figure 47. The different components of the sustainability test, performed by Kool et al. 2008, on New Mixed Farm. The score per component is positive, neutral or negative, compared to the reference - conventional livestock farming. Each component is positioned in the levels of scale on which it scores.

7.3.3 Landscape dialogue or co-design as a process formula

Table 10 summarises the key conclusions on the extent to which the various designs have been able to realise the character of landscape dialogue or co-design, as highlighted in Chapter 6.

With the hindsight gained from this publication, three long development lines can be sketched, in which the landscape dialogue is put into an overall national and even international perspective.

- In the Netherlands, Deltapark and Agrocentrum Westpoort paved the way, via a discussion framework, in which opponents and supporters, operating particularly on the national level, took up their position in the system innovation of agroparks. Furthermore, the designs instigated a number of international publications and media articles. An informal Community of Practice emerged from the Agrologistics Platform which, with the help of TransForum, set up two explicit co-design processes for regional agroparks; Biopark Terneuzen (without intensive livestock farming in the park), which is now being implemented; and New Mixed Farm (ultimately reduced to intensive livestock farming only – by necessity), which is still completing the permit procedure. Therefore, there are co-design processes at the regional/local level but also against the backdrop of a much wider debate, being held at the national and even international level.

- In China, two local co-design procedures were completed, both of which have led to a conceptual master plan and one of which (WAZ-Holland Park) is being partially implemented. Only after these designs were finished, was a start made on recruiting support at the national government level.
- In India, support was created first at the federal government level and in particular at the state authority level. The system innovations agroparks were presented to the Chief Ministers of both Andhra Pradesh and Gujarat, and received their personal commitment. Thereafter private enterprises were sought and found, to invest in the design for a conceptual master plan.

Obviously, the regimes of the Netherlands, China and India, where these three developments evolved, cannot be compared with one another. And yet it seems from the early projects in the Netherlands as well as the projects in China, that it is not sufficient to have the support of local authorities only. The later projects in the Netherlands, in particular Biopark Terneuzen and the IFFCO-Greenport Nellore project in India, show that the combination of top-down support of an otherwise bottom-up developed design is more successful. So, one is forced to conclude that an initiative that is only undertaken from the bottom up has less chance of succeeding than if there are two directed lines, one of which consists of obtaining top-down support and commitment, which is fleshed out – and put into practice on the basis of bottom-up designs of real-life projects.

A transdisciplinary knowledge base emerged from all projects, part of which has been laid out in this publication. Key transdisciplinary insights generated in the various projects are as follows:

- From Deltapark: the importance of a well-thought-out communication plan in the anchoring of system innovations.
- From Agrocentrum Westpoort: the expansion of the industrial ecology between different agrosectors with the other industries – such as waste processing and energy production – which are also present in the port.
- From New Mixed Farm: the importance of careful communication with the local authorities and citizens, and the obscure role of some NGOs and how to deal with them in the KENGi process.

Table 10. Conclusions concerning the landscape dialogue in the various designs from Chapter 6.

Deltapark	Drawing-board design on the basis of which a wide-ranging discussion was held on the pros and cons of agroparks. No landscape dialogue as process.
Agrocentrum Westpoort	Some key criteria of the landscape dialogue were not been met. Compromises had to be sought; there was no free space, in which a creative process between the KENGi parties could take place. There was a large group of professional amateurs around the table but too many of them were tied to the opinions of their company or organisation.
New Mixed Farm	The breeding ground for setting up a network around this project was the Regional Dialogue North Limburg, which had all the hallmarks of co-design. The New Mixed Farm project can be characterised as the implementation process of one of the projects of the regional dialogue, but itself resumed the character of a landscape dialogue at the local level of scale.
WAZ-Holland Park	The design came about as a result of research by design with knowledge institutions, government and enterprises from the Netherlands and China.
Biopark Terneuzen	The work process in Biopark Terneuzen can definitely be seen as a form of co-design together with the Agrocentrum Westpoort and New Mixed Farm projects. The researchers involved iteratively took the inventions, which arose from previous projects, off the shelf in the design phase, weighed them up in a transdisciplinary environment and with the KENGi parties involved developed them into value propositions.
Greenport Shanghai	The landscape dialogue is applied in this project in an international setting. Starting from the criteria which the Master Plan had to fulfil in the Chinese planning system, the plan was extended with scenarios, which included options for implementation with combinations of companies, and beneath that, a fixed pattern of zoning and infrastructure. In the organisation of the enterprise network in particular, the open innovative character of the co-design was highlighted, in which the creation of trust between the participants was paramount. From their position as SMEs, the entrepreneurs had difficulty playing the different roles of innovator, support generator and decision-maker simultaneously. Only when the conceptual master plan had been completed, was a start made on creating support among the national authorities of China. The design phase in the project was completed prematurely. There was no design of powerscape or mindscape; instead negotiations on investments were started immediately without there being a basis for trust between the Dutch and Chinese partners.
IFFCO-Greenport Nellore	The work process in India began with the drumming up of support from the federal authorities and from the governments of the various states. This can be compared in a way to the part played by the debate on Deltapark in the Netherlands in creating support for or opposition to agroparks. The design process of the actual IFFCO-Greenport Nellore project, like that of Greenport Shanghai, was set up as a landscape dialogue, extended with various scenarios. However, the scenarios were not worked out in parallel with each other but in an iterative process, in which different entrepreneurs got the chance to input their wishes.

- From three previously mentioned processes and Biopark Terneuzen together: the dominant role of intensive livestock farming and in particular intensive pig farming in the social debate, and the acceleration of the design process that can be achieved by consciously excluding intensive livestock farming.
- From WAZ-Holland Park: the realisation that, as long as it is well designed, an agropark can function as a key tourist attraction and has great recreational potential.
- From Greenport Shanghai: The design of the demo-trade-production trio.
- From (the preparation of) IFFCO-Greenport Nellore: insight into the importance of a local strategic partner for international projects, in which the Dutch network operates.

This knowledge base is being supported by the members of an informal Community of Practice. I will come back to the future prospects of this Community of Practice in Section 7.5. It is also clear from the examples that co-design goes hand in hand with taking opportunities (*Kairos*) – which cannot be planned in advance. The KENGi network of agropark designers operating in the Community of Practice can be regarded as being in a much better position to see and use these opportunities.

This publication and its theories can be used to further develop that learning capacity. On a level of abstraction higher than the various designs that have been discussed, this book combines as a whole an inductive with a deductive approach, as indicated in Figure 10, in Section 5.1. In the inductive approach I have used working hypotheses, derived from various theories (resource use efficiency theory, landscape theory and co-design) as a guiding principle for testing the content and process of the designs. In the ongoing projects the working hypotheses are being used to intervene in the design and in the process.

In the deductive approach these working hypotheses, insofar as they are confirmed by the examples, are used in this last chapter to enhance these theories. This emergent theory has been brought into the scientific debate via this publication and once more given rise to an enhancement of the theoretical principles and conclusions.

In each subsequent design and during the further implementation and operationalisation of the current designs, this process can be used and extended time and again with subsequent iterations. That is one of the tasks in the monitoring and evaluation of the learning process. This way the emergent theories can be used again with each successive design as the basis for the co-design process. Others can obviously be added where relevant.

As soon as the designs become operational, the obvious next step is to measure the results. These three theories can again be used as a starting point for this purpose.

7.4 Greenport Holland

In its report *'Plankgas voor het glas'* the Council for Rural Areas recommended no longer looking for expansion locations for glass horticulture in West Netherlands in the inner ring of the Groene Hart area, but instead in West Brabant and on the Zeeland and South Holland islands, around Schiphol and in Flevoland[331]. The reasoning behind this recommendation is that in the coming decades, new markets will emerge, due to technological developments within agrologistics and in particular the development of fresh transport via sea containers, both in food horticulture and ornamental horticulture, which can not only be accommodated by the expected increase in productivity in the existing glass crops, but for which an expansion of existing ground for cultivation is being discussion:

> 'Expansion of the European Union and the increased demand for glass horticulture products is expected to create new growth opportunities. The advent of Coolboxx sea transport and short sea lines create new export options to the United States of America, Spain, the German Ruhr area, England and the Baltic Sea region. (...) Starting with a cultivation area of more than 10,300 ha in 2003 (...), the Board expects that (...) by 2020 an area of between 12,500 and 14,000 ha will be needed.'[332]

But achieving this growth, in line with the vision of the Council, is an exercise that presupposes collaboration on the basis of a similar sounding vision between five provinces, one of which (South Holland) will have to leave a significant part of the anticipated expansion in the cultivation area of glass horticulture to the other four. The question is whether the national government will be content merely to facilitate this complex operation. The allocation of the Greenport label – which as I postulated in Section 4.3, is an operation without many ambitions in the National Policy Document on Spatial Planning and primarily confirms the current situation – seems insufficient at any rate for facilitating all the spatial reorganisation of glass horticulture. It is precisely the development perspective of glass horticulture and ornamental horticulture, including the long-term expansion options, which will have to be part of the spatial development policy concerning agriculture.

The Council for Rural Areas is consistent in that it also advises the LNV Minister to stop using state funds to subsidise the development of glass horticulture locations outside the Greenports, citing as the key reason that this dissipated development only creates satellite locations which are largely dependent for delivery and sales on the Greenports, and therefore create additional logistical movements. It would be better

---

[331] Raad voor het Landelijk Gebied (2005). *Plankgas voor glas? Advies over duurzame ontwikkeling van de glastuinbouw in Nederland.* Report RLG 05/2, Raad voor het Landelijk Gebied, The Hague, the Netherlands.

[332] *Ibid.*: 30.

to find locations for expanding the Dutch cultivation area right next to the existing Greenports[333].

But the Council for Rural Areas is clinging to a traditional perspective, which continues to start from a continuous development of monofunctional glass horticultural areas. In contrast, I am arguing in this publication for the development of glass horticulture, integrated with the other forms of industrial agriculture in the form of agroparks. The search for good locations must therefore include identifying optimal site conditions for glass horticulture. It concerns in any case the combination of intensive livestock farming and glass horticulture with production and processing which have been further industrialised – such as mushroom cultivation, closed fish breeding and all kinds of processing of agroproducts. Last but not least, this also applies to dairy farming. This sector is heading in the direction of industrial animal husbandry systems, which have disengaged their feed supply from their own land. There is still a regional link to land among these businesses, whereby they are transferring their preference from soil suitable for pasture to soil suitable for arable. But the import and export of raw feed and manure respectively, no longer occurs exclusively from and to the farmhouse plot and, like all other imports and exports, by road. An important added justification for state intervention in this sector is to be found mainly in the expected acceleration in the industrialisation of this sector, which will occur if the European quota policy is watered down or lifted in 2015.

State policy will, in other words, have to widen its focus from horticulture, now established in the rural allocation of five Greenports, to the potential combinations which glass horticulture and intensive livestock farming, dairy farming, other industrial primary production and related activities such as logistics and trade, together in agroparks, could deliver.

This is related in any case to state intervention in recent years in intensive livestock farming. Since 1998 this intervention has taken the form of reconstruction. In terms of Spatial Development Policy, the reconstruction areas were often seen as a textbook example of development areas, whereby the state specifies, 'to prevent a levelling of the spatial quality (...) which type of development needs to dominate in a certain area'[334]. Since then the reconstruction process has taken shape and zonings of agricultural development, recruitment and expansion areas have been established by the five provinces involved in the reconstruction areas. But on the matter of the whole reconstruction policy, there is an overriding feeling that the objectives have

---

333 *Ibid.*: 33.

334 Wetenschappelijke Raad voor het Regeringsbeleid (1998). *Ruimtelijke ontwikkelingspolitiek.* Wetenschappelijke raad voor het Regeringsbeleid, The Hague, the Netherlands: 151.

only partially been realised[335]. Here too the perspective of agroparks goes much further than the spatial concentration of intensive livestock farming in Agricultural Development Areas. It is also a new development prospect for this sector.

Starting from the surface area which is now taken up by glasshouses, stalls, processing companies, trade, storage and transport businesses, and all kinds of support activities, there should be space in the Netherlands, in the event of a complete conversion, for approximately 20 agroparks measuring between 1,000 ha and 5,000 ha. That number (or the average size) could rise by about 50% if the lion's share of the dairy farming sector found a place in the agroparks. It is also relevant that, for various reasons, the Netherlands is an ideal country for such an integrated form of metropolitan foodclusters. Firstly, because of its existing export position, founded on even greater progress in terms of productivity and quality, and on an increasingly expanding trade and distribution network. This export position of domestic products also means (together with the general position as 'gateway to Europe') that the Netherlands is a major importer of products that are sent on to other European countries. Then there is the high-quality transport infrastructure, knowledge infrastructure, a favourable climate, and last but not least, a very innovative stock of agricultural entrepreneurs. All this leads to permanent expectation of growth, as the result of further expansion and intensification of sales markets and increasing control of a global network.

With such future prospects, what would be the ideal location of 20 to 30 (perhaps even 40) agroparks each totalling an average 2,000 ha to 3,000 ha? There are a number of criteria which have more or less weight according to whether an agropark gets a specific focus in terms of production and processing activities.

- Proximity of major population concentrations, not only in the Netherlands, but also across the border (Rhine-Ruhr area and Flemish Diamond). The emphasis of production and processing is on processed fresh products which arrive ready-to-eat in the shops, and must then be sold quickly. Major population concentrations are also a source of labour and can provide waste that can be used in the agropark for climate-neutral energy generation, while they can sell an energy surplus, by making intelligent use of day-night discrepancies. Within the Netherlands, this criterion is not all that distinctive.
- Proximity to sea ports or inland ports, suitable for transfer of agroproducts (focus on livestock farming for bulk import of feed, or on overseas export of fruit and vegetables, or import and export of huge quantities of frozen produce).
- Proximity of main railway line for goods transport (Betuwe route, IJzeren Rijn; if possible also use of HSL-Zuid and -Oost), for the necessary shift from road to rail.

335 Boonstra F.F.G., W.W. Kuindersma, H.H. Bleumink, S.S.d. Boer and A.A.M.E. Groot (2007). *Van varkenspest tot integrale gebiedsontwikkeling: Evaluatie van de reconstructie zandgebieden*. Report 1441. Alterra, Wageningen, the Netherlands.

- Link to existing major, modern glass horticulture concentrations (as closest current approach to agroparks).
- Environment with relatively intensive land-dependent activities, the products of which can be concentrated, processed and distributed in the agropark (open-ground horticulture, bulb cultivation, ornamental horticulture, dairy farming in clay meadow areas which will only make the move to non-land-based production later).
- Proximity of knowledge institutions and creative centres.
- Proximity of sources of waste heat and CO_2.
- To make transition easier: proximity of current concentration areas (mainly of intensive livestock farming, but also old glasshouse areas) which are not viable in the long term either financially or because of their effect on the environment.
- Links to other urban elements of the same size and scale: major industrial terrains, ports and airports, major new-build regions, big, intensively designed recreational areas – in landscapes which can tolerate such a large scale (for example, reclaimed land, young sea clay areas, recently reclaimed sand areas, peat colonies).

The proximity of port facilities and existing (mainly glass horticulture) concentrations, seem to be the most important and most distinctive criteria on this list.

Figure 48[336] shows locations, with sea and inland ports, where the transfer of agroproducts is being discussed. Schiphol is also indicated because the airport plays a key role in the import and export of agroproducts by air – the *raison d'être* for the Aalsmeer complex. The map also shows existing glass horticultural areas and concentrations of intensive livestock farming as well as the locations of the agropark projects which have been described in this publication.

This map should certainly not be regarded as a hierarchical reflection or as part of the agricultural main structure in the discussion of the role of the government in system innovation (see Section 7.2). It merely indicates the best physical locations, particularly as concerns accessibility of areas, for developing agroparks. The tumultuous development of AgriportA7 in the Wieringermeer, from concentrated glass horticulture towards an agropark, but remote from water or rail infrastructure, shows that it can also occur in other places.

None of this is a plea for a new national project, like the Mainports of the Fourth National Policy Document on Spatial Planning or the major housing projects of subsequent documents. But it is necessary for the national government to adopt the much more radical ambition of agroparks as a clear and explicit policy principle and to actually apply the necessary tools such as subsidy regulations, extra supportive legislation (like the Reconstruction Act), and in particular an effective space for

[336] This map was composed by Herman Agricola, Alterra, Wageningen UR on the basis of the previously mentioned criteria.

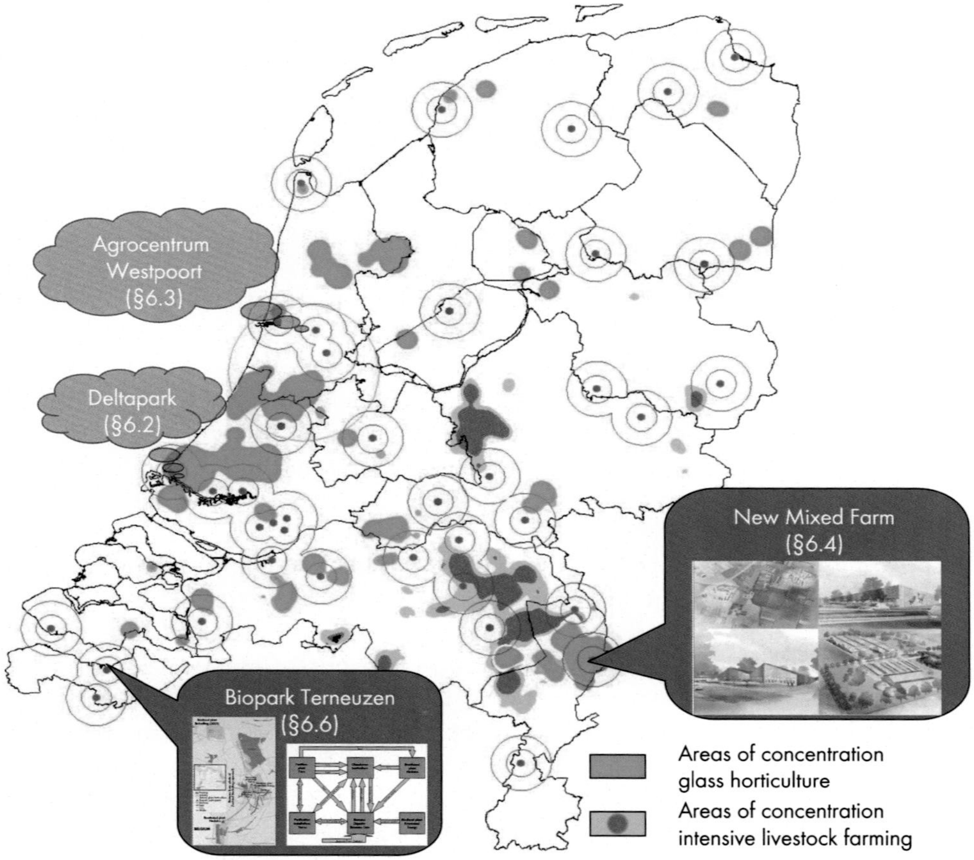

Figure 48. Potential locations for agroparks in relation to existing concentrations of glass horticulture and intensive livestock farming and the location of the project examples in this book.

experimentation for innovative projects. The regional development of the projects can then be taken over by collaborating provinces and cities. This all culminates in a final generic conclusion on the Greenport policy.

System innovation agroparks can be far better facilitated by provincial and local governments, if the national government makes them into directive starting points in terms of spatial development policy and experimentation space. The Greenport concept is a first step in this process, but it must embrace all Dutch agrofood.

7.5 The knowledge infrastructure of Greenport Holland

An informal knowledge infrastructure has emerged from the agropark projects in the previous chapter. This was structured via the Agrologistics Community of Practice in the period between 2002 and 2006. This Community of Practice has not been active since 2006, but the participants are continuing to work in various combinations and in different projects, where the knowledge infrastructure, albeit informally, is becoming more defined. In mid-2009 there were six active knowledge networks – six expeditions, the participants of which partly share the same history described in this publication.

- Together with several agricultural entrepreneurs and with the help of a consultancy, staff at Green Space and Agrocluster Innovation Network are busy developing business plans for various locations. They are deliberately working in the shelter of publicity and have discarded the KENGi principle[337].
- As the next step in the development of Biopark Terneuzen, Biopark Terneuzen, Hogeschool Zeeland and Ghent Bio-energy Valley have set up a collaboration which will lead to a trial biorefinery and a training institute for process operators in this industry. The project is being supported by an EU-Region subsidy[338].
- The New Mixed Farm Steering Group is still supervising the entrepreneurs, who are gradually making progress setting up a similar agropark. In the same region, the Streamlining Greenport Venlo initiative is also ongoing, with its focus on glass horticulture (Greenport Venlo) and is operating at the logistical hub. KnowHouse is playing the key role here in projects supported by TransForum.
- The Greenport Shanghai Steering Group is an informal network of province and city directors, knowledge institutions, KENGi brokers and some entrepreneurs, who are acting as the contact point for Chinese organisations, trying to implement Greenport Shanghai. The basis for Greenport Shanghai was laid from a collaboration

---

337 The most recent report on this action can be seen on the innovation network website: http://www.innovatienetwerk.org/nl/bibliotheek/nieuws/218/Juni2008PlannenAgrocentrumgepresenteerd: 'On 18 June 2008, the plans for the Agrocentrum were presented at a conference on mega-stalls. The CLM report commissioned by Friends of the Earth Netherlands was published on the same day. Conclusion: "Mega-stalls as they are currently foreseen in the country, contribute nothing to any of these developments towards sustainability. Therefore, mega-stalls are no more sustainable than smaller, conventional livestock farming businesses. Agroproduction parks, with a combination of horizontal and vertical integration, seem to offer better environmental prospects than mega-stalls". Various articles on the Agrocentrum appeared in the press as a result of this conference.'

338 The following appears on the Zeeland Seaports website: http://www.bioparkterneuzen.com/cms/publish/content/showpage.asp?pageid=1236: '19-12-2008: Bio Base Europe is granted € 21 million to become the first open innovation and education center for the biobased economy in Europe. On December 12, Europe, Flanders and The Netherlands have joined forces within the framework of an Interreg IV project and allocated € 21 million to Bio Base Europe. Bio Base Europe is the largest Interreg project ever granted to the Dutch-Flemish border region. Bio Base Europe will build research and training facilities for biobased activities, in order to speed up the development of a sustainable biobased economy in Europe.'

between the network around New Mixed Farm, Wageningen-UR and TransForum. TransForum has been coordinating these activities since 2008.

- Wageningen Agroparken is a network of staff from Wageningen-UR that is acquiring new agropark projects in India but also in other countries, in a strategic collaboration with Yes Bank active in India, in Agropark IFFCO Kisan SEZ Nellore and from there with KENGi partners from an international network.
- TransForum is organising a global knowledge network, the Platform on Innovation of Metropolitan Agriculture, which builds on the existing network of the *International Food and Agribusiness Management Association* (IAMA).

Little effort will be required to activate the latter knowledge network as an international Community of Practice, which will take a structured approach to organising the knowledge infrastructure around metropolitan foodclusters. The Agrologistics Community of Practice was the forerunner. Now it is time to connect and activate the network again with the current projects at the basis of the infrastructure, whereby the international expansion is the first item under discussion.

A number of functionalities need to be added to this knowledge infrastructure, simply because the agropark projects have since developed in the direction of implementation and operationalisation – in other words – from invention to system innovation. From the projects analysed in Chapter 6 and in the first sections of Chapter 7 of this publication, the functionalities are as follows.

- The Community of Practice can use the combined inductive-deductive approach developed in this book (see Section 5.1 and 7.3) as a starting point for its work method. The design teams, which will work in co-design on developing agroparks, must be a transdisciplinary mix of scientists, entrepreneurs, local citizens and government staff and NGOs.
- Sustainable development is the core foundation in this knowledge infrastructure. It is becoming the quality hallmark of the production method and the resulting products. This means that the Community of Practice occupied with developing metropolitan foodclusters must hold a permanent discussion, in which sustainable development is enhanced with new scientific truths, adapted to changing political insights and must also connect with the feeling of quality and beauty of individually engaged participants. Section 7.3 gives an initial impetus for a test on sustainable development.
- The design teams, occupied with designing agroparks, must simultaneously be able to incorporate matterscape, powerscape and mindscape and switch between the various aspects in the iterative design process. But the design teams must also be able to work simultaneously at the local scale level of rural transformation centres and agroparks, and from there be able to switch to the global level of Intelligent Agrologistics Networks.
- The scientists have multiple roles in this transdisciplinarity. It is their task to bring in explicit knowledge from fundamental, strategic and applied research. But

they are also responsible for controlling the quality of the knowledge as a whole, by means of the *ex-ante* and *ex-post* testing of the new designs. By monitoring and evaluating the process, they improve the design process, and are sometimes themselves the KENGi broker, establishing the learning experiences and thereby offering new educational material for the continuous education of the design teams.
- This Community of Practice is a work form in which trust forms the basis for collaboration[339], in addition to the principle that the knowledge used is as scientifically rigorous as possible. Designing for sustainable development is not a value-free science but a political discourse at the cutting edge. Successes, in the sense of design assignments and permit allocation for implementation, or an operational agropark, are certainly not obtained simply on the power of the arguments but on the basis of successful communication campaigns, in which, as the examples described demonstrate, motives other than sustainable development may play an important role. The Metropolitan foodclusters Community of Practice must be able to play its role here or it will be quickly marginalised.

Of the projects evaluated in this publication, it appears that this multiple role of the knowledge institutions is divided over several organisations in a number of projects. In two projects this role is assumed entirely by one knowledge institution (Wageningen-UR) and this latter also plays the role of KENGi broker.

Given the specific situation, in which the KENGi parties around metropolitan foodclusters currently find themselves in the Netherlands, Wageningen-UR should assume responsibility for the long-term development of a knowledge infrastructure, which facilitates the Metropolitan foodclusters Community of Practice. The knowledge infrastructure which manifests itself in this work has been described by others as that of the third space in innovation policy and is connected to the so-called third-generation university.

[339] Kersten P. and R. Kranendonk (2002). *Cop op Alterra; 'use the world around as a laerning resource and be a learning resource for the world'.* Report 546, Alterra, Wageningen, the Netherlands. See also Wetenschappelijke Raad voor het Regeringsbeleid (2008). *Innovatie vernieuwd. Opening in viervoud.* Amsterdam University Press, Amsterdam, the Netherlands.

7.5.1 The third-generation university[340] as knowledge manager of the Metropolitan foodclusters Community of Practice

The role of Wageningen-UR as knowledge manager of the Metropolitan foodclusters Community of Practice described above is an extension of the strategic and applied research functions. It is research by design (co-design), whereby the generation of appealing, cohesive, creative concepts is as important as the evaluation of those concepts via quantitative analytical research.

By using the knowledge that the Community of Practice generates, the knowledge institution becomes a knowledge enterprise that generates funding with its knowledge, which can be used for new knowledge development. The use of knowledge can occur via the re-application of previously produced knowledge in a new customised design. The use of the design knowledge on agroparks, as described in the examples from the previous chapter, is an example of this. But the knowledge can also be used in training and education of (existing) personnel from these self-same agroparks and in the training of the (future) scientific personnel of the knowledge institution itself.

Once it is operational, an agropark can become a foundation for the creation of a knowledge value chain (Figure 49). It is a practical laboratory, in which much can be learned in situ about agroproduction because of the highly regulated production conditions. This knowledge benefits not only Wageningen-UR but all partners in the consortium, and forms the basis for the establishment of training and education programmes for personnel at the agropark. This structural collaboration in a knowledge value chain can be laid down in a joint venture. The implementation and exploitation of an agropark requires knowledge transfer and capacity building. Education and training in the countries where an agropark is built, must result in the feedback of knowledge on sustainable development and exploitation, adapted to the local environment.

The continuous participation of globally operating knowledge institutions like Wageningen-UR in close collaboration with local institutions will not only create value for these institutions themselves, but also, because the knowledge is developed further on the basis of monitoring and evaluation in the agroparks, the latter will act as laboratories. This knowledge concerns not only the hardware knowledge of the agroproduction process, but also the orgware of the internal organisation and the

[340] Much of this section comes from the book by Wissema J.G. (2009). *Towards the third generation university. Managing the university in transition.* Edward Elgar Publishing, Cheltenham, UK, and is further inspired by article of Rabbinge R. and M.A. Slingeland (2009). Change in knowledge infrastructure: The third generation university. In: K.J. Poppe, C. Termeer and M. Slingerland (eds.) *Transitions towards sustainable agriculture, food chains and peri-urban areas.* Wageningen Academic Publishers, Wageningen, the Netherlands, pp. 51-62.

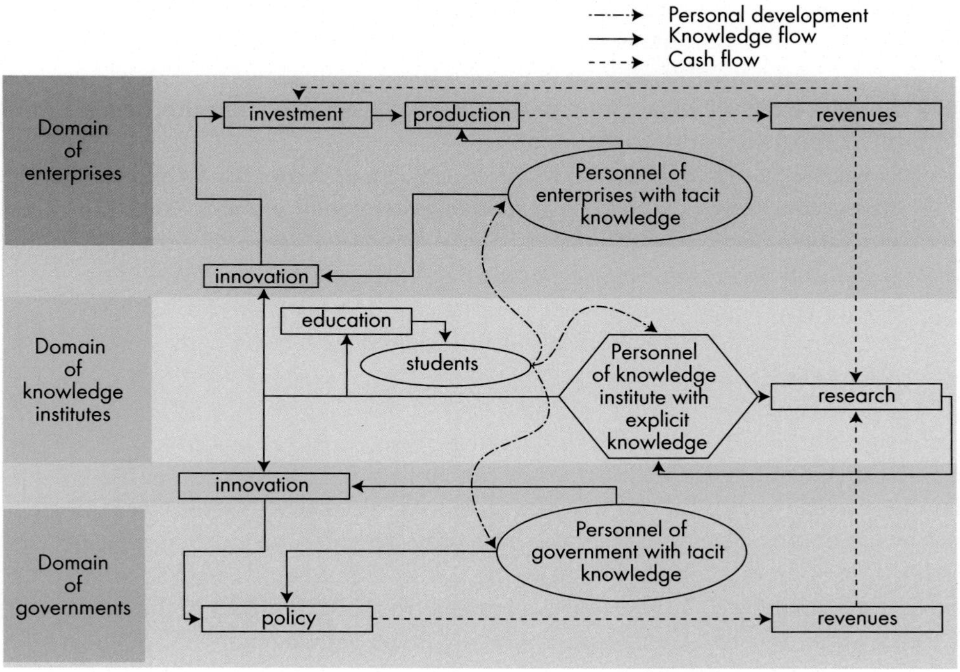

Figure 49. The metropolitan foodclusters knowledge value chain.

relations with governments and other NGOs in the outside world. The knowledge will be handed down to students – some of whom will be the future managers of agroparks – other lecturers or researchers in a knowledge institution and also to others working for the government involved in the same area. In this way the knowledge value chain can be developed into a network of agroparks. In the innovation process, researchers bring in explicit knowledge, but the tacit knowledge of entrepreneurs, government personnel and others is equally important. This is transdisciplinary knowledge development.

7.5.2 Roles in the metropolitan foodclusters knowledge value chain

The metropolitan foodclusters knowledge value chain is basically aimed at generating system innovations that come about in a process of open innovation. The practical examples in this book show that, ultimately, it is about processes that are activated at a regional or local level of scale. This applies to agroparks themselves, which were regarded as a landscape, but also to their wider surroundings in the form of the Intelligent Agrologistics Network. It also applies to the spatial reorganisation, which emerges as a possibility from the further concentration of industrial agriculture components in spatial clusters. That is an argument for allowing the knowledge value chain to be connected firstly to this same regional level of scale as well. This conclusion is closely connected to a more generally formulated conclusion of the WRR[341]. KENGi networks emerge in each of these regions and have strong mutual links within the region while being more loosely connected with other regions. The WRR refers to small worlds in this context:

> 'Small world networks are networks that have a great density locally with strong connections, but which are combined remotely with less dense and weaker connections with other, similarly structured networks elsewhere. (...) Locally this provides a good basis for collaboration and joint capacity to absorb knowledge from elsewhere, while cognitive distance and flexibility are introduced by the weaker, more flexible connections with sources elsewhere.'[342]

According to the WRR, universities are the perfect institutions to connect these regional small worlds in a knowledge network, and to link them up to similar systems elsewhere in the world. The development of the Agrologistics Community of Practice and the continuing development of concrete projects with other partners in a growing network, demonstrates precisely that process, with TransForum and Wageningen-UR as the spiders in the web. KnowHouse operates as the KENGi broker in the small world of North Limburg and Van de Bunt in that of Ghent-Terneuzen.

But according to the WRR, what is necessary for the organising and managing of this infrastructure is an organisation that simultaneously has a close connection with the various KENGi parties but is not one of them. The WRR talks of a third space:

> 'These are virtual or physical organisations for interaction between university and enterprise, which are partially connected to and partially removed from universities and enterprises, in order to protect exploration from too much commercial pressure[343]. (...) A third space may also be necessary to allow for interdisciplinary research which often never gets off the ground within universities.'[344]

---

341 Wetenschappelijke Raad voor het Regeringsbeleid (2008). *Innovatie vernieuwd. Opening in viervoud.* Amsterdam University Press, Amsterdam, the Netherlands. 'It is not just R&D that offers prospects for regions. In addition to science and technology, testing/prototyping, commercialisation, production, supply and distribution are also important and every region has its own competencies and potential in various different areas. These considerations, in addition to the need to choose and implement at precisely the right moment the policy that matches the different development stages of regional clusters, lead the Council to the (...) recommendation (...) to strive in the regional innovation policy towards reticence in design and control by the central government and to allow for sufficient space for regional initiatives and specificities.'

342 *Ibid.*: 98.

343 *Ibid.*: 70.

344 *Ibid.*: 73.

In fact, in recent years TransForum has acted as an informal pioneer of just such a third space via the investments in and management of the innovation process in the New Mixed Farm, Biopark Terneuzen and Greenport Shanghai projects. Wageningen-UR did the same in the ongoing projects in India, but effectively will have to accommodate these activities in a separate third space.

With the formulation of this process-like ambition for a third space, the cycle of content and process management has been completed once more. In the first four chapters of this book I sketched in time and space the context which led to the emergence of metropolitan foodclusters in the network society of the Netherlands. Chapter 5 is the theoretical and methodological basis for the description in Chapter 6 of the research by design on the basis of seven actual agropark designs, whereby time and space are developed via the use of the production-ecological resource use efficiency theory and the landscape theory and whereby the process by which these projects have come to fruition is described and analysed. This leads to the conclusions on content and process which are linked back to the theoretical and methodological principles in the last chapter, and result in the proposal to develop Greenport Holland in the form of a number of agroparks working together in an agrologistics network. The better this network evolves in the Netherlands and North West Europe, the more likely it is to be the support system for knowledge export (exploitation). The projects in China and India demonstrate the first examples of this. A knowledge infrastructure is a component of that – as described in this final section – and it is vital that the knowledge manager in this process – Wageningen-UR – as a third-generation university sets up a third space together with the other KENGi parties which can powerfully stimulate the development of these promising knowledge value chains for metropolitan foodclusters.

References

Alleblas J.T.W. (1996). *Vier kassengebieden in europa; visie op ruimtelijke kwaliteit.* Landbouw-Economisch Instituut. Staring-Centrum, The Hague, the Netherlands. [In Dutch].

Anonymous (2004). *Agrocentrum westpoort. De haalbaarheid verkend.* Gemeentelijk Havenbedrijf Amsterdam, Amsterdam, the Netherlands. [In Dutch].

Anonymous (2005). Nieuw gemengd bedrijf Horst: 'Neem alle belanghebbenden mee in het proces'. *Syscope: kwartaalblad van systeeminnovatieprogramma's* 3: 6-7. [In Dutch].

Asbeek Brusse W., H. van Dalen and B. Wissink (2002). *Stad en land in een nieuwe geografie. Maatschappelijke veranderingen en ruimtelijke dynamiek.* Wetenschappelijke Raad voor het Regeringsbeleid, The Hague, the Netherlands. [In Dutch].

Bakema, A. H., R. G. W. Dood, G. J. Manschot, C. W. M. v. d. Pol, J. M. T. Stam, I. d. Fries & M. P. Wijermans (1999). *Eindrapport; pilotproject meervoudig ruimtegebruik Zuid-West Nederland.* Ministerie van Verkeer en Waterstaat, Directoriaal-Generaal Rijkswaterstaat, The Hague, The Netherlands/RIVM, Bilthoven, the Netherlands; Overlegorgaan voor Vastgoedinformatie; DLO, Wageningen, the Netherlands. [In Dutch].

Baldock D., G. Beaufoy, F. Brouwer and F. Godeschalk (1996). *Farming at the margins; abandonment or redeployment of agricultural land in Europe.* Institute for European Environmental Policy; Agricultural Economics Research Institute, The Hague, the Netherlands.

Beaverstock J.V., P.J. Taylor and R.G. Smith (1999). A roster of world cities. *Cities* 16: 445-458.

Berkhout P. and C. van Bruchem (2004). *Landbouw-Economisch Bericht 2004.* Report PR.04.01. LEI, The Hague, the Netherlands.

Bethe F. and E.C.A. Bolsius (1995). *Marginalisation of agricultural land in the Netherlands, Denmark and Germany.* National Physical Planning Agency, The Hague, the Netherlands / Kopenhagen, Denmark / Bonn, Germany.

Bieleman J.J. (1992). *Geschiedenis van de landbouw in Nederland, 1500-1950. Veranderingen en verscheidenheid.* Boom, Meppel, the Netherlands, 423 pp. [In Dutch].

Bijpost S. and R. Overdevest (2005). *Beoordeling haalbaarheidskansen agrocentrum vanuit economisch perspectief.* Ballast Nedam. Bouw Speciale Projecten, Nieuwegein, the Netherlands. [In Dutch].

Blom G., M. Brinkhuijsen, W.B. Harms, M. van Mansfeld, I. Mulders, P.J.A.M. Smeets, L. Xiuzhen and E. Zuidema (2002). *Mianzi for all: Shanghai international wetland park 2002. Landscape ecological design, arranging concepts and principles for the chongming east headland.* Alterra 2002, WUR Wageningen-Bureau Zuidema Rotterdam, the Netherlands / Shanghai Industrial Investment Holdings Co; Shanghai, China PR.

Boekema F., M. Gijzen, F. Timmer and J. Dagevos (2008). Biopark Terneuzen. Een innovatief en duurzaam cluster. *Geografie* 17: 17-20. [In Dutch].

Bolsius E.C.A. (1993). *Pigs in space.* Ministry of Housing, Spatial Planning and the Envirionment, The Hague, the Netherlands, 63 pp.

Boone K., K. de Bont, K.J. van Calker, A. van der Knijff and H. Leneman (2007). *Duurzame landbouw in beeld. Resultaten van de Nederlandse land- en tuinbouw op het gebied van people, planet en profit.* Report 2.07.09. LEI, The Hague, the Netherlands. [In Dutch].

Boonstra F.F.G., W.W. Kuindersma, H.H. Bleumink, S.S.d. Boer and A.A.M.E. Groot (2007). *Van varkenspest tot integrale gebiedsontwikkeling: evaluatie van de reconstructie zandgebieden.* Report 1441. Alterra, Wageningen, the Netherlands. [In Dutch].

Breure A.S.H., P.J.A.M. Smeets and J. Broeze (2005). *Agrocentrum westpoort: Utopie of innovatie? Reflecties en leerpunten rond een systeeminnovatief project.* Report 1394. Alterra, Wageningen, the Netherlands. [In Dutch].

Broeze J., E. Annevelink and M. Vollebregt (2007). *Onderzoek biomassa en energie biopark Terneuzen.* Report AFSG 848, Agrotechnology and Food Sciences Group, Wageningen, the Netherlands. [In Dutch].

Broeze J., I.A.J.M. Eijk, K.H. de Greef, P.W.G. Groot Koerkamp, J.A. Stegeman and J.G. de Wilt (2003). *Animal care. Diergezondheid en dierwelzijn in ruimtelijke clusters.* Report 03.2.028, InnovatieNetwerk Groene Ruimte en Agrocluster, The Hague, the Netherlands. [In Dutch].

Broeze J., S. Schlatmann, M. Timmerman, A. Veeken, L. Bisschops, D. Kragić, J. van Doorn and A. Boersma (2006). *Uitwerking ontwerp bioenergiecentrale ngb bij het integraal project transforum agro & groen: nieuw gemengd bedrijf.* Agrotechnology & Food Sciences Group, Wageningen, the Netherlands. [In Dutch].

Broeze J., A.E. Simons, P.J.A.M. Smeets, J.K.M. te Boekhorst, J.H.M. Metz, P.W.G. Groot Koerkamp, T. van Oosten-Snoek and N. Dielemans (2000). Deltapark: Een havengebonden agroproductiepark. In: De Wilt J.G., H.J. van Oosten and L. Sterrenberg (2000). *Agroproductieparken perspectieven en dilemma's.* Innovatienetwerk Groen Ruimte en Agrocluster, The Hague, the Netherlands.

Broeze J., M.G.N. Van Steekelenburg and P.J.A.M. Smeets (2005). *Agrocentrum amsterdam. Ontwerpen voor agroparken in havengebieden.* Report 05.2.106. InnovatieNetwerk Groene Ruimte en Agrocluster, Utrecht, the Netherlands. [In Dutch].

Brouwer F.M., C.J.A.M. de Bont, H. Leneman and H.A.B. van der Meulen (2004). *Duurzame landbouw in beeld*, Landbouweconomisch Instituut, the Hague, the Netherlands. [In Dutch].

Buijs S., P.J.A.M. Smeets, H. Guozheng, B. Xinmin and M. Van Mansfeld (2007). *Masterplan Greenport Shanghai agropark. Knowledge report 1. Planning methodology.* Alterra WUR-rapport 1391 sub. 2, Wageningen University and Research Centre, Wageningen the Netherlands.

Buijs S.C. (1990). De stedebouwkundige ontwikkeling van Jakarta. In: Rijksplanologische Dienst (ed.) *Ruimtelijke verkenningen 1990.* Ministerie VROM, The Hague, the Netherlands. [In Dutch].

Castells M. (1996). *The information age: economy, society and culture. Volume 2: the power of identity*. Blackwell, Oxford, UK.

Castells M. (2000a). *The information age: economy, society and culture. Volume 3: end of millennium.* Blackwell, Oxford, UK.

Castells M. (2000b). *The information age: economy, society and culture. Volume 1: the rise of the network society.* Blackwell, Oxford, UK.

CBS (2006). *De Nederlandse economie 2003*. Centraal Bureau voor de Statistiek, Voorburg, the Netherlands. [In Dutch].

College van Rijksadviseurs (2007). *Advies megastallen*. College van Rijksadviseurs, The Hague, the Netherlands. [In Dutch].

Cornelis A. (1999). *De vertraagde tijd*. Stichting Essence, Amsterdam, the Netherlands, 174 pp. [In Dutch].

Covey S.R. (2000). *De zeven eigenschappen van effectief leiderschap*. Uitgeverij Contact, Amsterdam, the Netherlands. [In Dutch].

Daalder A. and J. Koopman (2004). *Verguld en verguisd. Agroparken in de media*. Report 04.2.075. InnovatieNetwerk Groene Ruimte en Agrocluster, The Hague, the Netherlands. [In Dutch].

Dammers E., F. Verwest, B. Staffhorst and W. Verschoor (2004). *Ontwikkelingsplanologie. Lessen uit en voor de praktijk*. Ruimtelijk Planbureau, NAi Uitgevers, The Hague, the Netherlands. [In Dutch].

Darwin J. (2007). *After Tamerlane*. Allen Lane (Penguin Books), London, UK.

De Geyter X., G. Bekaert, L. de Boeck and V. Patteeuw (2002). *After-sprawl: onderzoek naar de hedendaagse stad*. NAi Uitgevers, Rotterdam, the Netherlands, 255 pp. [In Dutch].

De Graaff M. (2008). Agrocentrum varkens in zicht. *Agrarisch Dagblad*. [In Dutch].

De Jonge J. (2009). *Landscape architecture between politics and science. An integrative perspective on landscape planning and design in the network society*. Wageningen University and Research Centre, Wageningen, the Netherlands, 233 pp. [In Dutch].

De Koning G.H.J., H. Van Keulen, R. Rabbinge and H. Janssen (1995). Determination of input and output coefficients of cropping systems in the European community. *Agricultural Systems* 48: 485-502.

De Nijs T. and E. Beukers (eds.) (2002). *Geschiedenis van Holland (4 dln)*. Verloren, Hilversum, the Netherlands. [In Dutch].

De Wilt J.G. and T. Dobbelaar (2005). *Agroparken. Het concept, de ontvangst, de praktijk*. InnovatieNetwerk Groene Ruimte en Agrocluster, Utrecht, the Netherlands: 22.

De Wilt, J.G., H.J. Van Oosten and L. Sterrenberg (2000). *Agroproductieparken perspectieven en dilemma's*. Innovatienetwerk Groen Ruimte en Agrocluster, The Hague, the Netherlands. [In Dutch].

De Wit C.T. (1992). Resource use efficiency in agriculture. *Agricultural Systems* 40: 125-151.

De Wit C.T. (1993). Tussen twee vuren. In: Themagroep Landbouw-Milieu (ed.) *Intensivering of extensivering*. Studium Generale. Landbouwuniversiteit Wageningen., Wageningen, the Netherlands, pp. 3-16. [In Dutch].

De Wit C.T., H.H. Huisman and R.R. Rabbinge (1987). Agriculture and its environment: Are there other ways? *Agricultural Systems* 23: 211-236.

Diamond J. (2000). *Zwaarden, paarden en ziektekiemen*. Uitgeverij Het Spectrum, Utrecht, the Netherlands. [In Dutch].

Dirkx G.H.P., M. Jacobs, J.M. De Jonge, J.F. Jonkhof, J.A. Klijn, A. Schotman, P.J.A.M. Smeets, J.T.C.M. Sprangers, M. Van den Top, H. Wolfert and E. Vermeer (2001). Kubieke landschappen kennen geen grenzen. In: *Jaarboek alterra 2000*, Alterra, Wageningen, the Netherlands. [In Dutch].

Driessen P.P.J. (1995). *Koersen tussen rijk en provincie: Evaluatie van de doorwerking van het koersenbeleid voor het landelijk gebied naar het provinciaal ruimtelijk beleid*. Universiteit Utrecht, Utrecht, the Netherlands, 201 pp. [In Dutch].

Dumont M.J., R. Groot, R. Schröder, P.J.A.M. Smeets and H. Smit (2003). *Nieuwe bruggen naar de toekomst. Weergave van een speurtocht naar nieuwe perspectieven voor het gelders landelijk gebied*. Report 674, Alterra, Wageningen, the Netherlands. [In Dutch].

Economist (2004). Rags and riches. Survey on fashion. *The Economist* 370.

European Commission (1995). *Europe 2000+*. Office for Official Publications of the European Commision, Luxembourg, Luxembourg.

European Commission (1996). *Prospects for the development of the central and capital cities and regions*. Office for Official Publications of the European Commision, Luxembourg, Luxembourg.

Florida R. (2002). *The rise of the creative class*. Basic Books, New York, NY, USA.

Frosch R.A. and N.E. Gallopoulos (1989). Strategies for manufacturing. *Scientific American* 261: 144-152.

Gemeente Horst aan de Maas (2007). *Informatiedocument gebiedsvisie Witveldweg*. Gemeente Horst aan de Maas, Horst aan de Maas, the Netherlands. [In Dutch].

Gemeente Horst aan de Maas (2008). Persbericht bij het verschijnen van de duurzaamheidstoets nieuw gemengd bedrijf. Gemeente Horst aan de Maas, Horst aan de Maas, the Netherlands. [In Dutch].

Gies E., J. Van Os, T. Hermans and R. Olde Loohuis (2007). *Megastallen in beeld*. Report 1581, Alterra, Wageningen, the Netherlands. [In Dutch].

Gijzen M., F. Timmer, J. Dagevos and F. Boekema (2009). Biopark Terneuzen: Een duurzaam en innovatief voorbeeld voor zuidwest Nederland. In: Smulders H., M. Gijzen and F. Boekema (eds.) *Agribusiness clusters: Bouwstenen van de regionale biobased economy*. Shaker Publishing, Maastricht, the Netherlands, pp. 37-48. [In Dutch].

Glendining M.J., A.G. Dailey, A.G. Williams, F.K. Van Evert, K.W.T. Goulding and A.P. Whitmore (2009). Is it possible to increase the sustainability of arable and ruminant agriculture by reducing inputs? *Agricultural Systems* 99: 117-125.

Goedman J., D. Langendijk, E. Opdam, S. Reinhard, I.d. Vries and M. Wijermans (2002). *Zee en land meervoudig benut; beknopt projectverslag*. Alterra, Wageningen, the Netherlands. [In Dutch].

Gordijn F. (2004). *Communities of Practice als Managementinstrument: Over de meerwaarde en het faciliteren van CoP's*. MSc Thesis Communication & Innovation, Wageningen University, Wageningen, the Netherlands. [In Dutch].

Groen T., J.W. Vasbinder and E. Van de Linde (2006). *Innoveren. Begrippen, praktijk, perspectieven*. Spectrum, Utrecht, the Netherlands. [In Dutch].

Groot A.M.E. and P.J.A.M. Smeets (2006). Transitie en transitiemanagement. In: O. Oenema, J.W.H. van der Kolk and A.M.E. Groot (eds.) *Landbouw en milieu in transitie*. Wettelijke Onderzoekstaken Natuur & Milieu, Wageningen, the Netherlands. [In Dutch].

Hall P. and K. Pain (2006). *The polycentric metropolis. Learning from mega-city regions in europe*. Earthscan / James & James, London, UK, 256 pp.

Harvey D. (1989). *The condition of postmodernity: An enquiry into the origins of cultural change.* Blackwell, Oxford, UK, 378 pp.

Hemel Z. (2008). Middelpunt zoekende krachten. *Stedebouw & Ruimtelijke Ordening* 89: 28-34. [In Dutch].

Hoes A., B. Regeer and J. Bunders (2008). Transformers in knowledge production. Building science-practice collaborations. *Action Learning: Research and Practice* 5: 207-220.

Ibn Khaldun (1967). *The muqaddimah.* Princeton University Press, Princeton, NJ, USA.

Imber C. (2002). *The Ottoman empire.* Palgrave (Macmillan), Basingstoke, UK.

Innovatienetwerk Groene Ruimte en Agrocluster (2000). *Initiëren van systeeminnovaties.* Report 00.3.002. Innovatienetwerk Groene Ruimte en Agrocluster, The Hague, the Netherlands. [In Dutch].

Innovatieplatform (2005). Creativiteit. De gewichtloze brandstof van de economie. Available at: http://www.innovatieplatform.nl/assets/binaries/documenten/2005/creatieve_industrie/rapportcreatieveindustrie2.pdf. Accessed October 5, 2005. [In Dutch].

Israel J.I. (1996). *De republiek 1477-1806.* Van Wijnen, Franeker, the Netherlands. 1368 pp. [In Dutch].

Jacobs J. (1984). *Cities and the wealth of nations.* Vintage, New York, NY, USA.

Jacobs M. (2002). *Landschap3.* Expertisecentrum Landschapsbeleving, Alterra, Wageningen, the Netherlands, 68 pp. [In Dutch].

Jacobs M. (2004). Metropolitan matterscape, powerscape and mindscape. In: Tress G., B. Tress, W.B. Harms, P.J.A.M. Smeets and A. Van der Valk (eds.) *Planning metropolitan landscapes. Concepts, demands, approaches, Delta series.* Wageningen University, Wageningen, the Netherlands, pp. 26-39.

Jacobs M. (2006). *The production of mindscapes. A comprehensive theory of landscape experience.* Wageningen University and Research Centre, Wageningen, the Netherlands, 268 pp.

James L. (1994). *The rise and fall of the british empire.* Little, Brown and Company, London, UK.

Kamphuis H.W. (1991). De vierde nota extra. Koersbepaling landelijke gebieden. *Landschap* 8: 47-58. [In Dutch].

Kamphuis H.W., P.L. Dauvelier, J. Groen, H.C. Jacobs and G.J. Wijchers (1991). *Platteland op weg naar 2015.* Rijksplanologische Dienst, The Hague, the Netherlands. [In Dutch].

Kersten P. and R. Kranendonk (2002). *Cop op Alterra; "use the world around as a laerning resource and be a learning resource for the world".* Report 546. Alterra, Wageningen, the Netherlands.

KnowHouse (2006). Ontwerp voor nieuw gemengd bedrijf. Folder met artist impressions gebaseerd op ontwerpen van TRZIN, Amsterdam. Quintrix, TRZIN, KnowHouse, Horst, the Netherlands. [In Dutch].

Kool A., I. Eijck and H. Blonk (2008). *Nieuw gemengd bedrijf. Duurzaam en innovatief?* Blonk Milieu Advies, SPF Gezonde Varkens, Gouda, the Netherlands. [In Dutch].

Kranendonk R., F. Gordijn, P. Kersten and P.J.A.M. Smeets (2003). *Cop agrologistiek; verslag van werkatelier (6-7 November, Venraij).* Alterra/WING, Wageningen, the Netherlands. [In Dutch].

Kranendonk R., P. Kersten and P. Smeets (2005a). *Cop agrologistiek. Verslag van Cop bijeenkomst (14 December, kasteel Groeneveld Baarn).* Alterra, Wageningen, the Netherlands. [In Dutch].

Kranendonk R., P. Kersten, P. Smeets and F. Gordijn (2004). *Cop agrologistiek; verslag van werkatelier (7-8 April, Zaandam).* WING Proces Consultancy, Alterra, Wageningen, the Netherlands. [In Dutch].

Kranendonk R., P. Kersten and P.J.A.M. Smeets (2006). *Cop agrologistiek verslag van de Copbijeenkomst (14 Juni 2006, living tomorrow Amsterdam).* Alterra/WING, Wageningen, the Netherlands. [In Dutch].

Kranendonk R.P., P.H. Kersten, P. Smeets and F. Gordijn (2005b). *Cop agrologistiek: Verslag van masterclass (12 Januari 2005, Den Bosch).* Alterra/WING, Wageningen, the Netherlands. [In Dutch].

Leenstra F.R., E.K. Visser, M.A.W. Ruis, K.H. de Greef, A.P. Bos, I.D.E. van Dixhoorn and H. Hopster (2007). *Ongerief bij rundvee, varkens, pluimvee, nertsen en paarden. Inventarisatie en prioritering en mogelijke oplossingsrichtingen.* Report 71, Animal Sciences Group, Wageningen UR, Wageningen, the Netherlands. [In Dutch].

Leeuwis C.C., R.R. Smits, J.J. Grin, L.L.W.A. Klerkx, B.B.C. Mierlo and A.A. Kuipers (2006). *Equivocations on the post privatization dynamics in agricultural innovation systems. The design of an innovation-enhancing environment.* Transforum working papers. TransForum, Zoetermeer, the Netherlands.

Lesger C. (2001). *Handel in Amsterdam ten tijde van de opstand.* Verloren, Hilversum, the Netherlands. [In Dutch].

Makaske B. (2008). De kwetsbaarheid van Delta's. Zeven plagen in een geologisch perspectief. *Geografie* 17: 50-55. [In Dutch].

Marcuse H. (1970). *De een-dimensionale mens: Studies over de ideologie van de hoog-industriële samenleving.* Paul Brand, Bussum, the Netherlands, 279 pp. [In Dutch].

McNeill W. (1963/1991). *The rise of the west.* The University of Chicago Press, Chicago, IL, USA.

McNeill W.H. (1996). *De excentriciteit van het wiel en andere wereldhistorische essays.* Uitgeverij Bert Bakker, Amsterdam, the Netherlands. [In Dutch].

Meijers E. (2007). *Synergy in polycentric urban regions. Complementarity, organising capacity and critical mass.* Delft University of Technology, Delft, the Netherlands, 196 pp.

Ministerie van Landbouw Natuurbeheer en Visserij (2004). *Het Nederlandse agrocluster in kaart.* Ministerie van Landbouw, Natuur en Voedselkwaliteit, The Hague, the Netherlands, 36 pp.

Ministerie van Volkshuisvesting Ruimtelijke Ordening en Milieu (1992). *Vierde nota over de ruimtelijke ordening extra.* SDU Uitgevers, The Hague, the Netherlands.

Ministerie van Volkshuisvesting Ruimtelijke Ordening en Milieu (1995). *Milieu, ruimte en wonen; tijd voor duurzaamheid.* VROM, The Hague, the Netherlands, 91 pp.

Ministerie van Volkshuisvesting Ruimtelijke Ordening en Milieu (2004). *Nota ruimte, ruimte voor ontwikkeling.* SDU Uitgevers, The Hague, the Netherlands.

Ministerie van Volkshuisvesting Ruimtelijke Ordening en Milieubeheer (2001). *Ruimte maken, ruimte delen: Vijfde nota over de ruimtelijke ordening 2000/2020 vastgesteld door de ministerraad op 15 december 2000, The Hague.* SDU, The Hague, the Netherlands.

Mumford L. (1961/1989). *The city in history.* Harcourt, San Diego, CA, USA.

Nonaka I. and H. Takeuchi (1995). *The knowledge-creating company: How Japanese companies create the dynamics of innovation.* Oxford University Press, New York, NY, USA.

OECD (2007). *Oecd review of agricultural policies China.* OECD, Parijs, France, 235 pp.

Oosterberg W. and C. Van Drimmelen (2006). *Rode Delta's.* Ministerie van Verkeer en Waterstaat, The Hague, the Netherlands.

Penning de Vries F.W.T., R. Rabbinge and J.J.R. Groot (1997). Potential and attainable food production and food security in different regions. *Philosophical transactions of the Royal Society of London. Series B, Biological sciences* 352: 917-928.

Pijlman F.T.A. (2005). Strong increase in total delta-thc in cannabis preparations sold in dutch coffee shops. *Addiction biology* 10: 171-180.

Platform Agrologistiek (2004). Ministers Veerman en Peijs zetten greenports op de kaart. In: Agrologistiek in uitvoering 2004, 5 oktober 2004, Amsterdam. Ministerie van LNV, The Hague, the Netherlands, pp. 4. [In Dutch].

Porter M.M.E. (1998). Clusters and the new economics of competition. *Harvard business review* 76: 77.

Projectbureau Biopark Terneuzen (2007). Biopark Terneuzen. Position paper. Available at: http://www.bioparkterneuzen.com/cms/publish/content/downloaddocument.asp?document_id=198. Accessed 2 June 2007.

Projectgroep Landelijke Gebieden & Europa (1995). *De toekomst van het landelijk gebied. Discussienota eurokompas '95.* Ministerie van Volkshuisvesting, Ruimtelijke Ordening en Natuurbeheer, The Hague, the Netherlands. [In Dutch].

Projectgroep Landelijke Gebieden & Europa (1997). *Landelijke gebieden in europa: Eindrapport.* Ministerie van Volkshuisvesting, Ruimtelijke Ordening en Natuurbeheer, The Hague, the Netherlands. [In Dutch].

Raad Landelijk Gebied (2008). *Het megabedrijf gewogen. Advies over het megabedrijf in de intensieve veehouderij.* Report 08/03, Raad Landelijk Gebied, Amersfoort, the Netherlands. [In Dutch].

Raad voor het Landelijk Gebied (2005). *Plankgas voor glas? Advies over duurzame ontwikkeling van de glastuinbouw in nederland.* Report RLG 05/2, Raad voor het Landelijk Gebied, The Hague, the Netherlands. [In Dutch].

Raad voor het Landelijk Gebied (2007). *Samen of apart? Advies over de wenselijkheid van een agrarische hoofdstructuur op rijksniveau.* Report RLG 07/7, Raad voor het Landelijk Gebied, The Hague, the Netherlands. [In Dutch].

Rabbinge R. (2000a). The future role of agriculture. In: Boekestein A., P. Diederen, W. Jongen, R. Rabbinge and H. Rutten, (eds.) *Towards an Agenda for Agricultural Research in Europe.* Wageningen Pers, Wageningen, the Netherlands, pp. 161-168.

Rabbinge R. (2000b). World food production, food security and sustainable land use. In: El Obeid A.E., S.R. Johnson, J.H. H. and L.C. Smith (eds.) *Food security: New solutions for the twenty-first century. Symposium honoring the tenth anniversary of the world food prize.* Wiley & Sons, New York, NY, USA, pp. 218-235.

Rabbinge R. (2006). Ruimtelijke ontwikkelingspolitiek. In: Aarts N., R. During and P. Van der Jagt (eds.) *Te koop en andere ideeën over de inrichting van Nederland.* Wageningen UR, Wageningen, the Netherlands, pp. 195-200. [In Dutch].

Rabbinge R. and M.A. Slingeland (2009). Change in knowledge infrastructure: The third generation university. In: K.J. Poppe, C. Termeer and M. Slingerland (eds.) *Transitions towards sustainable agriculture, food chains and peri-urban areas.* Wageningen Academic Publishers, Wageningen, the Netherlands, pp. 51-62.

Rabbinge R. and H.C. Van Latesteijn (1992). Long-term options for land use in the european community. *Agricultural Systems* 40: 195-210.

Rabbinge R., H.C. van Latesteijn and P.J.A.M. Smeets (1996). Planning consequences of long term land-use scenario's in the european union. In: Jongman R.H.G. (ed.) *Ecological and landscape consequences of land use change in europe.* European Centre for Nature Conservation, Tilburg, The Netherlands.

Rabbinge R.R. and P.S. Bindraban (2005). Poverty, agriculture and biodiversity. In: Riggs J.A. (ed.) *Conserving biodiversity.* The Aspen Institute, Washington, DC, USA, pp. 65-77.

RECLUS (1989). *Les villes europeénnes,* DATAR. RECLUS, Montpellier, France.

Regeer B. (2007). *Leren van biopark Terneuzen. Communicatie van kennis in context,* Afdeling Wetenschapscommunicatie, Athena Instituut, Amsterdam, the Netherlands. [In Dutch].

Rienks W.A., W. van Eck, B.S. Elbersen, K. Hulsteijn, W.J.H. Meulenkamp and K.R. de Poel (2003). *Melkveehouderij op schaal. Nieuwe concepten voor grootschalige melkveehouderij.* Report 03.2.051, InnovatieNetwerk Groene Ruimte en Agrocluster, The Hague, the Netherlands. [In Dutch].

Rohlen T.P. (2000). *Hong Kong and the Pearl River Delta: "one country, two systems" in the emerging metropolitan context.* A/PARC, Stanford University, Stanford, CA, USA.

Rotmans J. (2003). *Transitiemanagement: Sleutel voor een duurzame samenleving.* Koninklijke van Gorcum., Assen, the Netherlands, 243 pp. [In Dutch].

Ruijs M.N.A., A. Van der Knijff, J. Van der Lugt and C.E. Reijnders (2007). *Position paper glastuinbouw biopark Terneuzen. Deelrapport 2: kansen voor glastuinbouw(complex) in biopark Terneuzen,* Projectcode 4057200. Landbouweconomisch Instituut, The Hague, the Netherlands. [In Dutch].

Sassen S. (2001). *The global city. New York, London, Tokyo.* Princeton University, Princeton, NJ, USA, 447 pp.

Schot J. (2005). Transities: Veranderen met het verleden en de toekomst. *De Eerste Verdieping* 1(1). [In Dutch].

Shi P., J. Wang, M. Yang, Y. Wang, Y. Ding, L. Zhuo and J. Zhou (2002). Integrated risk management of flood disaster in metropolitan regions of China. To balance flood disaster magnitude and vulnerability in metropolitan regions. In: Second Annual IIASA-DPRI Meeting, 29-31 July, 2002, IIASA, Laxenburg, Austria, pp. 16.

Slicher van Bath B. (1960/1980). *De agrarische geschiedenis van West-Europa 500-1850.* Uitgeverij Het Spectrum, Utrecht, the Netherlands. [In Dutch].

Sloterdijk P. (2006). *Het kristalpaleis. Een filosofie van de globalisering.* Uitgeverij Boom/SUN, Amsterdam, the Netherlands. [In Dutch].

Smeets P.J.A.M. (2009). Transforum en de innovatie van de kennis-infrastructuur in de nederlandse landbouw In: Smulders H., M. Gijzen and F. Boekema (eds.) *Agribusiness clusters: bouwstenen van de regionale biobased economy.* Shaker Publishing, Maastricht, the Netherlands, pp. 83-94. [In Dutch].

Smeets P.J.A.M. and M.J.M. Van Mansfeld (2002). The landscape dialogue: interactive planning as a way to sustainable land use in metropolitan areas. Cases from Northwestern Europe. In: *The International Engineering Consultancy Forum on Sustainable development of Shanghai.* Shanghai Investment Consulting Corporation, Shanghai, P.R. China.

Smeets P.J.A.M., S. Buijs, M. Van Mansfeld, A.E. Simons, K. Chakravarthi, R. Poosapati, R. Olde Loohuis, H. Jansen, J. Broeze, H. Soethoudt, P. Bartels and G. Koneti (in press). *Agropark iffco kisan sez nellore. Conceptual masterplan.* Alterra, Agrotechnology & Food, Wageningen, the Netherlands / Yes Bank, New Dehli, India.

Smeets P.J.A.M., W.B. Harms, M.J.M. Van Mansfeld, A.W.C. Van Susteren and M.G.N. Van Steekelenburg (2004a). Metropolitan delta landscapes. In: Tress G., B. Tress, W.B. Harms, P.J.A.M. Smeets and A. Van der Valk (eds.) *Planning metropolitan landscapes. Concepts, demands, approaches,* Delta series, Wageningen University, Wageningen, the Netherlands, pp. 103-114.

Smeets P.J.A.M., M. Van Mansfeld, R. Olde Loohuis, M. Van Steekelenburg, P. Krant, F. Langers, J. Broeze, W. De Graaff, R. Van Haeff, P. Hamminga, B. Harms, E. Moens, R. Van de Waart, L. Wassink and J. De Wilt (2004b). *Masterplan WAZ-Holland Park. Design for an eco-agricultural sightseeing park in Wujin polder, Changzhou, China,* Alterra, Wageningen, the Netherlands.

Smeets P.J.A.M., M.J.M. Van Mansfeld, C. Zhang, R. Olde Loohuis, J. Broeze, S. Buijs, E. Moens, H. Van Latesteijn, M. Van Steekelenburg, L. Stumpel, W. Bruinsma, T. Van Megen, S. Mager, P. Christiaens and H. Heijer (2007). *Master plan Greenport Shanghai agropark.* Report 1391, Alterra, Wageningen UR, Wageningen, the Netherlands.

Steenbekkers A., C. Simon and V. Veldheer (2006). *Thuis op het platteland. De leefsituatie van platteland en stad vergeleken.* Sociaal en Cultureel Planbureau, The Hague, the Netherlands. [In Dutch].

Stichting Onderzoek Wereldvoedselvoorziening van de Vrije Universiteit Amsterdam (2009). Population density. SOW VU, Amsterdam, the Netherlands.

Taylor P.J., ed., (2003). *European cities in the world network, The European metropolis 1920-2000.* Eramus Universiteit, Rotterdam, the Netherlands.

Taylor P.J. (2004). *World city network: A global urban analysis.* Routledge, London, UK, 241 pp.

Termeer C. (2006). *Vitale verschillen. Over publiek leiderschap en maatschappelijke innovatie. Oratie, 7 december 2006.* Wageningen Universiteit en Researchcentrum, Wageningen, , the Netherlands. 48 pp. [In Dutch].

Termeer C.J.A.M. (2008). Barriers for new modes of horizontal governance. A sensemaking perspective. In: Proceedings Twelfth Annual Conference of the international Research Society for Public Management Queensland University of Technology, Brisbane, Australia.

Termeer C.J.A.M. and B. Kessener (2007). Revitalizing stagnated policy processes: Using the configuration approach for research and interventions. *Journal of Applied Behavioral Science* 43: 256-272.

Thomas H. (2003). *Rivers of gold.* Phoenix, Londen, UK.

Timmer F., M. Gijzen, J. Dagevos and F. Boekema (2007). *Kanaalzone: broedplaats van de biobased economy.* Radboud Universiteit, Nijmegen, the Netherlands. [In Dutch].

Tress B., G. Tress and G. Fry (2003a). Potential and limitations of interdisciplinary and transdisciplinary landscape studies. In: Tress B., G. Tress and G. Fry (eds.) *Interdisciplinary and transdisciplinary landscape studies: Potentials and limitations.* Alterra, Wageningen, the Netherlands, pp. 182-192.

Tress G., B. Tress and M. Bloemmen (eds.) (2003b). *From tacit to explicit knowledge in integrative and participatory research,* Delta series 3. Alterra, Wageningen, the Netherlands, 147 pp.

UNFPH (2008). *State of world population. Unleashing the potential of urban growth.* Available at: http://www.unfpa.org/swp/2007/english/introduction.html. Accessed 22 December 2008.

United Nations Department of Economic and Social Affairs (2008). *World urbanization prospects. The 2007 revision.* United Nations, UNDESA, Population Division, New York, NY, USA.

Van de Klundert A.F., A.G.J. Dietvorst and J. van Os (1994). *Back to the future; nieuwe functies voor landelijke gebieden in Europa.* Report 354. Staring Centrum, Instituut voor Onderzoek van het Landelijk Gebied (SC-DLO), Wageningen, the Netherlands. [In Dutch].

Van de Ven G.W.J., N. De Ridder, H. Van Keulen and M.K. Van Ittersum (2003). Concepts in production ecology for analysis and design of animal and plant-animal production systems. *Agricultural Systems* 76: 507-525.

Van den Broeck J., M. Barendrecht, P. De Boe, F. D'hondt, P. Govaerts, P. Janssens, R. Kragt, M. Van Ginderen and W. Zonneveld (1996). *Ruimte voor samenwerking; tweede Benelux structuurschets.* Brussel, Belgium, 185 pp. [In Dutch].

Van der Woud A. (2007). *Een nieuwe wereld.* Uitgeverij Bert Bakker, Amsterdam, the Netherlands. [In Dutch]..

Van Duinhoven G. (2004). De vergeelde koersen. *Landwerk* 5 (2): 26-28. [In Dutch].

Van Eck W., R. Groot, K. Hulsteijn, P.J.A.M. Smeets and M.G.N. Van Steekelenburg (eds.) (2002a). *Voorbeelden van agribusinessparken.* Report 594, Alterra, Wageningen, the Netherlands. [In Dutch].

Van Eck W., A. Van den Ham, A.J. Reinard, R. Leopold and K.R. de Poel (2002b). *Ruimte voor landbouw; uitwerking van vier ontwikkelingsrichtingen,* Report 530, Alterra, Wageningen, the Netherlands. [In Dutch].

Van Eck W., B. Van der Ploeg, K.R. De Poel and B.W. Zaalmink (1996). *Koeien en koersen; ruimtelijke kwaliteit van melkveehouderijsystemen in 2025.* Report 431, Staring Centrum, Instituut voor Onderzoek van het Landelijk Gebied, Wageningen, the Netherlands. [In Dutch].

Van Eck W., A. Wintjes and G.J. Noij (1997). Landbouw op de kaart. In: *Jaarboek 1997 van het Staring Centrum.* Staring Centrum, Wageningen, the Netherlands, pp. 4-20. [In Dutch].

Van Gendt S., G. De Groot and C. Boendermaker NIBConsult B.V. (2003). *Globaal Businessplan van een Agro-center.* InnovatieNetwerk Groene Ruimte en Agrocluster, The Hague, the Netherlands. [In Dutch].

Van Ittersum M.K. and R. Rabbinge (1997). Concepts in production ecology for analysis and quantification of agricultural input-output combinations. *Field Crops Research* 52: 197-208.

Van Ittersum M.K., R. Rabbinge and H.C. Van Latesteijn (1998). Exploratory land use studies and their role in strategic policy making. *Agricultural Systems* 58: 309-330.

Van Mansfeld M., M. Pleijte, J. De Jonge and H. Smit (2003a). De regiodialoog als methode voor vernieuwende gebiedsontwikkeling. De casus Noord-Limburg. *Bestuurskunde* 12: 262-273. [In Dutch].

Van Mansfeld M., A. Wintjes, J. De Jonge, M. Pleijte and P.J.A.M. Smeets (2003b). *Regiodialoog: Naar een systeeminnovatie in de praktijk.* Report 808, Alterra, Innonet, WISI, Wageningen, the Netherlands. [In Dutch].

Van Ravesteyn N. and D. Evers (2004). *Unseen Europe: A survey of eu politics and its impact on spatial development in the Netherlands.* NAi Publishers, The Hague, the Netherlands, 157 pp.

Van Susteren A.W.C. (2005). *Metropolitan world atlas.* O10 Publishers, Rotterdam, the Netherlands. .

Van Weel P. (2003). *Ontwerpen van geïntegreerde concepten voor agrarische productie in het kader van een agro-eco park in Horst aan de Maas.* Report PPO 588, Praktijkonderzoek Plant & Omgeving B.V. Sector Glastuinbouw, Wageningen, the Netherlands. [In Dutch].

Veerman, C. (2006). *Landbouw verbindend voor Europa. Van vrijheid in gebondenheid naar vrijheid in verbondenheid.* Ministerie van Landbouw, Natuur en Voedselkwaliteit, The Hague, the Netherlands. [In Dutch].

Veldkamp A., A.C. Van Altvorst, R. Eweg, E. Jacobsen, A. Van Kleef, H. Van Latesteijn, S. Mager, H. Mommaas, P.J.A.M. Smeets, L. Spaans and H. Van Trijp (2008). Triggering transitions towards sustainable development of dutch agriculture: Transforum's approach. *Agronomy for Sustainable Development* 29: 87-96.

Verkennis A. and T. Groenewegen (eds.) (1997). Ontwikkelingen in de regio Randstad-Rijn/Ruhr. In: Zonneveld, W. and F. Evers (eds.) *Van delta naar Europees achterland.* NIROV-Europlan, The Hague, the Netherlands.

Wallerstein I. (1974). *The modern world system. Capitalist agriculture and the origins of the european world-economy in the sixteenth century.* Academic Press, New York, NY, USA.

Wallerstein I. (1979). *The capitalist world-economy.* Cambridge Universty Press, Cambridge, UK, 305 pp.

Wallerstein I. (1980). *The modern world system ii. Mercantilism and the consolidation of the European world economy 1600-1750.* Academic Press, New York, NY, USA, 370 pp.

Wallerstein I. (1989). *The modern world system iii. The second era of great expansion of the capitalistic world-economy, 1730-1840s.* Academic Press, New York, NY, USA, 372 pp.

Wenger E. and W. Snyder (2000). Communities of practice: the organizational frontier. *Harvard Business Review* January-February 2000: 139-145.

Wesseling H.L. (2003). *Europa's koloniale eeuw.* Uitgeverij Bert Bakker, Amsterdam, the Netherlands. [In Dutch].

Wetenschappelijke Raad voor het Regeringsbeleid (1992). *Ground for choices; four perspectives for the rural areas in the European community.* Wetenschappelijke raad voor het Regeringsbeleid, The Hague, the Netherlands.

Wetenschappelijke Raad voor het Regeringsbeleid (1994). *Duurzame risico's: Een blijvend gegeven.* Report 44, Wetenschappelijke Raad voor het Regeringsbeleid, The Hague, the Netherlands. [In Dutch].

Wetenschappelijke Raad voor het Regeringsbeleid (1998). *Ruimtelijke ontwikkelingspolitiek.* Wetenschappelijke Raad voor het Regeringsbeleid, The Hague, the Netherlands. [In Dutch].

Wetenschappelijke Raad voor het Regeringsbeleid (2008). *Innovatie vernieuwd. Opening in viervoud.* Amsterdam University Press, Amsterdam, the Netherlands. [In Dutch].

Wetzels W., A.W.N. van Dril and B.W. Daniëls (2007). *Kenschets van de Nederlandse glastuinbouw.* Report ECN-E--07-095, Energiecentrum Nederland, Petten, the Netherlands. [In Dutch].

Wissema J.G. (2009). *Towards the third generation university. Managing the university in transition.* Edward Elgar Publishing, Cheltenham, UK.

About the author

Peter Smeets was born on 8 May 1953 in Roermond, the Netherlands, where he also attended secondary school. In 1978 he graduated as an ecologist from the Catholic University of Nijmegen.

Between 1978 and 1986 he worked as a researcher in the Environmental Sciences department at the above university, as well as at the National Institute for Research in Forestry and Landscape ('De Dorschkamp') in Wageningen. The core activity here was research into nature and landscape conservation by agricultural enterprises.

From 1986 to 1989 Peter worked at the Forestry and Landscape Management department of the Ministry of Agriculture, Nature Conservation and Fisheries as a project leader in area-based studies which provided the landscape-ecological basis for land reallotment projects.

From 1989 to 1996 he was a senior policy advisor at the National Physical Planning Agency, working on plans for river areas, and as leader for the project on Rural Areas and Europe between 1992 and 1996.

In 1996 he was appointed to the Agricultural Research Unit as head of the Landscape and Spatial Use department, which later merged with teaching chair groups from Wageningen University and developed into the Alterra Landscape Centre in 2004, with 180 staff, including 140 scientists. The Landscape Centre initiates interdisciplinary research and planning projects aimed at the relationship between agriculture and space, at urban-rural problems, infra-landscapes and at management and perception issues.

In 2004 he was seconded to TransForum, the BSIK programme on the transition to sustainable agriculture. Thereafter, he has also worked on a thesis about agropark designs, an area in which he has been involved since 2000.

Since 2000 he has been active, often as a project leader, in the Netherlands, China and India, as a designer of agroparks together with other scientists, entrepreneurs, government officials and NGOs.

Keyword index

A

Action Research 42, 119
advanced producer services (APS) 20, 74, 75, 76, 82, 100
AF+ model 213, 221, 245
Agglomerationsräume 88, 91
agricultural
- main structure 113
- productivity 96, 110, 125, 275
- society 48
agriculture 38, 53, 57, 71
- demand-driven 261
- in the network society 93, 108, 115
- supply-driven 261
AgriportA7 291
Agrocentrum Westpoort 150, 161
agrologistics 99, 240, 273
- platform 168, 178, 193
agropark 39, 40, 44, 99, 105, 118, 130, 145
agro-systems 124
ammonia emissions 183
Amsterdam 69, 70, 72, 84, 89, 161, 174
- Port Authority 161, 170
animal welfare 148, 154, 161, 168, 187, 231, 266
antibiotics 185
APS – *See:* advanced producer services
arable farming 100
arcadian notion of farming 54
Asia 38, 77
Athena Institute 192, 218, 220
automobile 70

B

Ballast Nedam 166, 167, 170, 174, 176
Beijing 79
Biopark Terneuzen 150, 210
Blonk Milieu Advies 264, 283
blue banana 67
BSIK – *See:* Investment in Knowledge Infrastructure

C

cannabis cultivation 104
capital cities 66
Castells 47, 49, 50, 51, 52, 56, 83, 93
central cities 65, 66, 68
centralised government 71
Central Processing Unit (CPU) 200, 202, 204, 206, 227, 245
Changzhou 200
China 71, 85, 205, 224, 229, 231
Chongming 224
chronos 148
citizens 56
climate change 97
co-design 42, 43, 135, 138, 191, 206, 219, 284, 296
- theory 120
co-digestion 180
Combined Heat and Power 216
Common Agricultural Policy 106
communication 156, 160, 168, 177, 190, 217, 251, 285
communicative self-steering 54, 95, 96, 128, 143, 265, 268
Community of Practice 138, 149, 158, 162, 168, 172, 192, 207, 220, 285, 287, 293, 295, 296
composting 180
consolidation centre 241, 242
consumer 56
- responsiveness 273
cooling 253
Cornelis 54
cost-benefit analysis 166
countryside 20, 38, 40, 62, 65, 91, 111, 116, 117, 156, 190, 193
CPU – *See:* Central Processing Unit
Crystal Palace 48, 50, 51, 57, 83, 93, 94, 143, 196, 266
currencies 135, 141

D

dairy farming 100, 289
De Jonge 138, 148
De Wit 43, 121
deductive approach 287
Delta metropolis 65, 84, 85, 86, 87, 88, 89, 109
Deltapark 146, 150, 151
democratic system 55
demonstration park 226
Demo>Trade>Processing>Production principle 226, 234, 238, 261
design 131, 136, 137, 161, 171, 177, 200, 210, 224, 240, 282
– aesthetics 280
– criteria 132
– process 43, 132
diversification 103
Dongtan 224

E

environmental
– movement 56, 57
– pollution 126
Europe 38, 66, 67, 71, 85, 88, 101, 145
European
– policy 61, 106, 107
– quota policy 289
– Union 94, 105, 124, 137, 288
expedition 41
– agroparks 41, 58, 96, 147

F

Fifth National Policy Document on Spatial Planning 60, 106
food 21, 39, 41, 47, 53, 65, 71, 85, 86, 91, 93, 103, 105, 109, 110, 132, 134, 145, 239, 240, 242, 277, 288
– processing 242
– safety 156, 230
– shortages 47, 95
fossil fuel 49, 97, 98, 126, 132, 137, 184, 253, 274, 283
Fourth National Policy Document on Spatial Planning 113
the fourth world 52, 83
Freshpark Venlo 243

G

glacial time 56
glass horticulture 100, 124
Global Age 48, 50, 62, 72, 76, 78, 82, 93, 94, 95
global economy 83, 84, 94
globalisation 37, 38, 41, 42, 47, 72, 85, 93, 94, 96, 99
Greece 65
green cokes 214
greenhouse gas emissions 184
Greenport 107, 115, 193, 232, 240, 247, 260, 288
– Holland 288, 293
– India 251
– Shanghai 150, 224
Ground for choices 105, 124, 274, 275
Grubbenvorst 190, 193, 198

H

hardware 129
Heros 212
Holland 67
hope 95
horizontal integration 117
Horst aan de Maas 177, 187, 199
hunger 47, 132

I

IAN – *See:* Intelligent Agrologistics Networks
IFFCO – *See:* Indian Farmers Fertilisers Co-operation
– -Greenport Nellore 150, 240, 245
image 49, 138, 147, 160, 167, 171, 173, 176, 179, 189, 195, 198, 200, 209, 246, 257, 260, 262, 265
import replacement 72, 78

India 71, 242
Indian Farmers Fertilisers Co-operation (IFFCO) 247, 251
inductive approach 287, 294
industrial
- cities 69
- ecology 40, 126, 153, 180, 213, 216, 219, 227, 261, 285
- revolution 69
- society 48
industrialisation 97, 98
- of agriculture 38
information technology revolution 49
infrastructure 69
inhibiting context 55, 56, 59, 62, 105, 157, 196
innovation 72, 104, 105, 133, 135, 140, 146, 179, 190, 191, 199, 206, 209, 219, 221, 233, 234, 251, 253, 262, 280, 297
- network 146, 151, 154, 155, 157, 160, 161, 204, 293
Intelligent Agrologistics Networks (IAN) 40, 241, 242, 250, 255, 261, 273
interdisciplinarity 42, 121, 129
invention 42, 133, 140, 145, 151, 156, 157, 174, 191, 206, 233, 251, 255, 262, 286
Investment in Knowledge Infrastructure (BSIK) 235

J

Jacobs 128
justice 127
just landscape 127

K

kairos 138, 148, 287
KENGi 44, 74, 135, 142, 171, 186, 206, 215, 230, 248, 258, 264, 268, 275, 279, 283
- brokers 44, 160, 172, 185, 191, 200, 207, 219, 238, 248, 250, 252, 254, 265, 293, 295, 298
KnowHouse 178, 185, 189, 226, 227, 232, 234, 235, 265, 293, 298
knowledge 49, 275
- cycle 140
- enterprise 296
- infrastructure 293
- institutions 141
- management 148
- networks 293
- value chain 231, 250, 296

L

landscape 127, 129
- dialogue 43, 127, 129, 135, 136, 252, 284
- theory 43
- visions 136
learning histories 218
Limburg 146, 177, 226, 235
livestock farming 124
- intensive 100, 116, 161
London 70, 72, 79, 84, 86, 89

M

manure processing 180, 183
matterscape 43, 128, 147, 153, 162, 180, 204, 212, 222, 227, 241
mega-city region 73, 78
mega-stables 117
mega-trend 93, 94, 95, 96, 98, 99, 103, 125
methane emissions 184
metropolitan
- areas 9, 39, 74, 76, 78, 93, 132, 231
- foodclusters 41, 91, 108, 112, 271, 276, 290, 294, 297
- foodclusters Community of Practice 296
mindscape 43, 128, 148, 154, 168, 187, 205, 216, 222, 231, 249, 265
mode 2 learning 119, 138, 172, 189, 192, 218, 219, 276, 277, 287
MRSA 185

multilateral institutions 94
myth of the negative image 173

N

National Policy Document on Spatial Planning 107, 115
nation state 21, 57, 58, 72, 73, 86, 87, 94, 95, 105, 115, 143
Nedalco 212
Negative Space 58
the Netherlands 70
network
- cities 82, 83
- enterprise 49
- society 37, 38, 47, 48, 49, 50, 51, 52, 71, 82, 84, 93, 94, 105, 241, 271, 276, 277

new mixed farms 145, 150, 177
New York 79, 84, 89
NGOs 56, 161, 267, 268, 280, 285
NIB-Capital 154, 166, 167
North American 77
North West Europe 38, 45, 59, 76, 88, 90, 91, 100, 106, 209, 231, 257, 266, 299
Northwest European Delta 38
- Metropolis 88, 90, 111

O

odour emission 184
Oostland 116
open innovation 235, 297
orgware 129

P

participative planning 43, 276
particulate matter 185
pig flat 155, 156, 160, 170, 173
pluri-activity 101
polycentric
- mega-city region 20, 73, 76, 85
- metropolis 37

population growth 91
port locations 162
powerscape 43, 128, 148, 154, 166, 185, 204, 215, 222, 246

R

Rabbinge 61, 96, 124
railway 70
Randstad 89
reconstruction 289
research by design 42, 119, 147
resource use efficiency theory 43, 98, 120, 144, 261, 272, 275
risk
- analysis 154
- assessment 249

river deltas 84, 85, 87
Rome 66
Rosendaal Energy 212
Rotterdam 153
RTC – *See:* Rural Transformation Centre
Rural Areas and Europe 106
Rural Transformation Centre (RTC) 241, 243, 244, 250

S

scenarios 211, 227, 228, 252
Scientific Council for Government Policy (WRR) 59, 60, 61, 62, 97, 106, 124, 233, 272, 278, 298
Shanghai 79, 84, 89, 224, 234
- Industrial Investment Company (SIIC) 224, 228, 232

SIIC – *See:* Shanghai Industrial Investment Company
Sloterdijk 48, 50, 53, 55, 57, 62, 85, 95
small and medium-sized entreprises (SMEs) 200, 205, 234, 236, 280
small farmers 71, 253
SMEs – *See:* small and medium-sized entreprises
software 129
space 274

– of flows 38, 52, 53, 72, 109, 130, 159, 174, 196, 221, 277
– of place 38, 52, 83, 109, 127, 130, 196, 276
– pump 40, 91, 279
spatial
– clustering 125
– development policy 58, 62, 63, 95, 114, 116, 137, 275, 278, 289
– planning 58, 95
stigma 155
subsidies 235, 238, 249, 255
suburbanisation 70
Sustainability Test 182
sustainable development 40, 57, 58, 91, 123, 134, 159, 175, 179, 197, 221, 262, 283, 294
system innovation 40, 42, 43, 133, 135, 139, 141, 147, 159, 161, 190, 192, 196, 197, 221, 248, 252, 253, 262, 264, 268, 279, 280, 281, 282

T
tacit knowledge 62
Termeer 119, 194, 197
Terneuzen 211
third-generation university 295
third space 295, 298
three-dimensional landscape 43, 128
– theory 120
time 51
timeless time 51
Tokyo 79
trade 226
trading cities 65, 66, 67, 86
transdisciplinarity 42, 44, 121, 137, 139, 141, 143, 148, 158, 173, 192, 197, 207, 219, 238, 248, 251, 265, 275, 279, 282, 285, 294, 297
TransForum 108, 141, 217, 225, 227, 235, 252, 294
transition 133
transportation 125
true landscape 127
trust 236, 240, 258, 268, 295
trustful landscape 127
trustfulness 127
truth 127

U
University of Ghent 216
urbanisation 37, 60, 65, 66, 69, 93
urban sprawl 87

V
Van de Bunt 211, 217, 222, 265, 298
Veerman 145, 178, 185, 187, 189, 194, 197, 199
vertical integration 117
veterinary risks 184
virtuality 49

W
Wageningen-UR 108, 141, 148, 178, 200, 202, 225, 240, 247, 250, 254, 262, 265, 294, 295, 296, 298, 299
WAZ-Holland Park 150, 200
Westland 116
work process 149
world economy 20, 72, 82
WRR – *See:* Scientific Council for Government Policy

Y
Yara 211
Yes Bank 240, 246, 251, 254, 255, 265, 294

Z
Zeeland 146, 211
– Seaport 211, 215, 217

Printed in the United States
by Baker & Taylor Publisher Services